GEOMORPHOLOGY

The Orange River gorge downstream of the Augrabies Falls, South Africa.
(Photograph © Paul Bishop. Reproduced with permission.)
This bedrock gorge has been cut by headward propagation of the Augrabies Falls knick point in response to Cenozoic uplift.
The prominent and extensive planar erosion surface that flanks the gorge and into which the gorge has cut is an excellent
example of the major surfaces that characterize many ancient landscapes in the Gondwanan continents. On the skyline is a
series of residual hills that rise above the erosion surface.

Geomorphology: Critical Concepts in Geography

General Editor David J.A. Evans

Volume VII

Landscape Evolution

Edited by Paul Bishop

Routledge
Taylor & Francis Group

LONDON AND NEW YORK

First published 2004
by Routledge
2 Park Square, Milton Park, Abingdon, OXON, OX14 4RN

Simultaneously published in the USA and Canada
by Routledge
711 Third Avenue, New York, NY 10017

Transferred to Digital Printing 2006

Routledge is an imprint of the Taylor & Francis Group

British Library Cataloguing in Publication Data
A catalogue record for this book is available from the British Library

Library of Congress Cataloging in Publication Data
A catalog record for this book has been requested

ISBN 10: 0-415-27608-X (Set)
ISBN 10: 0-415-27615-2 (Volume VII)
ISBN 13: 978-0-415-27608-5 (Set)
ISBN 13: 978-0-415-27615-3 (Volume VII)

Publisher's Note
References within each chapter are as they appear in the original complete work

CONTENTS

CONTENTS

ACKNOWLEDGEMENTS

The publishers would like to thank the following for permission to reprint their material:

Blackwell Publishing for permission to reprint W.M. Davis, 'The geographical cycle', *Geographical Journal* 14 (1899): 481–504.

Blackwell Publishing for permission to reprint D.L. Linton, 'The problem of tors', *Geographical Journal* 121 (1955): 470–87.

J.T. Hack, 'Interpretation of erosional topography in humid temperate regions', *American Journal of Science* 258-A (1960): 80–97. Reprinted by permission of *American Journal of Science*.

Blackwell Publishing for permission to reprint C.D. Ollier, 'Evolutionary geomorphology of Australia and Papua–New Guinea', *Transactions of the Institute of British Geographers* 4 (1979): 516–39.

C.R. Twidale, 'On the survival of paleoforms', *American Journal of Science* 276 (1976): 77–95. Reprinted by permission of *American Journal of Science*.

From A.R. Gilchrist, M.A. Summerfield and H.A.P. Cockburn, 'Landscape dissection, isostatic uplift, and the morphologic development of orogens', *Geology* 22 (1994): 963–6. Reproduced with permission of the publisher, the Geological Society of America, Boulder, Colorado, USA. Copyright © 1994 Geological Society of America.

D. Griggs, 'A theory of mountain-building', *American Journal of Science* 237 (1939): 611–50. Reprinted by permission of *American Journal of Science*.

The University of Chicago Press for permission to reprint P.E. Potter, 'Significance and origin of big rivers', *Journal of Geology* 86 (1978): 13–33.

S.A. Schumm and R.W. Lichty, 'Time, space, and causality in geomorphology', *American Journal of Science* 263 (1965): 110–19. Reprinted by permission of *American Journal of Science*.

Disclaimer

The publishers have made every effort to contact authors/copyright holders of works reprinted in *Geomorphology: Critical Concepts in Geography*. This has not been possible in every case, however, and we would welcome correspondence from those individuals/companies who we have been unable to trace.

Chronological Table of reprinted articles and chapters

Date	Author	Article/chapter	References	Chap.
1899	W.M. Davis	The geographical cycle	*Geographical Journal* 14: 481–504	1
1939	D. Griggs	A theory of mountain-building	*American Journal of Science* 237: 611–50	9
1948	J.H. Mackin	Concept of the graded river	*Bulletin of the Geological Society of America* 59: 463–511	6
1953	L.C. King	Canons of landscape evolution	*Bulletin of the Geological Society of America* 64: 721–51	12
1955	D.L. Linton	The problem of tors	*Geographical Journal* 121: 470–87	3
1955	S.W. Wooldridge and D.L. Linton	The Mid-Tertiary cycle of erosion and the summit peneplain	S.W. Wooldridge and D.L. Linton, *Structure, Surface and Drainage in South-East England*, London: George Philip and Son, pp. 34–44	2
1960	J.T. Hack	Interpretation of erosional topography in humid temperate regions	*American Journal of Science* 258-A: 80–97	4
1965	S.A. Schumm and R.W. Lichty	Time, space, and causality in geomorphology	*American Journal of Science* 263: 110–19	11
1976	C.R. Twidale	On the survival of paleoforms	*American Journal of Science* 276: 77–95	7
1978	P.E. Potter	Significance and origin of big rivers	*Journal of Geology* 86: 13–33	10
1979	C.D. Ollier	Evolutionary geomorphology of Australia and Papua–New Guinea	*Transactions of the Institute of British Geographers* 4: 516–39	5
1994	A.R. Gilchrist, M.A. Summerfield and H.A.P. Cockburn	Landscape dissection, isostatic uplift, and the morphologic development of orogens	*Geology* 22: 963–6	8

GENERAL EDITOR'S PREFACE

All areas of research benefit from periodic exercises in reflection – a time for assessing where we are and how we have arrived, who set the ball rolling and whether it still has momentum, and even whether or not we are re-inventing our own metaphorical wheels. Such reflection is especially valuable in these times of accelerating productivity and expanding volumes of literary output, because no geomorphologist could ever hope to keep abreast of every advance in all aspects of the discipline. For the academic this has just as much to do with shrinking library budgets as it has to do with the erosion of reading and processing time. Additionally, when caught up in the frenzy of pushing back research frontiers we can lose sight of the origins of the critical concepts in our science. We all use our relatively small set of core references when compiling papers for submission to journals but these necessarily mostly originate in the more modern literature and often focus on the novel application of a critical concept rather than its initial airing. In other words, we could all name a 'top ten' in our area of expertise but would they be specifically examples of excellent practice rather than benchmarks or breakthroughs? From beginning to end I have found the exercise of selection very much like repeatedly choosing the best ever England football team – 'So we've got Gordon Banks in goal . . .'

The seven volumes in this series contain reproductions of papers that the individual editors regarded as the initiators of critical concepts. It can, and undoubtedly will, be argued by the readership that someone somewhere thought of concept 'x' before author 'y' in Volume 'z'. We have hopefully covered these eventualities in our individual assessments by attempting to integrate the chosen paper with what came before and what has developed since its publication. Furthermore, we have all deliberated long and hard over our selections in the knowledge that our own volume must be objective in its reflection of the development of a subject area. Nonetheless, despite wide consultation with each other and with the wider geomorphic community, the compilations are ultimately personal and therefore reflect the sources of our inspiration. We hope that they coincide with those of other geomorphologists. As few libraries now house collections comprehensive enough to cover both the subject and age range represented in these seven volumes, we hope that they become a valuable reference source for both students and practitioners.

Whilst compiling this series I have been most ably assisted by Rebekah Taylor at Routledge, who possesses admirable levels of patience and the power of gentle persuasion necessary to get academics to deliver book manuscripts in times when their employers would frankly prefer to see them doing other things. I am glad that we persevered, for the exercise was informative and rewarding. Thanks also to Les Hill in Geography and Geomatics at the University of Glasgow for his continued excellent support in the reproduction of photographs that in many cases at the outset did not leave much margin for improvement! Finally, I would also like to thank each of the editors for their knowledgeable insight and hard labour in compiling the critical concept papers in geomorphology.

David J.A. Evans
University of Glasgow

INTRODUCTION

Paul Bishop

Chorley *et al.* (1964: 446) observed in their overview of pre-Davisian geomorphology that:

> all geomorphologists would do well to remember that by the 1870s a large storehouse of quantitative and geometric data already existed. Thereafter any propounder of theories of landscape evolution would be wise to consult and accept the findings of physicists and engineers.

This comment highlights, perhaps unintentionally, the fact that practitioners of landscape evolution studies have generally been either those who have consulted and accepted the findings of physicists and engineers, or those who have not. It is difficult, however, to criticize the latter, the non-consulters who have ignored Chorley *et al.*'s (1964) admonition, because the steps required to synthesize the findings of physicists and engineers to produce theory in landscape evolution are formidable (cf. Church 1996). Thus, many of the critical concepts in landscape evolution are grand, or even 'heroic', in scale. These are the big ideas on which our discipline was founded and which still underpin much geomorphology, concepts such as base-level, equilibrium, grade, the cycle of erosion and so on. Moreover, many of these concepts date from the very dawn of 'modern' science – Domenico Gugliemini's late seventeenth-century description of the concave-up equilibrium fluvial long profile, for example, springs to mind (Chorley *et al.* 1964: 87) – and these concepts were often both expounded and progressively developed in the great scientific works of the eighteenth and nineteenth century. Many of these grand works went through multiple editions incorporating re-statement and gradual refinement of these concepts. It is thus not as straightforward to choose a series of papers that embodies the critical concepts of landscape evolution as it might be for more focused areas of geomorphology. Some will no doubt disagree with my choices of the key papers, and other reviews in this area should also be consulted (e.g. Thomas and Summerfield 1987; Summerfield 2000a). It is also worth noting that it is often difficult to identify the key paper that embodies a critical concept in landscape evolution. Moreover, for several concepts I have used a paper that enables exploration of the concept and associated concepts. A good example of this is Linton's 'The problem of tors', which is used to explore not only the origin of tors but the issue of the erosive power of cold- and warm-based ice. It should also be noted that it is impossible to reference all papers relevant to critical concepts in landscape evolution, not least because there is a multitude of views on the appropriate timescale for landscape evolution.

The following of the development of ideas in landscape evolution is more than manageable because of the superb charting of this development by Richard Chorley, Robert Beckinsale and Antony Dunn (Chorley *et al.* 1964, 1973; Beckinsale and Chorley 1991). Chorley *et al.* (1964) have provided an exhaustive treatment of the late eighteenth-century contributions of James Hutton. Hutton was a giant not least because he realized the great depth of time available in Earth history for landscapes to evolve, and because of his emphasis on the centrality of slow non-catastrophic fluvial processes in this landscape evolution. It is not difficult to see in Hutton's words the roots of much of what William Morris Davis would say a century later:

> The natural operations of this globe, by which the size and shape of our land are changed, are so slow as to be altogether imperceptible . . . We have but to consider the mountains as formed by the hollowing out of the valleys, and the valleys as hollowed out by the attrition of hard materials coming from the mountains . . . Our solid earth is everywhere wasted, where exposed to the day. The summits of the mountains are necessarily degraded. The solid weighty materials of those mountains are everywhere urged through the valleys by the force of running water.
>
> (Hutton 1795, vol 2: 295, 401, 561)

Such landscape evolution was clearly demonstrated in many situations. Thus, in Charles Lyells' words:

> Desmarest [an early nineteenth-century worker], after a careful examination of the Auvergne, pointed out first the most recent volcanos which had their craters still entire, and their streams of lava conforming to the level of the present river-courses. He then showed that there were others of an intermediate epoch, whose craters were nearly effaced, and whose lavas were less intimately connected with the present valleys; and, lastly, that there were volcanic rocks still more ancient, without any discernible craters or scoriae.
>
> (Lyell 1830: 59)

Likewise, in developing and explaining Hutton's ideas, John Playfair emphasized the efficacy of fluvial erosion (Chorley *et al.* 1964: 60). Indeed, Playfair's law of accordant tributary junctions was a powerful demonstration of the central role of fluvial erosion in landscape evolution, as well as of drainage basin order and adjustment, concepts that have resonated down the decades from Powell to Davis to Strahler to Leopold and Wolman to Hack.

If Davis is a Colossus standing astride the gateways of the discipline of landscape evolution, he is standing – to mix the metaphor somewhat – on the shoulders of the giants who had preceded him. In the late eighteenth century, the French hydrologist Du Buat wrote of Infancy, Youth, Middle Course ('serious and wise') and Old Age of a river system, terminology remarkably prescient of the Davisian scheme (Chorley *et al.* 1964: 88). The idea of landform 'age' was certainly not new with Davis, and Chorley *et al.* (1964: 455–6) have highlighted several nineteenth-century workers who expressed a relationship between a landform and its age. Another very important immediate precursor was Major John Wesley Powell whose great contributions to the critical concepts in landscape evolution include antecedent valleys, superimposed valleys and consequent valleys (Powell 1875). He also coined the term 'base level' for the fundamental control on vertical erosion, sometimes expressing it in terms reminiscent of Playfair's law of tributary junctions:

> There is a limit to the effect of these conditions [of denudation], for it should be observed that no valley can be eroded below the level of the principal stream, which carries away the products of surface degradation . . . We may consider the level of the sea to be a grand base level . . . but we may also have, for local or temporary purposes, other base levels of erosion, which are the levels of the beds of the principal streams which carry away the products of erosion.
> (Powell 1875: 163, 203)

Thus was the stage set for William Morris Davis, with whose name many of the critical concepts in landscape evolution are linked. One of Davis's key contributions was to integrate much of the insight that preceded him into the cycle of erosion.

Davis's cycle of erosion

W.M. Davis (1899) The geographical cycle. *Geographical Journal* 14: 481–504

Chorley *et al.* (1973) devoted the whole of their 874-page book to the commanding figure of Davis (see also Summerfield 2000a). One of Davis's key contributions, the geographical cycle or the cycle of erosion, came quite early in his career. The cycle is the sequential development of landforms towards base level as a result of sub-aerial denudation. It is triggered by a short, sharp uplift event that is followed by tectonic quiescence and slow, steady sub-aerial denudation. The cycle's emphasis on the operation of uniform processes thus embodies an extreme form of Lyellian uniformitarianism (cf. Chorley *et al.* 1964).

Many critical concepts in landscape evolution are associated with the Davisian cycle, including the notion of a cycle itself, and the cycle's emphasis on the primacy of sub-aerial denudation (as opposed to marine denudation – see Davis 1896). Base-levelling, slope decline and the 'age' terms that have come to be associated with the different stages of the cycles, namely, youth, maturity and old age, are also important contributions, as are the concepts of the peneplain and the monadnock. The concept of the peneplain, the low relief erosional surface that characterizes the geographical cycle's end stage ('old age') is a key concept in geomorphology, not least because uplifted peneplains are perhaps the most frequently identified element in reconstructions of regional landscape evolution. Key characteristics of a peneplain are its low relief, its surficial cover of deep and well-weathered ancient soils and regolith, and the way in which its surface morphology truncates or bevels geological structure. In the Davisian cycle, peneplains are the end stage of prolonged base-levelling and are formed close to sea-level. The identification of such low-relief surfaces at high elevations therefore implies long-lasting sub-aerial denudation (to form the peneplain) followed by uplift (to elevate the peneplain to the higher elevation). Uplifted peneplains have been identified almost countless times in landscape evolution studies and used to infer tectonic stability and activity.

Davis's anthropomorphization of landscape evolution in a life cycle reflected to some extent the nineteenth-century's embracing of Darwinism (but see Stoddart 1966). More formally, Davis envisaged landscape evolution as a function of his 'trio of geographic controls': 'Structure, Process and Stage'. The concept of 'Structure' embodies lithology and geological structure, and the ways in which they influence landscape evolution; the term 'Process' covers the weathering and erosion responsible for that evolution. For Davis, the third geographic control, 'Stage' or the passage of time, was the principal control that comes to dominate in landscape evolution, so that irrespective of structure, landforms are a function of their stage in the cycle. A landscape that exhibits the old age morphology of the peneplain must have passed through all the preceding stages in the cycle, with successive stages being indicated by progressively lower slope angles. In this way, landscape morphology was a direct measure of its age in terms of the geographic cycle. The Davisian viewpoint of progressive slope decline with the passage of the cycle contrasts with other schemes that envisage landscape evolution as being accompanied by retreat of slopes at a constant angle (see p. 7).

Early critics of the geographical cycle included Shaler (1898) and Tarr (1898); Chorley (1965), Flemal (1971) and Bishop (1980) are three of the many who have summarized these and subsequent criticisms. The early criticisms were, in effect, brushed aside by Davis (1899b). His subsequent papers (e.g. Davis 1902) sought to cement the position of the geographical cycle as the key to understanding landscape evolution. The cycle came to be so central to American geology more widely because landform morphology became a key clue to Earth history. Theories of orogenesis, for example, were judged on their consistency with the geographical cycle: the principal tectonic or orogenic cycles had to be episodic because uplifted peneplains 'proved' that the Earth evolved via cycles of uplift and stability (for peneplanation) followed by another episode of uplift (see Bishop 2004 for further discussion of this). So, much Anglophone landscape history in the first half of the twentieth century (particularly in the USA and eastern Australia – see Young 1977, 1978 and Bishop 1982 for the latter) was dominated by the Davisian approach building on the critical concepts of short periods of uplift, slope lowering and peneplanation to a base-level, and the linking of river channel morphology and stage within the geographical cycle.

The classic mid-1950s work of Wooldridge and Linton (1955) in this area (see pp. 5–8) is a relatively minor revision and 're-touching' of their 1930s ideas (Wooldridge and Linton 1939) and is built squarely on a Davisian foundation. Elsewhere in the research community, however, dissatisfaction with the application of the Davisian critical concepts had already become widespread by the mid-1950s, with Davis coming to be caricatured by some 'as an old duffer with a butterfly-catcher's sort of interest in scenery' (Mackin 1963: 136). The dissatisfaction was embodied in Strahler's (1952) call for radical change and the embracing of a whole new approach and set of underpinning concepts: 'If geomorphology is to achieve full stature ... operating upon the frontier of research into fundamental principles and laws of earth science, it must turn to the physical and engineering sciences and mathematics for vitality which it now lacks' (Strahler 1952: 924; see also Strahler 1992). Strahler would no doubt have agreed that Davis's conceptualization of landscape morphology and evolution in terms of structure, process and time (or stage) 'succinctly expressed the fundamental truth of geomorphology' (Brown 1960: 1), but equally he would have argued that clearer articulation of these three concepts into testable mechanisms that controlled landscape evolution was also required.

Denudation chronology

S.W. Wooldridge and D.L. Linton (1955) The Mid-Tertiary cycle of erosion and the summit peneplain. *Structure, Surface and Drainage in South-East England.* London: George Philip, pp. 34–44.

This chapter from a classic book incorporates many of the critical concepts associated with the reconstruction of the erosional development of the Earth's surface and so is representative of the considerable body of work devoted to such denudation chronology in the first half of the twentieth century. Wooldridge and Linton noted that early British workers tended to favour a marine origin for uplifted erosional plains (e.g. Ramsay 1846; Jukes 1862), and that this viewpoint survived William Morris Davis's arguments to the contrary (Davis 1895). George (1966), for example, was a relatively recent proponent of a marine origin for uplifted erosion surfaces.

Wooldridge and Linton's (1955) work demonstrates very clearly the ways in which uplifted erosion surfaces were reconstructed, often by projections of surfaces across rather long distances. They also present detailed argument for the ages of surfaces, dating the surface by the way in which it locally passes beneath dated marine deposits. This argument depends of course on confidence in the long-distance projections on which the reconstructions of the surface are based. Wooldridge and Linton also highlight the way in which different parts of a surface may have marine or sub-aerial origins depending on the palaeogeography. They thereby avoid the artificial polarization of the debate between marine and sub-aerial origins for planation surfaces (as well as highlighting the requirement that planation, be it marine or sub-aerial, must be consistent with the known palaeogeography). Sissons (1967) makes a similar point in his comment on the palaeogeographic difficulties of George's hypothesis (1966) of a marine origin for the Scottish upland surfaces. Despite their balanced treatment of the origins of surfaces, Wooldridge and Linton (1955) do seem generally to prefer a sub-aerial origin for the erosion surfaces of south-east England, citing, for example, the local presence of sub-aerial residual soils of the type to be expected on a peneplain and the lack of marine erosional features. Widespread marine incursion *is* invoked to explain aspects of the development of the regional drainage system.

D.K.C. Jones has been a key dissenter from the Wooldridge and Linton interpretation and has provided a very useful summary of the development of ideas in this area (Jones 1999a; see also Jones 1999b). The differences in detail of the Wooldridge and Linton and Jones interpretations of the evolution of southern England are not of as overriding concern to us here as the difference in approach. Thus, the characteristically Davisian tectonic stability required to produce extensive erosion surfaces is not particularly in evidence in southern England in Cenozoic times. There is little evidence for the widespread marine incursion invoked by Wooldridge and Linton, and Jones (1999a) showed clearly how this marine incursion came to be suggested as regionally significant on the basis of supposed links between regional marine transgressions and the development of drainage networks that are discordant with structure. This link had been stressed to Linton by Douglas Johnson, one of Davis's most faithful disciples and key advocates. One of Jones's

(1999a) key criticisms of Wooldridge and Linton is their over-reliance on morphological evidence and he is particularly critical of their treatment of soils. Indeed, Jones (1999a) makes strong argument that greater weight be placed on the evidence available from 'materials' (soils and sediments).

Brown (1960) followed Wooldridge and Linton (1955) in basing his interpretation of the evolution of Wales on the Davisian approach. This approach was clearly being put to one side in the 1950s and 1960s, both by those who acted on Strahler's (1952) call for more process-based work, as well as by those who preferred a dynamic equilibrium-based approach to landscape evolution (see pp. 9–11). The fact remains, however, that low relief upland surfaces are clearly present in the landscape and challenge us to be explained. As Brown remarked at the end of the 1970s: 'the High Plateau, the Middle and Low Peneplains [in Wales] are still there and have not been eliminated either physically or mentally since I first mapped them in the 1950s' (Brown 1979: 459). The origin of these high elevation, low relief surfaces remains unresolved and even enigmatic. Lester King is widely identified with a viewpoint opposed to that of Davis, namely, that such low relief surfaces form close to base-level by pediplanation involving parallel retreat of slopes (rather than Davisian downwasting). His detailed writings, however, argue that both peneplanation (in the Davisian sense) and pediplanation may operate in landscape evolution, depending on the combination of climate and rock characteristics (and hence dominant processes) (King 1962, chapter V). It is true, nonetheless, that King did emphasize parallel slope retreat and pediplanation as the dominant landscape process – as did Penck (1953) – an emphasis in King's case that perhaps grew out of his considerable familiarity with semi-arid Gondwanan landscapes (see also pp. 21–22).

Etchplanation has latterly received more attention as a mechanism to explain low relief surfaces (Büdel 1982; Thomas 1989a, 1989b). Etchplanation involves weathering of the landsurface, perhaps deeply, and the removal of that weathered regolith to give a low relief surface that more-or-less mimics the weathering front between regolith and relatively unweathered bedrock (Mabbutt 1961). Etchplanation has been widely invoked as a powerful mechanism of relief lowering, not least because many low relief surfaces bear considerable evidence of deep weathering. Fairbridge and Finkl (1980) invoked etchplanation to explain low relief surfaces in stable cratonic areas, and the mechanism has also been widely used to explain deeply weathered, low relief surfaces in other settings (e.g. northern Britain: Hall 1991; Scandinavia: Lidmar-Bergström 1995, 1999; central and northern Europe: Migon and Lidmar-Bergström 2001). Indeed, Jones (1999a) has invoked etchplanation as a key process in his re-interpretation of Wooldridge and Linton's (1955) history of southern England.

A key, but widely neglected, issue related to the origin of high elevation, low relief surfaces is the elevation of their formation. Implicit in many, and probably most, treatments of high elevation surfaces is the concept that they must have formed (by whichever mechanism that is invoked) at low elevations, close to sea-level, and then have been subsequently uplifted. Young (1977) has argued, however, that high elevation, low relief surfaces may be graded to resistant lithologies which may have controlled the level of planation. Thus, it is argued, planation can occur at any elevation and a high elevation, low relief surface cannot necessarily be interpreted as indicating planation followed by uplift. This notion undoubtedly introduces a further complexity to studies of landscape evolution, but it may be that this complexity may not be so relevant when dealing with areally extensive upland surfaces, especially when local resistant base-levels to which the upland surface is graded cannot be identified. Moreover, apparent grading of an upland surface to a perched base-level does not preclude planation to that base-level at a lower elevation and subsequent uplift of the whole ensemble.

Tors

D.L. Linton (1955) The problem of tors. *Geographical Journal* 121: 470–87

Tors are a relatively minor landform but this paper is included for several reasons. Linton's first important contribution was to attempt to provide an overarching explanation for tors in terms of a two-stage process involving initial (deep) weathering along joint lines and subsequent removal of the interstitial weathered material to expose the tors. The common association between tors and low relief (upland) surfaces, and Linton's obvious reliance on the Davisian approach (as well as Davis's suggestion that low relief surfaces are likely to be characterized by deeply weathered residuum), help to explain the likely origin of Linton's interpretation. Hitherto, tors had been explained by local effects such as variable rock resistance to weathering and/or development of the details of the tor structure after the rocks had emerged and been exposed to atmospheric processes (e.g. Palmer 1955). Tors were also explained by periglacial processes (e.g. Common 1954), such explanations appearing even after the publication of Linton's two-stage explanation (e.g. Pullan 1959), but this explanation appears not to have found favour (e.g. Bird 1967). Haines-Young and Petch (1986: 162) highlight some of the methodological issues associated with theories of tor formation.

Linton's second major contribution to landscape evolution was that his two-stage evolutionary sequence for tor formation implies formerly warmer and more humid climates for the deep weathering (or 'rotting') phase. This implication was partly by analogy with the deep weathering observed in lower latitudes and partly by deduction that for tors now found at high elevations in middle to high latitudes, a snow and ice cover would not be conducive to deep rotting of rock. Thus, the tors in the Scottish Highlands have been linked to warmer and wetter pre-glacial climates (Hall 1985, 1986).

The common presence of tors on upland surfaces, such as the Scottish Highlands (e.g. Hall 1991) and the Scandinavian uplands (e.g. Olvo *et al.* 1999), raises a third important issue: the relationship between tors and glaciation. Linton commented: 'it seems necessary to regard parts of these uplands, with their tors . . . as essentially unglaciated during the last Ice Age, though doubtless deeply covered by névé' (Linton 1955: 479). In other words, how could tors, formed in part by deep pre-glacial bedrock rotting, survive the subsequent glaciation(s)? Linton resolved this apparent contradiction by implying that the northern Britain tors were not covered by the British ice sheet. It is clear, however, that the British ice sheet did cover the tors of northern Britain, and Sissons (1967) therefore reiterated Linton's observation that tors are absent from the most intensely glaciated west (presumably having been removed by intense glacial erosion) and by further observing that those tors that are present tend to be in locations that were sheltered from the advancing ice sheet. These explanations are not entirely satisfactory, however, and indeed limited glacial erosion of the northern shields had been argued since the nineteenth century, principally on the basis of the occurrence of remnants of pre-glacial saprolites (Lidmar-Bergström 1997). Ice sheets may be minimally erosive because they are locally frozen to their substrate (Hall and Sugden 1987; Dyke 1993; Kleman 1994). Such cold-based ice protects pre-glacial features, such as periglacial landforms (e.g. Kleman and Borgström 1994) and pre-glacial weathering features (e.g. Dyke 1983; Dyke *et al.* 1992). The identification of minimally erosive cold-based ice has revolutionized glacial studies (see Volume III of this series) in that landform palimpsests are being recognized beneath the veil of the most recent glacial features (e.g. Hättestrand and Stroeven 2002; Johansson *et al.* 2001; Stroeven *et al.* 2002).

Dynamic equilibrium

J.T. Hack (1960) Interpretation of erosional topography in humid temperate regions. *American Journal of Science* 258-A: 80–97

This important paper introduced (or at least synthesized and re-emphasized) several critical concepts in landscape evolution. Hack took a classic area of Davisian analysis, the Appalachians, and used ideas of Davis's contemporary Grove Karl Gilbert (1877) to argue that its evolution may be interpreted in terms of a dynamic equilibrium between erosional energy and rock resistance (see also Hack 1979). In other words, once an equilibrium is established between erosional energy and rock resistance, this equilibrium landscape persists and, unless the energy applied to the system changes, the landscape downwastes with a constant morphology, resulting in time-independent landforms. Whereas Davis had argued that morphology changes with the passage of time, and the influence of lithology declines as the geographical cycle progresses, Hack argued for the opposite: the persistence of lithological influence and the constancy of morphology.

Thorne and Welford (1994) have argued that the concept of dynamic equilibrium has been used in various confusing (and confused) ways but still find value in the concept (see Kennedy 1994 for a counter viewpoint). Assessment of whether a landscape is evolving under conditions of a Hack-type dynamic equilibrium (or, indeed, according to a Davisian cycle) is a qualitative, or at best semi-quantitative, exercise, as Hack (1975) makes clear. Bishop et al. (1985), for example, argued that Hack-type dynamic equilibrium may be demonstrated in inland SE Australia by the apparent adjustment of fluvial long profiles to lithology and the approximate equivalence between rates of river incision and catchment-wide denudation (i.e. catchment-wide downwasting). It is now clear, however, that the long profiles are characterized by much disequilibrium, presumably precluding a Hack-type interpretation (Bishop and Goldrick 2000).

The reason for the inapplicability of a Hack-type explanation for low energy settings, such as the passive margin uplands of inland SE Australia, is presumably the insufficiency of erosional energy and sediment load to maintain equilibrium throughout the landscape. On the other hand, a type of dynamic equilibrium characterizes rapidly uplifting areas, such as the Southern Alps or the Himalayas, as has been demonstrated as a result of recent advances in age dating, including in low temperature thermochronology and isotopic systems such as cosmogenic isotope analysis. This requires steep valley-side slopes and river gradients, and high sediment loads, to generate the high incision rates that match uplift rates, as is found in the Southern Alps (Adams 1980, 1985) and the Himalayas (Burbank et al. 1996). Although not precisely the type of dynamic equilibrium envisaged by Hack ('steady-state' is commonly used – e.g. Whipple 2001; Willett and Brandon 2002), it does allow the differentiation of time-independent landforms and time-dependent landforms. Adams (1985) has persuasively argued that until uplift rates are sufficient to promote rapid incision and mass hillslope wasting, remnants of former high elevation low-relief surfaces may be preserved (as in eastern South Island, New Zealand, where the Pacific plate starts to be uplifted as it moves toward the Alpine Fault). These flat-topped hills and mountains are time-dependent landforms (they preserve evidence of their age and pre-uplift morphology) and contrast with the so-called 'spiky' mountains that characterize the higher uplift areas once the flat-topped remnants have been completely eroded. Within this zone of more rapid uplift close to the Alpine Fault, these spiky mountains have the same time-independent morphology irrespective of their distance from the fault and the zone of maximum uplift (Adams 1985). Burbank *et al.* (1996) have shown how slope morphology and process are closely adjusted to lithology in very rapidly uplifting areas in the Himalayas, but the adjustment is to the strength of the rock rather than to a more conventional measure of erodibility. Thus, tectonics, lithology and climate may interact to produce time-dependent and time-independent landscapes, apparently reconciling some of the apparent tensions between Davisian and Hack-type approaches to landscape evolution.

Evolutionary geomorphology

C.D. Ollier (1979) Evolutionary geomorphology of Australia and Papua–New Guinea. *Transactions of the Institute of British Geographers* 4: 516–39

The term 'evolutionary geomorphology' has been used in at least two distinct ways. Thornes (1983) used the term to describe a way of conceptualizing geomorphological processes and their interactions that does not rely on equilibrium concepts. It highlights thresholds and complex, even chaotic, behaviour of the geomorphological systems (cf. Phillips 2003). Thornes' signalling (1983) of the need to move away from simple equilibrium notions was largely concerned with shorter term processes and landforms, and is therefore of less relevance to our concern here with landscape evolution.

Ollier's (1979) paper, which requires only a short introduction here, argues that cyclical and dynamic equilibrium approaches to landscape evolution are not appropriate for many parts of the Earth's surface, such as less humid areas experiencing low rates of tectonic activity, and/or areas that have not experienced Quaternary glaciation. Settings such as these have been characterized by rates of denudation that would be considered low elsewhere and a longer time scale is therefore required for landscape evolution. Ollier (1979) has argued that the starting point for much landscape evolution in areas that experienced Quaternary glaciation is that glaciation, whereas this approach is inappropriate for areas outside the Quaternary glaciated area. In many of these latter areas, such as southern Africa and south-eastern Australia, the appropriate starting point for landscape evolution is likely to be the Permian glaciation, 250 million years ago.

This viewpoint means that cyclical approaches are inappropriate for such settings and that landscape evolution should be considered much more as a directional process (i.e. as 'evolutionary'). The counterpoising of 'directional' and 'cyclical' has been highlighted by Gould (1987) as providing complementary frameworks for thinking about Earth history. Thus, Ollier's (1979) notion of evolutionary geomorphology offers more than just another viewpoint on the appropriate theoretical framework for landscape evolution.

INTRODUCTION

The equilibrium fluvial long profile

J.H. Mackin (1948) Concept of the graded river. *Bulletin of the Geological Society of America* 59: 463–511

The concept of the concave-up equilibrium long profile had been identified and explained in the late seventeenth century by the Italian Domenico Gugliemini, and developed in some detail in the mid-nineteenth century by the French civil engineer Alexandre Surell (including the concepts of thalweg and grade) and then slightly later by the American James Dwight Dana (Chorley *et al.* 1964). The notion of an equilibrium between lithology and stream slope and hillslope was postulated by Dausse in the 1870s and this idea was extensively developed by Hack (1957, 1973, 1975), tying the concept to his concept of dynamic equilibrium.

There is an immense literature on the equilibrium long profile, with major twentieth-century statements or reviews by Davis (1902), Baulig (1926), Kesseli (1941), Mackin (1948), Woodford (1951), Dury (1965) and Knox (1975). Chorley (2000) reviewed Mackin's (1948) contribution, as well as the wider history of various approaches to the concave-up equilibrium long profile, emphasizing the ways in which Mackin integrated, often in a deductive way, the multitude of factors that determine long profile morphology.

The concave-up equilibrium (graded) stream long profile has been used in a wide range of geophysical, geomorphological and geological interpretations of landscape evolution in bedrock terrains. These include (with only one of many possible example references in each case) identification of neotectonic activity (Merritts *et al.* 1994); interpretations of post-breakup landscape evolution and tectonic history on passive continental margins (Pazzaglia and Gardner 1994), in other intraplate settings (McKeown *et al.* 1988) and in tectonic collision zones (Brookfield 1998); and reconstructions of other base level changes (Reed 1981). In the majority of cases, the existing long profile is assessed for departures from an equilibrium form of the long profile, with these departures (disequilibria) generally being interpreted in terms of base-level changes. Mackin (1948) stated that the graded longitudinal profile 'cannot be a simple mathematical curve in anything more than a loose or superficial sense' (p. 491), but all studies using long profiles in landscape evolution studies have relied on some mathematical form of the equilibrium long profile. This would seem to be especially important if assessment of base-level changes, for example, requires downstream projection of the long profile.

The first approach to the long profile equilibrium form involves no *a priori* assumption as to its form, and derivation of this equilibrium form by some form of curve fitting. Jones (1924) fitted power functions to his long profile data to formulate the equilibrium long profile form (see also Woodford 1951), and Hovius (2000) has reported a range of mathematical best-fit forms, including logarithmic, exponential and power function, as well as long profiles that are not well fit by any particular mathematical form. Recently, such curve fitting has been guided by use of the stream-power law (e.g. Whipple *et al.* 2000; Kirby and Whipple 2001). The second, and more common, approach has been to use Hack's semi-log relationship (1973) between downstream distance and channel elevation (the SL long profile form) as a theoretical model of the equilibrium long profile against which actual profiles have been assessed (e.g. Hack 1973, 1982; Keller 1977; Reed 1981; Seeber and Gornitz 1983; McKeown *et al.* 1988; Merritts and Vincent 1989; Fried and Smith 1992; Goldrick and Bishop 1995; Brookfield, 1998; Ramírez-Herrera 1998). Other possible *a priori* forms of the equilibrium long profile include the DS form (e.g. Brocard *et al.* 2003). The DS form includes Hack's SL form as a special case (Bishop and Goldrick 2000). The concave-up equilibrium long profile form is one of the longest known elements in landscape evolution, but its precise form and significance still require further elucidation, including with reference to the respective roles of climate and tectonics in long profile evolution (Pazzaglia *et al.* 1998; Roe *et al.* 2002).

Landscape antiquity

C.R. Twidale (1976) On the survival of paleoforms. *American Journal of Science* 276: 77–95

Ancient landform elements are commonly preserved in the present landscape, and these are not well accommodated by models of landscape evolution that incorporate landscape downwasting as a central element, such as the models of Davis (1899) and Hack (1975). Such remnants should be destroyed by the erosional downwasting central to such models, but we have already seen how Wooldridge and Linton (1955) demonstrated antiquity dating back into the Tertiary for elements of the SE England landscape. Other models of landscape evolution, such as the classic models incorporating parallel retreat of slopes (e.g. King 1962) and the model of landscape development under regimes of unequal activity (Crickmay 1975), do incorporate the preservation of ancient, relict landscapes. Twidale's paper is the first of several over the last quarter of a century which have developed this point, building on the contributions of the somewhat neglected Crickmay (1933, 1974, 1975) (e.g. Twidale 1991, 1998). Twidale's viewpoint has been considerably influenced by his many decades of research in Gondwanan landscapes where remnants of very ancient elements are preserved, and is in part a reaction to a widely held (mainly northern hemisphere) viewpoint that landscapes cannot be much older than the Tertiary and are probably no older than the Pleistocene (e.g. Thornbury 1964; Brown 1980; cf. Young 1983).

Crickmay's (1974, 1975) Hypothesis of Unequal Activity was built on his observation that fluvial incision is a far more effective agent of landscape fashioning than hillslope processes. This observation, which Crickmay took as self-evident (Crickmay 1975), means that rivers must deepen their valleys and that landscape must increase in relief through time. Twidale (1991) has argued that the increase in relief in many settings over the last 60 to 100 Ma is a clear expression of this principle. Twidale has not argued that increasing relief should characterize all settings, but this very observation implies the corollary, namely, that the grand landscape evolution schemes of Davis and Hack cannot likewise be expected to apply to all settings. We have already made a similar point in relation to the time-dependent and time-independent landforms of New Zealand's South Island.

More generally, this paper points to the importance in landscape evolution studies of preserved remnants of ancient landscapes. Except in rapidly eroding terrains, landscapes are likely to be composite of elements of different age, and as these elements get progressively older, they become progressively less likely to be eroded (i.e. more likely to persist). As Brunsden's tenth proposition has expressed it: 'The ability of a landscape to resist impulses of change tends to increase with time' (Brunsden 1990: 27). It remains unclear, however, whether it is possible for landscapes to have survived at the Earth's surface since, for example, the Cambrian (Stewart *et al.* 1986; Ollier *et al.* 1988). Even at the vanishingly low rate of denudation of 0.14 m.Ma^{-1} reported from frigid, arid Antarctica (Summerfield *et al.* 1999), a land surface would still have been lowered 70 m in the *c.*500 Ma since the Cambrian and more than that at more usual erosion rates. This extreme persistence seems unlikely.

Ancient landscape elements may persist despite the apparently intense erosion of an ice sheet; Crickmay and Twidale offer another perspective on this persistence via their emphases on the spatial variability of erosional intensity, reinforced by positive feedbacks. Twidale also emphasized a range of other factors such as structural influences (see also Twidale 1998). Ancient landscapes may also be present at the Earth's surface because of re-emergence from beneath cover rocks that have been removed by erosion. (Geomorphologists have termed this process 'exhumation', a term which recently has also come to describe rock uplift by sub-aerial and tectonic denudational process – see discussions by Summerfield 1991: 371ff and Burbank and Anderson 2001: 131ff of distinctions between (bed)rock uplift and surface uplift.) These exhumed surfaces are, of course, unconformities and are most clearly indicated when surfaces retain remnants of the formerly existing cover which has been removed to expose the surface, especially when it is clear that these cover rocks must formerly have been more continuous. Such exhumed surfaces are commonly more or less planar (a common finding when the overlying sediments show that the surface is marine).

The exhumation interpretation is therefore commonly used to account for surfaces of low relief at the Earth's surface, and it must be noted that this explanation does not avoid the necessity to account for the initial formation of the low-relief surface, especially if the signs of marine origin are absent, but merely displaces this necessity back in time. There is a large literature on exhumed unconformities, and Twidale (1998) has provided many examples. Lidmar-Bergström (1996) argued that the land surface formed on Precambrian basement in Sweden is a mosaic of old, exhumed surfaces, including Precambrian, sub-Cambrian and sub-Mesozoic surfaces. Of course, it is not necessary that an exhumed surface be planar, and so surfaces with relief may equally be exhumed, as Lidmar-Bergström (1987), for example, has shown. Examples abound of such exhumation of ancient topography. Two examples are the considerable relief (more than 600 m) on the exhumed sub-Torridinian surface of the Lewisian Gneiss in NW Scotland (Sissons 1967: 9) and the exhumed Cretaceous valleys of northern Australia (Nott 1995).

Fixing the age of exhumed surfaces, a key issue in understanding landscape evolution, is not always straightforward (Watchman and Twidale 2002). The overlying sedimentary cover provides age constraints on the age of formation of the surface but this cannot logically be the age of subsequent re-exposure of the surface as an element in the 'modern' landscape. Likewise, long-distance correlations between sub-aerial landsurface and sediment-covered unconformity (e.g. King 1962) attempt to date the age of formation of the surface and not its re-exposure. Watchman and Twidale (2002) have exhaustively reviewed the ways of dating such surfaces.

New isotopic techniques are revolutionizing the dating of surfaces in two distinctive ways. Cosmogenic isotope analysis enables for the first time the determination of the age of exposure of a bedrock surface (e.g. Lal 1991). The cautionary notes of Watchman and Twidale (2002) on methodological issues associated with the use of cosmogenic isotope analysis for the absolute age dating of surfaces are welcome but mainly signal the need for a thoughtful, carefully structured and rigorous sampling programme. Related issues underpin the apparent tensions and even contradictions between thermochronological and geomorphological (or geological) approaches to dating land surfaces (Kohn and Bishop 1999; Hill 1999). Low temperature thermochronometers, such as (U-Th)/He analysis in apatite (Ehlers and Farley 2003), offer new ways of dating major, long wavelength topography by exploiting the way in which this topography deforms the thermal structure of the crust (e.g. House *et al.* 1998; Persano *et al.* 2002).

Denudational isostasy

A.R. Gilchrist, M.A. Summerfield and H.A.P. Cockburn (1994) Landscape dissection, isostatic uplift, and the morphologic development of orogens. *Geology* 22: 963–6

The early applications of the concept of isostasy in landscape evolution seem to have been largely related to the isostatic implications of glacial loading and unloading, and subsidence and rebound as a result of fluctuating Pleistocene lake levels (e.g. Gilbert 1890). Such applications remain fundamental in understanding medium- to long-term landscape evolution in formerly glaciated terrains (e.g. Nansen 1922; England 1983; Smith and Dawson 1983; Guilcher *et al.* 1986).

The concept of denudational isostasy is also important for landscape evolution, but has been somewhat neglected until recently. Denudational isostasy in areas that are in isostatic equilibrium means that, for every metre that is removed from the Earth's surface by denudation, the Earth's surface is uplifted by about 0.8 m (0.8 being the approximate ratio of the densities of the crust and sub-lithospheric mantle). Denudational isostasy means that the rate of surface lowering as envisaged in most grand schemes of landscape evolution is radically slower than was envisaged. If this denudational isostatic rebound effect was acknowledged, it was then generally assumed away. Thus, Davis either ignored it as a factor in landscape evolution or was antipathetic to it (Davis 1910).

A further important implication of the concept of denudational isostasy flows from the fact that denudation may not be uniform across a region. The eminent geophysicist Harold Jeffreys seems to have been the first to argue for the role of denudational isostasy in the uplift of mountain peaks (Jeffreys 1931). Within a few years, Wager (1937) had applied this idea to aspects of drainage development and mountain peak uplift in the Himalayas, and this idea presaged by more than half a century the recent fruitful exploration and development of this concept. Molnar and England (1990) suggested that the isostatic response to valley incision may lead to mountain peak uplift, and a pair of papers in 1994 – Montgomery (1994) and the paper by Gilchrist *et al.* (1994) reproduced here – used numerical modelling to explore the notion of mountain peak uplift as a result of localized valley incision. Such peak uplift depends essentially on the strength of the lithosphere, so that local erosional unloading is isostatically accommodated regionally (see also Pinter and Brandon 1997). Thus, whereas an overall erosional lowering and decrease in mean elevation must accompany denudation (because under conditions of isostatic equilibrium isostatic recovery can only ever be about 80 per cent of that which was lost by denudation), elevations on uneroded mountain peaks may increase due to *regional* isostatic response to *local* incision. In effect, denudational isostasy means that surface uplift may not necessarily be tectonically induced, but may be the isostatic reponse to denudation (Small and Anderson 1995). This effect is limited, however, and Gilchrist, Summerfield and Cockburn (1994) show that it does not seem capable of generating the mountain peak altitudes envisaged by Molnar and England (1990) (cf. Whipple *et al.* 1999).

King (1955) and Pugh (1955) argued that the crustal unloading associated with escarpment retreat on continental margins such as southern Africa would lead to isostatic uplift of the margin. This uplift, they suggested, would be episodic and would account for stepped pediplain surfaces between the continental margin highlands and the sea. Denudational isostasy is generally understood to be a continuous response to unloading (except, perhaps, if the isostatic uplift is accommodated by faulting) and such episodicity in pediplanation would not generally be expected. Nonetheless, the concept of denudational unloading of continental margins has been extensively explored in recent decades. Low temperature thermochronological data from passive continental margins (using apatite fission track analysis and uranium-thorium/helium analysis in apatite) indicate kilometre-scale denudation along continental margins during continental break-up and shortly afterwards (Dumitru *et al.* 1991; Brown *et al.* 2002; Persano *et al.* 2002). The flexural uplift of these continental margins may not be as was suggested by King (1955) and Pugh (1955), but the differential erosion/ unloading associated with the concentration of denudation along the coastal plain of high elevation passive margins means that denudational flexural uplift should characterize such settings. This expectation has apparently been confirmed by several numerical models (e.g. Gilchrist and Summerfield 1990; Gilchrist, Kooi and Beaumont 1994; Kooi and Beaumont 1994; Tucker and Slingerland 1994). Such flexural rebound is an important element in maintaining passive margin relief and exhibits a powerful influence on landscape evolution and the development of drainage systems. Numerical models indicate that the escarpment lip, the topographic expression of which is enhanced by flexural rebound, most commonly evolves to become the principal drainage divide between seaward and inland river systems. The divide is, in effect, pinned as a result of the flexural rebound. Most high elevation passive margin escarpment lips are in fact the major continental drainage divide, apparently confirming the importance of this isostatic rebound mechanism.

The plate tectonic revolution

D. Griggs (1939) A theory of mountain-building. *American Journal of Science* 237: 611–50

The plate tectonic revolution in the 1960s and 1970s was no less revolutionary for understanding landscape evolution than it was for the earth sciences more generally. This is because plate tectonics is generally considered to provide an overall integration of tectonics and topography in a way that had been attempted, unsuccessfully, by Davis. One of the key advances has been to provide an explanation for the development of mountains, although there remain dissenters (e.g. Ollier and Pain 2000). Key to this is the notion of isostasy. Beckinsale and Chorley (1991) have summarized the development of the concept of isostasy, a term coined in the late nineteenth century by the American physiographer Clarence Edward Dutton to describe the hydrostatic balance or equilibrium in the Earth's crust, whereby the crust floats in the deeper sub-crustal substrate (Dutton 1889; see also Summerfield 1991).

Mountains have high elevation because they are supported isostatically on deep crustal roots; the mountain belt is associated with a gravity anomaly that indicates downfolding of lighter crust into a denser substrate. How this deep root of crustal material is formed is therefore a key issue in understanding the development of Earth surface topography. Griggs's (1939) paper made the crucial link between the deep crustal roots and sub-crustal convection currents. The notion of sub-crustal convection had been developed by Holmes (1928–9), and Bull (1931) had suggested that convection currents may split continents (e.g. Africa). Bull's comment that 'this is in agreement with the statement that the rift valleys and faults of Africa indicate a tension outwards' (1931: 496) clearly foreshadowed later plate tectonic interpretations of the East African Rift.

Much of the Earth's macrogeomorphology can now be satisfactorily interpreted using a plate tectonics-based overarching framework for landscape evolution (cf. Summerfield's chapters [1991] on Global morphology and tectonics, Landforms and tectonics of plate boundaries, Landforms and tectonics of plate interiors; see also Ollier 1985). More recent contributions elaborate the detail of these settings and various interactions (see various chapters in Summerfield 2000b).

Numerical models that couple tectonics and surface processes over tens of millions of years have revolutionized research into the links between tectonics and topography in a plate tectonic framework, especially, but not exclusively, for high elevation passive continental margins (e.g. Kooi and Beaumont 1994; Tucker and Slingerland 1994; Beaumont et al. 1996, 2000; van der Beek and Braun 1999; van der Beek et al. 2002). The numerical models suggest that surface uplift triggers increases in orographic precipitation, which, coupled with glaciation if surface elevations become sufficient, leads to enhanced erosion, which results in turn in enhanced rock uplift as a result of denudational isostasy (e.g. Beaumont et al. 2000). In effect, the crust is 'sucked' through the Earth's surface, perturbing the pressure and temperature conditions at depth in the crust and thereby playing a critical role in rock metamorphism. Thus, the phrase, 'the geomorphology of metamorphism', has recently been used (Zeitler et al. 2001), and the comment of Hoffman and Grotzinger (1993) elegantly captures the same point: 'Savour the irony should those orogens most alluring to hard-rock geologists owe their metamorphic muscles to the drumbeat of tiny raindrops' (p. 198).

A major implication of the interplay between tectonics and erosion is that surficial processes, including fluvial and glacial erosion, and even meteorological processes, are now seen to be capable of 'driving' tectonics, and these interactions can be assessed using a range of field- and laboratory-based tests from numerical models (Beaumont *et al.* 1992, 2000) and physical models (Bonnet and Crave 2003). For example, these models have suggested (with ways of testing) that the side of the orogen that experiences the maximum orographic precipitation is important in terms of rates and styles of denudation, crustal exhumation (rock uplift) and metamorphism (Beaumont *et al.* 1992, 2000). Thus, plate tectonics has provided a mechanism that elegantly integrates deep Earth processes and surficial geomorphology, and so landscape can again be used to test tectonic models, as it was in Davis's time (cf. Chamberlin 1909: 689: '[Davisian] base-leveling has come to play such a large part in our science, for it is clear that the doctrine of base-leveling is specifically inconsistent with the doctrine of perpetual deformation, for the very conditions prerequisite to the accomplishment of base-leveling involve a high degree of stability through a long period').

A particular insight into landscape evolution is provided by the plate tectonic interpretation of oceanic islands. Darwin (1842) had interpreted coral reefs as growing upwards from the subsiding basement of an oceanic volcano, an interpretation also favoured by Davis (1928) (Chorley *et al.* 1973, chapter 25; Watts 2000). Where these islands are found in linear chains, such as the Hawaiian Islands in the Pacific Ocean, the island geomorphology pointed to a younging in the island chain from one end to the other (in a southeasterly direction in the case of the Hawaiian chain). Wilson (1963) showed that oceanic islands, such as the Hawaiian Islands, were explicable in terms of the sea-floor being carried by mantle convection over a volcanic source erupting at the sea-floor from the mantle, and building up the volcano from the sea-floor. A further breakthrough was provided by radiometric dating, when McDougall (1964) provided a suite of radiometric ages which confirmed the age progression in the Hawaiian Islands. McDougall's contribution also heralded another breakthrough because absolute age dating revolutionized understanding of rates of landscape evolution. Ruxton and McDougall (1967) provided one of the earliest examples of this approach, and this type of study has proliferated (e.g. Bishop 1985; Young and McDougall 1993; Nott *et al.* 1996).

The numerical modelling revolution that has been so important in exploring the links between tectonics and landscape evolution in a plate tectonic context has brought landscape evolution studies full circle. Thus, Kooi and Beaumont (1996) and Gilchrist and Summerfield (1994) have used numerical models to test the classical models of landscape evolution.

Antiquity of drainage systems

P.E. Potter (1978) Significance and origin of big rivers. *Journal of Geology* 86: 13–33

Potter's paper provided several critical insights in landscape evolution studies. The first is that big rivers have long histories. This insight, built mainly on Potter's case study of the Mississippi, which has occupied its present place on the North American continent for approximately one-sixteenth of Earth history, harks back to the classic concept of stream antecedence, first proposed by Powell (1875). Once established, large streams with high stream power can readily match any uplift across their course. This is an important concept in its own right, and also provides a key framework for understanding landscape evolution in many formerly glaciated settings. For example, pre-glacial fluvial origins have been proposed for the inter-island open waterways of the Canadian Arctic archipelago (Dyke *et al.* 1992) and for the glaciated valleys of the Scottish Highlands (Sissons 1967; Hall 1991). The extent to which glacial systems deepen and widen these pre-existing fluvial valleys continues to be debated. Montgomery (2002) has provided a useful review and investigation of this issue, concluding on the basis of data from NW USA that pre-existing fluvial valleys are significantly deepened and widened by glacial activity.

The antiquity of major fluvial systems can be elucidated from ancient preserved channels that have escaped erosion (e.g. the Cenozoic basalt-filled channels of the SE Australian highlands – e.g. Bishop and Goldrick 2000), and from the rivers' continental deposits and their long-lived deltas (e.g. the Orange River: Dingle and Hendey 1984, Rust and Summerfield 1990; the Mississippi: Potter 1978; the Amazon: Driscoll and Karner 1994, Potter 1997; the N Britain systems draining to the North Sea: Morton 1979, Hall 1991; the systems of E Australia: Veevers 1984).

A major contribution of Potter (1978) is to highlight the plate tectonic setting conducive to the development of the world's major long-lived river systems that deliver the bulk of the sediment to the oceans (cf. Milliman and Meade 1983). The world's 28 largest rivers drain to passive continental margins, which have the world's 25 largest deltas. The ideal setting for a big river system is a large asymmetric continent that couples a distant high mountain range which is generally related to plate convergence (e.g. the Rockies, the Himalayas or the Andes) and is responsible for the bulk of the river's sediment load (i.e. the Andes for the Amazon – Gibbs 1967). The stable intraplate setting crossed by the river for the bulk of its length permits the river to be long-lived, and the broad continental shelf that is characteristic of a passive margin furnishes the platform on which the delta can be built (Potter 1978).

Time, space and causality in geomorphology

S.A. Schumm and R.W. Lichty (1965) Time, space, and causality in geomorphology. *American Journal of Science* 263: 110–19

As Schumm and Lichty pointed out in the introduction to this paper: 'a study of [geomorphic] process must attempt to relate causality to the evolution of the system' (1965: 110). This paper was an attempt to reconcile the process emphasis in geomorphology (cf. Strahler 1952) with the historical geomorphological approach that had dominated the first half of the twentieth century (Kennedy 1997). The issue remains an enduring one (e.g. Sugden *et al.* 1997). A key contribution of Schumm and Lichty was that they considered both time and space, voicing concern at the apparent neglect of the time dimension in both process studies and the time-independent approach of Hack (1960). The major contribution, therefore, was to identify different time spans relevant to understanding geomorphological processes and landscape evolution (cyclic time, graded time and steady time) and the ways in which the status of the various drainage basin and river variables relevant to geomorphological processes and landscape evolution varied according to the time span.

Detailed 'realist' based approaches to understanding geomorphological processes which aim at explanation rather than functional morphological relationships (e.g. Richards 1990, 1994; Rhoads 1994; Richards *et al.* 1997) indicate that river variables may not be dependent at one timescale and independent at another in the way that Schumm and Lichty (1965) argued (Lane and Richards 1997). Contingency means that prior landscape history may shape subsequent landscape history, highlighting yet again the methodological challenges of moving between scales in geomorphological research (e.g. Douglas 1982; Starkel 1982; Church 1996; Sugden *et al.* 1997). Indeed, the ways in which detailed local case studies can be generalized to inform landscape evolution remain to be clarified (Church 1996; Lane and Richards 1997). Moreover, appropriate procedures for generalizing from individual studies of place-specific landscape evolution to understandings of landscape evolution more generally also await clarification; see Bishop (1998) for a fuller discussion of this point.

Haines-Young and Petch (1986) highlighted further important methodological issues surrounding Schumm and Lichty's treatment (1965) of a variable's change between dependent and independent status, in particular the scheme's status as a trigger for the development of theory, noting that it does not constitute scientific theory. It is also noteworthy here, in the context of the issues of landscape antiquity and appropriate timescales for understanding landscape evolution, that Schumm and Lichty's suggestion that 'geologic time' corresponds to the Pleistocene is too conservative and restrictive for understanding landscape evolution in many areas.

Canons of landscape evolution

L.C. King (1953) Canons of landscape evolution. *Bulletin of the Geological Society of America* 64: 721–51

Brunsden (1990) pointed out that syntheses of overarching principles of landscape evolution date back to Hutton's 'Laws'; King's 'Canons' provide a mid-twentieth-century example. The importance of this paper is twofold: it embodies a statement of King's view of landscape evolution and it presents his 50 canons of landscape evolution.

Like Davis (and unlike Penck), King believed that landscape evolution was cyclical, but King disagreed with Davis (and agreed with Penck) on the mode of slope evolution, arguing that parallel retreat of slopes (rather than slope decline) was the principal mode of slope evolution. Parallel slope retreat leads to the formation of a pediment at the foot of the slope, and coalescing pediments form a pediplain (cf. Davis's peneplain). Unlike the peneplain, the pediment must have a range of ages (i.e. be diachronous) because it has formed progressively by slope retreat.

Slope processes and morphology were a research focus in the 1970s; Young (1972) has provided an introduction to the extensive literature on slope form and formation (see also Carson and Kirkby 1972; Parsons 1988). Numerical modelling approaches to slope processes are covered by Armstrong (1987) and Kirkby (1985).

King highlighted other points of departure from Davis, noting in passing that Davis was close to an overarching view that would have integrated landscape evolution under arid and humid conditions. Instead, King argued, Davis was too tied to his 'humid' model (the geographical cycle of 'normal' landscape evolution), which dominated his own thinking and the discipline for 50 years.

King's actual canons of landscape evolution, the second reason for including this paper, are more a summary of 'facts' about landscape evolution than conceptual statements. Brunsden's (1990) 'ten commandments' are more akin to the latter, and are therefore of perhaps more value than King's statements. Brunsden's statements build on the important concepts of magnitude and frequency (see Wolman and Miller 1960 [and Haines-Young and Petch 1996 for a contrary view]) as well as earlier statements by Brunsden and Thornes (1979). Brunsden's 'ten commandments' paper should be consulted for its many references to a wide range of issues, concepts and phenomena relevant to the issue of landscape evolution. The treatment of time echoes the issues raised by Schumm and Lichty (1965), touching on related issues such as the very high magnitude, low frequency events that we have not treated in detail here. As well, the issues of climatic geomorphology, a major area of landscape evolution that has received relatively little attention in the anglophone literature (Tricart *et al.* 1972; Derbyshire 1973; Peltier 1975; Büdel 1982; see Twidale and Lageat 1994 for a contrary viewpoint) are to an extent implicit in Brunsden's treatment, especially in relation to tectonic influences. Brunsden attempted to establish a conceptual framework for landscape evolution, and this remains a major challenge for the earth sciences community.

References

Adams, J. (1980) Contemporary uplift and erosion of the Southern Alps, New Zealand. *Geological Society of America Bulletin, Part II*: 91, 1–114.

Adams, J. (1985) Large-scale tectonic geomorphology of the Southern Alps. In M. Morisawa and J.T. Hack (eds) *Tectonic Geomorphology*. London: Allen and Unwin, pp. 105–28.

Armstrong, A.C. (1987) Slopes, boundary conditions, and the development of convexo-concave forms – some numerical experiments. *Earth Surface Processes and Landforms* 12: 17–30.

Baulig, H. (1926) La notion de profil d'équilibre: histoire et critique. *Comptes Rendus du Congrès International de Géographie (1925)* 3: 51–63.

Beaumont, C., Fullsack, P. and Hamilton, J. (1992) Erosional control of active compressional orogens. In K.R. McClay (ed.) *Thrust Tectonics*. London: Chapman and Hall, pp. 1–18.

Beaumont, C., Kamp, P.J.J., Hamilton, J. and Fullsack, P. (1996) The continental collision zone, South Island, New Zealand: comparison of geodynamical models and observations. *Journal of Geophysical Research* 101: 3333–60.

Beaumont, C., Kooi, H. and Willett, S. (2000). Coupled tectonic-surface process models with applications to rifted margins and collisional orogens. In M.A. Summerfield (ed.) *Geomorphology and Global Tectonics*. Chichester: Wiley, pp. 29–55.

Beckinsale, R.P and Chorley, R.J. (1991) *The History of the Study of Landforms*. Vol. 3. London: Routledge.

Bird, J.B. (1967) *The Physiography of Arctic Canada*. Baltimore: The Johns Hopkins Press.

Bishop, P. (1980) Popper's principle of falsifiability and the irrefutability of the Davisian cycle. *Professional Geographer* 32: 310–15.

Bishop, P. (1982) Stability or change: a review of ideas on ancient drainage in eastern New South Wales. *Australian Geographer* 15: 219–30.

Bishop, P. (1985) Southeast Australian late Mesozoic and Cenozoic denudation rates: a test for late Tertiary increases in continental denudation. *Geology* 13: 479–82.

Bishop, P. (1998) Griffith Taylor and the SE Australian highlands: testability of models of longterm drainage history and landscape evolution. *Australian Geographer* 29: 7–29.

Bishop, P. (2004) Tectonic and related landforms. In R.J. Chorley *et al.* (eds) *The History of the Study of Landforms*. Vol. 4. London: Routledge. In press.

Bishop, P. and Goldrick, G. (2000) Geomorphological evolution of the East Australian continental margin. In M.A. Summerfield (ed.) *Geomorphology and Global Tectonics*. Chichester: John Wiley, pp. 227–55.

Bishop, P., Young, R.W. and McDougall, I. (1985) Stream profile change and long-term landscape evolution – Early Miocene and modern rivers of the east Australian highland crest, central New South Wales, Australia. *Journal of Geology* 93: 455–74.

Bonnet, S. and Crave, A. (2003) Landscape response to climate change: insights from experimental modeling and implications for tectonic versus climatic uplift of topography. *Geology* 31: 123–6.

Brocard, G., van der Beek, P., Bourlès, D.L., Siame, L.L. and Mugnier, J.-L. (2003) Long-term fluvial incision rates and postglacial river relaxation time in the French Western Alps from [10]Be dating of alluvial terraces with assessment of inheritance, soil development and wind ablation effects. *Earth and Planetary Science Letters* 209: 197–214.

Brookfield, M.E. (1998) The evolution of the great river systems of southern Asia during the Cenozoic India–Asia collision: rivers draining southwards. *Geomorphology* 22: 285–312.

Brown, E. (1960) *The Relief and Drainage of Wales*. Cardiff: University of Wales Press.

Brown, E. (1979) The shape of Britain. *Transactions of the Institute of British Geographers* (New Series) 4: 449–62.

Brown, E.H. (1980) Historical geomorphology – principles and practice. *Zeitschrift für Geomorphologie Supplement* 36: 9–15.

Brown, R.W., Summerfield, M.A. and Gleadow, A.J.W. (2002) Denudational history along a transect across the Drakensberg Escarpment of southern Africa derived from apatite fission track thermochronology. *Journal of Geophysical Research* 107, doi:10.1029/2001JB000745.

Brunsden, D. (1990) Tablets of stone: toward the ten commandments of geomorphology. *Zeitschrift für Geomorphologie Supplement* 79: 1–37.

Brunsden, D. and Thornes, J.B. (1979) Landscape sensitivity and change. *Transactions of the Institute of British Geographers* 4: 463–84.

Büdel, J. (1982) *Climatic Geomorphology*. Trans. L. Fischer and D. Busche. Princeton, NJ: Princeton University Press.

Bull, A.J. (1931) The convection current hypothesis of mountain building. *Geological Magazine* 68: 495–8.

Burbank, D.W. and Anderson, R.S. (2001) *Tectonic Geomorphology*. Malden, Mass.: Blackwell.

Burbank, D.W., Leland, J., Fielding, E., Anderson, R.S., Brozovic, N., Reid, M.R. and Duncan, C. (1996) Bedrock incision, rock uplift and threshold hillslopes in the northwestern Himalayas. *Nature* 379: 505–10.

Carson, M.A. and Kirkby, M.J. (1972) *Hillslope Form and Process*. Cambridge: Cambridge University Press.

Chamberlin, T.C. (1909) Diastrophism as the fundamental basis of correlation. *Journal of Geology* 17: 685–93.

Chorley, R.J. (1965) A re-evaluation of the geomorphic system of W.M. Davis. In R.J. Chorley and P. Haggett (eds) *Frontiers in Geographical Teaching*. London: Methuen, pp. 21–38.

Chorley, R.J. (2000) Classics in physical geography revisited: Mackin, J.H. 1948. Concept of the graded river. *Progress in Physical Geography* 24: 563–78.

Chorley, R.J., Beckinsale, R.P. and Dunn, A.J. (1973) *The History of the Study of Landforms*. Vol. 2: *The Life and Work of William Morris Davis*. London: Methuen.

Chorley, R.J., Dunn, A.J. and Beckinsale, R.P. (1964) *The History of the Study of Landforms*. Vol. 1: *Geomorphology before Davis*. London: Methuen.

Church, M. (1996) Space, time and the mountain – how do we order what we see? In B.L. Rhoads and C.E. Thorn (eds) *The Scientific Nature of Geomorphology*. Chichester: John Wiley, pp. 147–70.

Common, R. (1954) The geomorphology of the East Cheviot area. *Scottish Geographical Magazine* 70: 124–38.

Crickmay, C.H. (1933) The later stages of the Cycle of Erosion. *Geological Magazine* 70: 337–47.

Crickmay, C.H. (1974) *The Work of the River*. London: Macmillan.

Crickmay, C.H. (1975) The hypothesis of unequal activity. In W.N. Melhorn and R.C. Flemal (eds) *Theories of Landform Development*. Binghamton: State University of New York Press, pp. 103–9.

Darwin, C. (1842) *The Structure and Distribution of Coral Reefs*. London: Smith, Elder.

Davis, W.M. (1895) On the origin of certain English rivers. *Geographical Journal* 5: 128–46.

Davis, W.M. (1896) Plains of marine and sub-aerial denudation. *Bulletin of the Geological Society of America* 7: 377–98.

Davis, W.M. (1899a) The geographical cycle. *Geographical Journal* 14: 481–504.

Davis, W.M. (1899b) The peneplain. *American Geologist* 23: 207–39.

Davis, W.M. (1902) Base-level, grade and the peneplain. *Journal of Geology* 10: 77–111.

Davis, W.M. (1910) The theory of isostasy (Abstract). *Bulletin of the Geological Society of America* 21: 777.

Davis, W.M. (1928) *The Coral Reef Problem*. American Geographical Society Special Publication 9.

Derbyshire, E. (ed.) (1973) *Climatic Geomorphology*. London: Macmillan.

Dingle, R.V. and Hendey, Q.B. (1984) Late Mesozoic and Tertiary sediment supply to the Eastern Cape Basin (SE Atlantic) and palaeo-drainage systems in south-western Africa. *Marine Geology* 56: 13–26.

Douglas, I. (1982) The unfulfilled promise: earth surface processes as a key to landform evolution. *Earth Surface Processes and Landforms* 7: 101.

Driscoll, N.W. and Karner, G.D. (1994) Flexural deformation due to Amazon Fan loading: a feedback mechanism affecting sediment delivery to margins. *Geology* 22: 1015–18.

Dumitru, T.A., Hill, K.C., Coyle, D.A., Duddy, I.R., Foster, D.A., Gleadow, A.J.W., Green, P.F., Kohn, B.P., Laslett, G.M. and O'Sullivan, A.J. (1991) Fission track thermochronology: application to continental rifting of south-eastern Australia. *APEA Journal* 31: 131–42.

Dury, G.H. (1965) The concept of grade. In G.H. Dury (ed.) *Essays in Geomorphology*. London: Heinemann, pp. 211–34.

Dutton, C.E. (1889) On some of the greater problems of physical geology. *Bulletin of the Philosophical Society of Washington* 11: 51–64.

Dyke, A.S. (1983) Quaternary geology of Somerset Island, District of Franklin. *Geological Survey of Canada Memoir* 404.

Dyke, A.S. (1993) Landscapes of cold-centred Late Wisconsinan ice caps, Arctic Canada. *Progress in Physical Geography* 17: 223–47.

Dyke, A.S., Morris, T.F., Green, D.E.C. and England, J. (1992) Quaternary geology of Prince of Wales Island, Arctic Canada. *Geological Survey of Canada Memoir* 433.

Ehlers, T.A. and Farley, K.A. (2003) Apatite (U-Th)/He thermochronometry: methods and applications to problems in tectonic and surface processes. *Earth and Planetary Science Letters* 206: 1–14.

England, J. (1983) Isostatic adjustments in a full glacial sea. *Canadian Journal of Earth Sciences* 20: 895–917.

Fairbridge, R.W. and Finkl, C.W. (1980) Cratonic erosional unconformities and peneplains. *Journal of Geology* 88: 69–86.

Flemal, R.C. (1971) The attack on the Davisian system of geomorphology: a synopsis. *Journal of Geological Education* 19: 3–13.

Fried, A.W. and Smith, N. (1992) Timescales and the role of inheritance in long-term landscape evolution, northern New England, Australia. *Earth Surface Processes and Landforms* 17: 375–85.

George, T.N. (1966) Geomorphic evolution in Hebridean Scotland. *Scottish Journal of Geology* 2: 1–34.

Gibbs, R.J. (1967) The geochemistry of the Amazon River System. Part 1: Factors controlling the salinity and the composition and concentration of suspended solids. *Bulletin of the Geological Society of America* 78: 1023–32.

Gilbert, G.K. (1877) *The Geology of the Henry Mountains (Utah)*. U.S. Geographical and Geological Survey of the Rocky Mountain Region. Washington, DC: Government Printing Office.

Gilbert, G.K. (1890) Lake Bonneville. *U.S. Geological Survey Monograph* 1.

Gilchrist, A.R. and Summerfield, M.A. (1990) Differential denudation and flexural isostasy in formation of rifted-margin upwarps. *Nature* 346: 739–42.

Gilchrist, A.R. and Summerfield, M.A. (1994) Tectonic models of passive margin evolution and their implications for theories of long-term landscape evolution. In M.J. Kirkby (ed.) *Process Models and Theoretical Geomrophology*. Chichester: John Wiley, pp. 55–84.

Gilchrist, A.R., Kooi, H. and Beaumont, C. (1994) Post-Gondwana geomorphic evolution of southwestern Africa: Implication for the controls on landscape development from observations and numerical experiments. *Journal of Geophysical Research* 99: 12211–28.

Gilchrist, A.R., Summerfield, M.A. and Cockburn, H.A.P. (1994) Landscape dissection, isostatic uplift, and the morphological development of orogens. *Geology* 22: 963–6.

Goldrick, G. and Bishop, P. (1995) Distinguishing the roles of lithology and relative uplift in the steepening of bedrock river long profiles: an example from southeastern Australia. *Journal of Geology* 103: 227–31.

Gould, S.J. (1987) *Time's Arrow, Time's Cycle*. Cambridge, Mass.: Harvard University Press.

Griggs, D. (1939) A theory of mountain-building. *American Journal of Science* 237: 611–50.

Guilcher, A., Bodere, J.-C., Coude, A., Hansom, J.D., Moign, A. and Peulvast J.-P. (1986) Le probleme des strandflats en cinq pays de hautes latitudes. *Revue de Geologie Dynamique et de Geographie Physique* 27: 47–79. (English translation 'The strandflat problem in five high latitude countries' available in D.J.A. Evans (ed.) *Cold Climate Landforms*. Chichester: Wiley, pp. 351–93.

Hack, J.T. (1957) Studies of longitudinal stream profiles in Virginia and Maryland. *US Geological Survey Professional Paper* 294-B: 45–97.

Hack, J.T. (1960) Interpretation of erosional topography in humid temperate regions. *American Journal of Science* 258-A: 80–97.

Hack, J.T. (1973) Stream-profile analysis and stream-gradient index. *Journal of Research of the US Geological Survey* 1: 421–9.

Hack, J.T. (1975) Dynamic equilibrium and landscape development. In W.N. Melhorn and R.C. Flemal (eds) *Theories of Landform Development*. Binghamton: State University of New York Press, pp. 87–102.

Hack, J.T. (1979) Rock control and tectonism – their importance in shaping the Appalachian highlands. *US Geological Survey Professional Paper* 1126-B.

Hack, J.T. (1982) Physiographic divisions and differential uplift in the Piedmont and Blue Ridge. *US Geological Survey Professional Paper* 1265.

Haines-Young, R. and Petch. J. (1986) *Physical Geography: Its Nature and Methods*. London: Harper and Row.

Hall, A.M. (1985) Cenozoic weathering covers in Buchan, Scotland and their significance. *Nature* 315: 392–5.

Hall, A.M. (1986) Deep weathering patterns in north-east Scotland and their geomorphological significance. *Zeitschrift für Geomorphologie* 30: 407–22.

Hall, A.M. (1991) Pre-Quaternary landscape evolution in the Scottish Highlands. *Transactions of the Royal Society of Edinburgh: Earth Sciences* 82: 1–26.

Hall, A.M. and Sugden, D.E. (1987) Limited modification of mid-latitude landscapes by ice-sheets: the case of northeast Scotland. *Earth Surface Processes and Landforms* 12: 531–42.

Hättestrand, C. and Stroeven, A.P. (2002) A relict landscape in the centre of Fennoscandian glaciation: geomorphological evidence of minimal Quaternary glacial erosion. *Geomorphology* 44: 127–43.

Hill, S.M. (1999) Mesozoic regolith and palaeolandscape features in southeastern Australia: significance for interpretations of denudation and highland evolution. *Australian Journal of Earth Sciences* 46: 217–32.

Hoffman, P.F. and Grotzinger, J.P. (1993) Orographic precipitation, erosional unloading and tectonic style. *Geology* 21: 195–8.

Holmes, A. (1928–9) Radioactivity and earth movements. *Transactions of the Geological Society of Glasgow* 18: 559–606.

House, M.A., Wernicke, B.P. and Farley, K.A. (1998) Dating topography of the Sierra Nevada, California using apatite (U-Th)/He ages. *Nature* 396: 66–9.

Hovius, N. (2000) Macroscale process systems of mountain belt erosion. In M.A. Summerfield (ed.) *Geomorphology and Global Tectonics*. Chichester: John Wiley, pp. 77–105.

Hutton, J. (1795) *Theory of the Earth*. 2 vols. Edinburgh.

Jeffreys, H. (1931) On the mechanics of mountains. *Geological Magazine* 68: 435–42.

Johansson, M., Olvmo, M. and Lidmar-Bergström, K. (2001) Inherited landforms and glacial impact of different palaeosurfaces in southwest Sweden. *Geografiska Annaler* 83A: 67–89.

Jones, D.K.C. (1999a) Evolving models of the Tertiary evolutionary geomorphology of southern England, with special reference to the Chalklands. In B.J. Smith, W.B. Whalley and P.A. Warke (eds) *Uplift, Erosion and Stability: Perspectives on Long-term Landscape Development*. London: Geological Society, Special Publication 162, pp. 1–23.

Jones, D.K.C. (1999b) On the uplift and denudation of the Weald. In B.J. Smith, W.B. Whalley and P.A. Warke (eds) *Uplift, Erosion and Stability: Perspectives on*

Long-term Landscape Development. London: Geological Society, Special Publication 162, pp. 25–43.

Jones, O.T. (1924) Longitudinal profiles of the Upper Towy drainage system. *Quarterly Journal of the Geological Society* 80: 568–609.

Jukes, J.B. (1862) On the mode of formation of some of the river valleys in the south of Ireland. *Quarterly Journal of the Geological Society* 18: 378–400.

Keller, E.A. (1977) Adjustment of drainage to bedrock in regions of contrasting tectonic framework. *Geological Society of America Abstracts with Programs* 9: 1046.

Kennedy, B. (1994) Requiem for a dead concept. *Annals of the Association of American Geographers* 84: 702–5.

Kennedy, B.A. (1997) Classics in physical geography revisited: Schumm, S.A. and Lichty, R.W. 1965. Time, space and causality in geomorphology. *Progress in Physical Geography* 21: 419–23.

Kesseli, J.E. (1941) The concept of the graded river. *Journal of Geology* 49: 561–88.

King, L.C. (1953) Canons of landscape evolution. *Bulletin of the Geological Society of America* 64: 721–51.

King, L.C. (1955) Pediplanation and isostasy: an example from South Africa. *Quarterly Journal of the Geological Society of London* 111: 353–9.

King, L.C. (1962) *The Morphology of the Earth*. Edinburgh: Oliver and Boyd.

Kirby, E. and Whipple, K. (2001) Quantifying differential rock-uplift rates via stream profile analysis. *Geology* 29: 415–18.

Kirkby, M.J. (1985) A model for the evolution of regolith mantled slopes. In M.J. Woldenberg (ed.) *Models in Geomorphology*. Boston: Allen and Unwin, pp. 213–37.

Kleman, J. (1994) Preservation of landforms under ice sheets and ice caps. *Geomorphology* 9: 19–32.

Kleman, J. and Borgström, I. (1994) Glacial land forms indicative of a partly frozen bed. *Journal of Glaciology* 40: 255–64.

Knox, J.C. (1975) Concept of the graded stream. In W.N. Melhorn and R.C. Flemal (eds) *Theories of Landform Development*. Binghamton: State University of New York Press, pp. 169–98.

Kohn, B. and Bishop, P. (1999) Introduction to thematic issue on apatite fission track thermochronology and geomorphology. *Australian Journal of Earth Sciences* 46: 155–6.

Kooi, H. and Beaumont, C. (1994) Escarpment evolution on high-elevation rifted margins: insights derived from a surface process model that combines diffusion, advection, and reaction. *Journal of Geophysical Research* 99: 12191–209.

Kooi, H. and Beaumont, C. (1996) Large-scale geomorphology: classical concepts reconciled and integrated with contemporary ideas via a surface process model. *Journal of Geophysical Research* 101: 3361–86.

Lal, D. (1991) Cosmic ray labeling of erosion surfaces; in situ nuclide production rates and erosion models. *Earth and Planetary Science Letters* 104: 424–39.

Lane, S.N. and Richards, K.S. (1997) Linking river channel form and process: time, space and causality revisited. *Earth Surface Processes and Landforms* 22: 249–60.

Lidmar-Bergström, K. (1987) Exhumed Cretaceous landforms in south Sweden. *Zeitschrift für Geomorphologie* 72: 21–40.

Lidmar-Bergström, K. (1995) Relief and saprolites through time on the Baltic Shield. *Geomorphology* 12: 45–61.

Lidmar-Bergström, K. (1996) Long-term morphotectonic evolution in Sweden. *Geomorphology* 16: 33–59.

Lidmar-Bergström, K. (1997) A long-term perspective of glacial erosion. *Earth Surfaces Processes and Landforms* 22: 297–306.

Lidmar-Bergström, K. (1999) Uplift histories revealed by landforms of the Scandinavian domes. In B.J. Smith, W.B. Whalley and P.A. Warke (eds) *Uplift, Erosion and Stability: Perspectives on Long-term Landscape Development*. London: Geological Society. Special Publication 162, pp. 85–91.

Linton, D.L. (1955) The problem of tors. *Geographical Journal* 121: 470–87.

Lyell, C. (1830) *Principles of Geology*. Vol. 1. London: John Murray.

Mabbutt, J.A. (1961) 'Basal surface' or 'weathering front'. *Proceedings of the Geologists Association* 72: 357–8.

McDougall, I. (1964) Potassium-argon ages of lavas of the Hawaiian Islands. *Bulletin of the Geological Society of America* 75: 107–28.

McKeown, F.A., Jones-Cecil, M., Askew, B.L. and McGrath, M.B. (1988) Analysis of stream-profile data and inferred tectonic activity, Eastern Ozark Mountains region. *US Geological Survey Bulletin* 1807.

Mackin, J.H. (1948) Concept of the graded river. *Bulletin of the Geological Society of America* 59: 463–511.

Mackin, J.H. (1963) Rational and empirical methods of investigation in geology. In C.C. Albritton (ed.) *The Fabric of Geology*, Stanford: Freeman, Cooper, pp. 135–63.

Merritts, D. and Vincent, K.R. (1989) Geomorphic response of coastal streams to low, intermediate, and high rates of uplift, Mendocino triple junction region, northern California. *Geological Society of America Bulletin* 101: 1373–88.

Merritts, D.J., Vincent, K.R. and Wohl, E.E. (1994) Long river profiles, tectonism, and eustacy: a guide to interpreting fluvial terraces. *Journal of Geophysical Research* 99: 14031–50.

Migon, P. and Lidmar-Bergström, K. (2001) Weathering mantles and their significance for geomorphological evolution of central and northern Europe since the Mesozoic. *Earth-Science Reviews* 56: 285–324.

Milliman, J.D. and Meade, R.H. (1983) World-wide delivery of sediment to the oceans. *Journal of Geology* 91: 1–21.

Molnar, P. and England, P. (1990) Late Cenozoic uplift of mountain ranges and global climate change: chicken or egg? *Nature* 346: 29–34.

Montgomery, D.R. (1994) Valley incision and the uplift of mountain peaks. *Journal of Geophysical Research* 99: 13913–21.

Montgomery, D.R. (2002) Valley formation by fluvial and glacial erosion. *Geology* 30: 1047–50.

Morton, A.C. (1979) The provenance and distribution of the Palaeocene sands of the North Sea. *Journal of Petroleum Geology* 2: 11–21.

Nansen, F. (1922) The strandflat and isostasy. *Vidensk. Skr. I. Math., Naturv. Kl., Oslo* 11: 313pp.

Nott, J. (1995) The antiquity of landscapes on the North Australian craton and the implications for theories of long-term landscape evolution. *Journal of Geology* 103: 19–32.

Nott, J., Young, R. and McDougall, I. (1996) Wearing down, wearing back, and gorge extension in the long-term denudation of a highland mass: Quantitative evidence from the Shoalhaven catchment, southeast Australia. *Journal of Geology* 104: 224–32.

Ollier, C.D. (1979) Evolutionary geomorphology of Australia and Papua–New Guinea. *Transactions of the Institute of British Geographers* 4: 516–39.

Ollier, C.D. (1985) Morphotectonics of continental margins with great escarpments. In M. Morisawa and J.T. Hack (eds) *Tectonic Geomorphology*. London: Allen and Unwin, pp. 3–25.

Ollier, C.D., Gaunt, G.F.M. and Jurkowski, I. (1988) The Kimberley plateau, Western Australia: a Precambrian erosion surface. *Zeitschrift für Geomorphologie* 32: 239–46.

Ollier, C. and Pain, C. (2000) *The Origin of Mountains*. London: Routledge.

Olvo, M., Lidmar-Bergström, K. and Lindberg, G. (1999) The glacial impact on an exhumed sub-Mesozoic etch surface in southwestern Sweden. *Annals of Glaciology* 28: 153–60.

Palmer, J. (1956) Tor formation at the Bridestones in North east Yorkshire and its significance in relation to problems of valley-side development and regional glaciation. *Transactions of the Institute of British Geographers* 22: 55–71.

Parsons, A.J. (1988) *Hillslope Form*. London: Routledge.

Pazzaglia, F.J. and Gardner, T.W. (1994) Late Cenozoic flexural deformation of the middle U.S. passive margin. *Journal of Geophysical Research* 99: 12143–57.

Pazzaglia, F.J., Gardner, T.W. and Merritts, D.J. (1998) Bedrock fluvial incision and longitudinal profile development over geologic time svcales determoned by fluvial terraces. In K.J. Tinkler and E.E. Wohl (eds) *Rivers over Rock: Fluvial Processes in Bedrock Channels*. American Geophysical Union Geophysical Monograph 107: 207–35.

Peltier, L.C. (1975) The concept of climatic geomorphology. In W.N. Melhorn and R.C. Flemal (eds) *Theories of Landform Development*. London: Allen and Unwin, pp. 129–44.

Penck, W. (1953) *Morphological Analysis of Land Forms*. London: Macmillan.

Persano, C., Stuart, F.M., Bishop, P. and Barfod, D. (2002) Apatite (U-Th)/He age constraints on the development of the Great Escarpment on the southeastern Australian passive margin. *Earth and Planetary Science Letters* 200: 79–90.

Phillips, J.D. (2003) Sources of non-linearity and complexity in geomorphic systems. *Progress in Physical Geography* 27: 1–23.

Pinter, N. and Brandon, M.T. (1997) How erosion builds mountains. *Scientific American*, April: 74–9.

Potter, P.E. (1978) Significance and origin of big rivers. *Journal of Geology* 86: 13–33.

Potter, P.E. (1997) The Mesozoic and Cenozoic paleodrainage of South America: a natural history. *Journal of South American Earth Sciences* 10: 331–44.

Powell, J.W. (1875) *Exploration of the Colorado River of the West (1869–72)*. Washington.

Pugh, J.C. (1955) Isostatic readjustment in the theory of pediplanation. *Quarterly Journal of the Geological Society of London* 111: 361–9.

Pullan, R.A. (1959) Notes on periglacial phenomena, tors. *Scottish Geographical Magazine* 75: 51–5.

Ramírez-Herrera, M.T. (1998) Geomorphic assessment of active tectonics in the Acambay Graben, Mexican Volcanic Belt. *Earth Surface Processes and Landforms* 23: 317–32.

Ramsay, A. (1846) On the denudation of South Wales and the adjacent counties of England. In *Essays*. Vol. 1: *Memoirs of the Geological Survey*, pp. 326–8.

Reed, J.C. (1981) Disequilibrium profile of the Potomac River near Washington, DC – A result of lowered base level or Quaternary tectonics along the Fall Line? *Geology* 9: 445–50.

Rhoads, B.L. (1994) On being a 'real' geomorphologist. *Earth Surface Processes and Landforms* 19: 269–72.

Richards, K.S. (1990) 'Real' geomorphology. *Earth Surface Processes and Landforms* 15: 195–7.

Richards, K.S. (1994) 'Real' geomorphology revisited. *Earth Surface Processes and Landforms* 19: 277–81.

Richards, K.S., Brooks, S., Clifford, N., Harris, T. and Lane, S. (1997) Theory, measurement and testing in 'real' geomorphology and physical geography. In D.R. Stoddart (ed.) *Process and Form in Geomorphology*. London: Routledge, pp. 265–92.

Roe, G.H., Montgomery, D.R. and Hallet, B. (2002) Effects of orographic precipitation variations on the concavity of steady-state river profiles. *Geology* 30: 143–6.

Rust, D.J. and Summerfield, M.A. (1990) Isopach and bore-hole data as indicators of rifted margin evolution in southwestern Africa. *Marine and Petroleum Geology* 7: 277–87.

Ruxton, B.P. and McDougall, I. (1967) Denudation rates in northeast Papua from potassium-argon dating of lavas. *American Journal of Science* 265: 545–61.

Schumm, S.A. and Lichty, R.W. (1965) Time, space, and causality in geomorphology. *American Journal of Science* 263: 110–19.

Seeber, L. and Gornitz, V. (1983) River profiles along the Himalayan Arc as indicators of active tectonics. *Tectonophysics* 92: 335–67.

Shaler, N.S. (1899) Spacing of rivers with reference to the hypothesis of base-levelling. *Bulletin of the Geological Society of America* 19: 263–76.

Sissons, J.B. (1967) *The Evolution of Scotland's Scenery*. Edinburgh: Oliver and Boyd.

Small, E.E. and Anderson, R.S. (1995) Geomorphically driven Late Cenozoic rock uplift in the Sierra Nevada, California. *Science* 270: 277–80.

Smith, D.E. and Dawson, A.G. (eds) (1983) *Shorelines and Isostasy*. London: Academic, Press.

Starkel, L. (1982) The need for parallel studies on denudation chronology and present-day processes. *Earth Surface Processes and Landforms* 7: 301–2.

Stewart, A.J., Blake, D.H. and Ollier, C.D. (1986) Cambrian river terraces and ridge-tops in Central Australia: oldest persisting landforms. *Science* 23: 758–61.

Stoddart, D.R. (1966) Darwin's impact on geography. *Annals of the Association of American Geographers* 56: 683–98.

Strahler, A.N. (1952) Dynamic basis of geomorphology. *Bulletin of the Geological Society of America* 63: 923–38.

Strahler, A.N. (1992) Quantitative/dynamic geomorphology at Columbia 1945–60: a retrospective. *Progress in Physical Geography* 16: 65–84.

Stroeven, A.P., Fabel, D., Hättestrand, C. and Harbor, J. (2002) A relict landscape in the centre of Fennoscandian glaciation: cosmogenic radionuclide evidence of tors preserved through multiple glacial cycles. *Geomorphology* 44: 145–54.

Sugden, D.E., Summerfield, M.A. and Burt, T.P. (1997) Editorial: Linking short-term geomorphic processes to landscape evolution. *Earth Surface Processes and Landforms* 22: 193–4.

Summerfield, M.A. (1991) *Global Geomorphology*. Harlow: Longman.

Summerfield, M.A. (2000a) Geomorphology and global tectonics: introduction. In M.A. Summerfield (ed.) *Geomorphology and Global Tectonics*. Chichester: John Wiley, pp. 3–12.

Summerfield, M.A. (ed.) (2000b) *Geomorphology and Global Tectonics*. Chichester: John Wiley.

Summerfield, M.A., Sugden, D.E., Denton, G.H., Marchant, D.R., Cockburn, H.A.P. and Stuart, F.M. (1999) Cosmogenic isotope data support previous evidence of extremely low rates of denudation in the Dry Valleys region, southern Victoria Land, Antarctica. In B.J. Smith, W.B. Whalley and P.A. Warke (eds) *Uplift, Erosion and Stability: Perspectives on Long-term Landscape Development*. London: Geological Society, Special Publication 162, pp. 255–67.

Tarr, R.S. (1898) The peneplain. *American Geologist* 21: 351–70.

Thomas, M.F. (1989a) The role of etch processes in landform development. I. Etching concepts and their applications. *Zeitschrift für Geomorphologie* 33: 129–42.

Thomas, M.F. (1989b) The role of etch processes in landform development. II. Etching and the formation of relief. *Zeitschrift für Geomorphologie* 33: 257–74.

Thomas, M.F. and Summerfield, M.A. (1987) Long-term landform development: key themes and research problems. In V. Gardiner (ed.) *International Geomorphology 1986 Part II*. Chichester: Wiley, pp. 935–56.

Thorn, C.E. and Welford, M.R. (1994) The equilibrium concept in geomorphology. *Annals of the Association of American Geographers* 84: 666–96.

Thornes, J.B. (1983) Evolutionary geomorphology. *Geography* 68: 225–35.

Thornbury, W.D. (1964) *Principles of Geomorphology*. New York: Wiley.

Tricart, J., Cailleux, A. and Kiewietdejonge, C.J. (1972) *Introduction to Climatic Geomorphology*. Harlow: Longman.

Tucker, G.E. and Slingerland, R.L. (1994) Erosional dynamics, flexural isostasy, and long-lived escarpments: a numerical modeling study. *Journal of Geophysical Research* 99: 12229–43.

Twidale, C.R. (1976) On the survival of paleoforms. *American Journal of Science* 276: 77–95.

Twidale, C.R. (1991) A model of landscape evolution involving increased and increasing relief amplitude. *Zeitschrift für Geomorphologie* 35: 85–109.

Twidale, C.R. (1998) Antiquity of landforms: an 'extremely unlikely' concept vindicated. *Australian Journal of Earth Sciences* 45: 657–68.

Twidale, C.R. and Lageat, Y. (1994) Climatic geomorphology – a critique. *Progress in Physical Geography* 3: 319–34.

van der Beek, P. and Braun, J. (1999) Controls on post-mid-Cretaceous landscape evolution in the southeastern highlands of Australia: insights from numerical surface process models. *Journal of Geophysical Research* 104: 4945–66.

van der Beek, P., Summerfield, M.A., Braun, J., Brown, R.W. and Fleming, A. (2002) Modeling postbreakup landscape development and denudational history across the southeast African (Drakensberg Escarpment) margin. *Journal of Geophysical Research* 107, doi:10.1029/2001JB000744.

Veevers, J.J. (ed.) (1984) *Phanerozoic Earth History of Australia*. Oxford: Clarendon Press.

Wager, R.L. (1937) The Arun River drainage pattern and the rise of the Himalaya. *Geographical Journal* 89: 239–50.

Watchman, A.L. and Twidale, C.R. (2002) Relative and 'absolute' dating of land surfaces. *Earth-Science Reviews* 58: 1–49.

Watts, A.B. (2000) The growth and decay of oceanic islands. In M.A. Summerfield (ed.) *Geomorphology and Global Tectonics*. Chichester: Wiley, pp. 339–60.

Whipple, K.X. (2001) Fluvial landscape response time: how plausible is steady-state denudation? *American Journal of Science* 301: 313–25.

Whipple, K.X., Kirby, E. and Brocklehurst, S.H. (1999) Geomorphic limits to climate-induced increases in topographic relief. *Nature* 401: 39–43.

Whipple, K.X., Snyder, N.P. and Dollenmayer, K. (2000) Rates and processes of bedrock incision by the Upper Ukak River since the 1912 Novarupta ash flow in the Valley of Ten Thousand Smokes, Alaska. *Geology* 28: 835–8.

Willett, S.D. and Brandon, M.T. (2002) On steady states in mountain belts. *Geology* 30: 175–8.

Wilson, J.T. (1963) Evidence from islands on the spreading of ocean floors. *Nature* 197: 536–8.

Wolman, M.G. and Miller, J.P. (1960) Magnitude and frequency of forces in geomorphic processes. *Journal of Geology* 68: 54–74.

Woodford, A.O. (1951) Stream gradients and the Monterey Sea valley. *Bulletin of the Geological Society of America* 62: 799–852.

Wooldridge, S.W. and Linton, D.L. (1939) Structure, Surface and Drainage in South-East England. *Institute of British Geographers Publication* 10.

Wooldridge, S.W. and Linton, D.L. (1955) The Mid-Tertiary cycle of erosion and the summit peneplain. In *Structure, Surface and Drainage in South-East England*. London: George Philip, pp. 34–44.

Young, A. (1972) *Slopes*. London: Longman.

Young, R.W. (1977) Landscape development in the Shoalhaven River catchment of southeastern Australia. *Zeitschrift für Geomorphologie* 21: 262–83.

Young, R.W. (1978) The study of landform evolution in the Sydney region: A review. *Australian Geographer* 14: 71–93.

Young, R.W. (1983) The tempo of geomorphological change: evidence from southeastern Australia. *Journal of Geology* 91: 221–30.

Young, R.W. and McDougall, I. (1993) Long-term landscape evolution: early Miocene and modern rivers in southern New South Wales, Australia. *Journal of Geology* 101: 35–49.

Zeitler, P.K., Meltzer, A.S., Koons, P.O., Craw, D., Hallet, B., Chamberlain, C.P., Kidd, W.S.F., Park, S.K., Seeber, L., Bishop, M. and Shroder, J. (2001) Erosion, Himalayan geodynamics, and the geomorphology of metamorphism. *GSA Today* 11: 4–9.

1

THE GEOGRAPHICAL CYCLE

W. M. Davis

Source: *Geographical Journal* 14 (1899): 481–504.

The genetic classification of land-forms

All the varied forms of the lands are dependent upon—or, as the mathematician would say, are functions of—three variable quantities, which may be called structure, process, and time. In the beginning, when the forces of deformation and uplift determine the structure and attitude of a region, the form of its surface is in sympathy with its internal arrangement, and its height depends on the amount of uplift that it has suffered. If its rocks were unchangeable under the attack of external processes, its surface would remain unaltered until the forces of deformation and uplift acted again; and in this case structure would be alone in control of form. But no rocks are unchangeable; even the most resistant yield under the attack of the atmosphere, and their waste creeps and washes downhill as long as any hills remain; hence all forms, however high and however resistant, must be laid low, and thus destructive process gains rank equal to that of structure in determining the shape of a land-mass. Process cannot, however, complete its work instantly, and the amount of change from initial form is therefore a function of time. Time thus completes the trio of geographical controls, and is, of the three, the one of most frequent application and of most practical value in geographical description.

Structure is the foundation of all geographical classifications in which the trio of controls is recognized. The Alleghany plateau is a unit, a "region," because all through its great extent it is composed of widespread horizontal rock-layers. The Swiss Jura and the Pennsylvanian Appalachians are units, for they consist of corrugated strata. The Laurentian highlands of Canada are essentially a unit, for they consist of greatly disturbed crystalline rocks. These geographical units have, however, no such simplicity as mathematical units; each one has a certain variety. The strata of plateaus are not strictly horizontal, for they slant or roll gently, now this way, now that. The corrugations of the Jura or of the Appalachians are not all alike; they might, indeed, be more truly described as all different, yet they preserve their essential features with much constancy. The disordered rocks of the Laurentian highlands have so excessively complicated a structure as at present to defy description, unless item by item; yet, in spite of the free variations from a single structural pattern, it is legitimate and useful to look in a broad way at such a region, and to regard it as a structural unit. The forces by which structures and attitudes have been determined do not come within the scope of geographical inquiry, but the structures acquired by the action of these forces serve as the essential basis for the genetic classification of geographical forms. For the purpose of this article, it will suffice to recognize two great structural groups: first, the group of horizontal structures, including plains, plateaus, and their derivatives, for which no single name has been suggested; second, the group of disordered structures, including mountains and their derivatives, likewise without a single name. The second group may be more elaborately subdivided than the first.

The destructive processes are of great variety—the chemical action of air and water, and the mechanical action of wind, heat, and cold, of rain and snow, rivers and glaciers, waves and currents. But as most of the land surface of the Earth is acted on chiefly by weather changes and running water, these will be treated as forming a normal group of destructive processes; while the wind of arid deserts and the ice of frigid deserts will be considered as climatic modifications of the norm, and set apart for particular discussion; and a special chapter will be needed to explain the action of waves and currents on the shore-lines at the edge of the lands. The various processes by which destructive work is done are in their turn geographical features, and many of them are well recognized as such, as rivers, falls, and glaciers; but they are too commonly considered by geographers apart from the work that they do, this phase of their study being, for some unsatisfactory reason, given over to physical geology. There should be no such separation of agency and work in physical geography, although it is profitable to give separate consideration to the active agent and to the inert mass on which it works.

Time as an element in geographical terminology

The amount of change caused by destructive processes increases with the passage of time, but neither the amount nor the rate of change is a simple function of time. The amount of change is limited, in the first place, by the altitude of a region above the sea; for, however long the time, the normal destructive forces cannot wear a land surface below this ultimate baselevel of their action; and glacial and marine forces cannot wear down a land-mass indefinitely beneath sea-level. The rate of change under normal processes, which alone will be considered for the present, is at the very first relatively moderate; it then advances rather rapidly to a maximum, and next slowly decreases to an indefinitely postponed minimum.

Evidently a longer period must be required for the complete denudation of a resistant than of a weak land-mass, but no measure in terms of years or centuries can now be given to the period needed for the effective wearing down of highlands to featureless lowlands. All historic time is hardly more than a negligible fraction of so vast a duration. The best that can be done at present is to give a convenient name to this unmeasured part of eternity, and for this purpose nothing seems more appropriate than a *geographical cycle.*" When it is possible to establish a ratio between geographical and geological units, there will probably be found an approach to equality between the duration of an average cycle and that of Cretaceous or Tertiary time, as has been indicated by the studies of several geomorphologists.

"Theoretical" geography

It is evident that a scheme of geographical classification that is founded on structure, process, and time, must be deductive in a high degree. This is intentionally and avowedly the case in the present instance. As a consequence, the scheme gains a very "theoretical" flavour that is not relished by some geographers, whose work implies that geography, unlike all other sciences, should be developed by the use of only certain ones of the mental faculties, chiefly observation, description, and generalization. But nothing seems to me clearer than that geography has already suffered too long from the disuse of imagination, invention, deduction, and the various other mental faculties that contribute towards the attainment of a well-tested explanation. It is like walking on one foot, or looking with one eye, to exclude from geography the "theoretical" half of the brain-power, which other sciences call upon as well as the "practical" half. Indeed, it is only as a result of misunderstanding that an antipathy is implied between theory and practice, for in geography, as in all sound scientific work, the two advance most amiably and effectively together. Surely the fullest development of geography will not be reached until all the mental faculties that are in any way pertinent to its cultivation are well trained and exercised in geographical investigation.

All this may be stated in another way. One of the most effective aids to the appreciation of a subject is a correct explanation of the facts that it presents. Understanding thus comes to aid the memory. But a genetic classification of geographical forms is, in effect, an explanation of them; hence such a classification must be helpful to the travelling, studying, or teaching geographer, provided only that it is a true and natural classification. True and natural a genetic classification may certainly be, for the time is past when even geographers can look on the forms of lands as "ready made." Indeed, geographical definitions and descriptions are untrue and unnatural just so far as they give the impression that the forms of the lands are of unknown origin, not susceptible of rational explanation. From the very beginning of geography in the lower schools, the pupils should be possessed with the belief that geographical forms have meaning, and that the meaning or origin of so many forms is already so well assured that there is every reason to think that the meaning of all the others will be discovered in due time. The explorer of the Earth should be as fully convinced of this principle, and as well prepared to apply it, as the explorer of the sky is to carry physical principles to the furthest reach of his telescope, his spectroscope, and his camera. The preparation of route-maps and the determination of latitude, longitude, and altitude for the more important points is only the beginning of exploration, which has no end till all the facts of observation are carried forward to explanation.

It is important, however, to insist that the geographer needs to know the meaning, the explanation, the origin, of the forms that he looks at, simply because of the aid thus received when he attempts to observe and describe the forms carefully. It is necessary clearly to recognize this principle, and constantly to bear it in mind, if we would avoid the error of confounding the objects of geographical and geological study. The latter examines the changes of the past for their own sake, inasmuch as geology is concerned with the history of the Earth; the former examines the changes of the past only so far as they serve to illuminate the present, for geography is concerned essentially with the Earth as it now exists. Structure is a pertinent element of geographical study when, as nearly always, it influences form; no one would to-day attempt to describe the Weald without some reference to the resistent chalk layers that determine its rimming hills. Process is equally pertinent to our subject, for it has everywhere been influential in determining form to a greater or less degree, and it is everywhere in operation to-day. It is truly curious to find geographical text-books which accept the movement of winds, currents, and rivers as part of their responsibility, and yet which leave the weathering of the lands and the movement of land-waste entirely out of consideration. Time is certainly an important geographical element, for where the forces of uplift or deformation have lately (as the Earth views time) initiated a cycle of change, the destructive processes can have accomplished but little work, and the land-form is "young;" where more time has elapsed, the surface will have been more thoroughly carved, and the form thus becomes "mature;" and where so much time has passed that the originally uplifted surface is worn down to a lowland of small relief, standing but little above sea-level, the form deserves to be called "old." A whole series of forms must be in this way evolved in the life-history of a single region, and all the forms of such a series, however unlike they may seem at first sight, should be associated under the element of time, as merely expressing the different stages of development of a single structure. The larva, the pupa, and the imago of an insect; or the acorn, the full-grown oak, and the fallen old trunk, are no more naturally associated as representing the different phases in the life-history of a single organic species, than are the young mountain block, the maturely carved mountain-peaks and valleys, and the old mountain peneplain, as representing the different stages in the life-history of a single geographic group. Like land-forms, the agencies that work upon them change their behaviour and their appearance with the passage of time. A young land-form has young streams of torrential activity, while an old form would have old streams of deliberate or even of feeble current, as will be more fully set forth below.

The ideal geographical cycle

The sequence in the developmental changes of land-forms is, in its own way, as systematic as the sequence of changes found in the more evident development of organic forms. Indeed, it is chiefly for this reason that the study of the origin of land-forms—or geomorphogeny, as some call it—becomes a practical aid, helpful to the geographer at every turn. This will be made clearer by the specific consideration of an ideal case, and here a graphic form of expression will be found of assistance.

The base-line, $\alpha\omega$, of Fig. 1 represents the passage of time, while verticals above the base-line measure altitude above sea-level. At the epoch 1, let a region of whatever structure and form be uplifted, B representing the average altitude of its higher parts, and A that of its lower parts; thus AB measuring its average initial relief. The surface rocks are attacked by the weather. Rain falls on the weathered surface, and washes some of the loosened waste down the initial slopes to the trough-lines where two converging slopes meet; there the streams are formed, flowing in directions consequent upon the descent of the troughlines. The machinery of the destructive processes is thus put in motion, and the destructive development of the region is begun. The larger rivers, whose channels initially had an altitude, A, quickly deepen their valleys, and at the epoch 2 have reduced their main channels to a moderate altitude, represented by C. The higher parts of the interstream uplands, acted on only by the weather without the concentration of water in streams, waste away much more slowly, and at epoch 2 are reduced in height only to D. The relief of the surface has thus been increased from AB

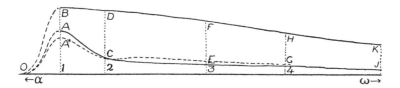

Figure 1

to CD. The main rivers then deepen their channels very slowly for the rest of their life, as shown by the curve CEGJ; and the wasting of the uplands, much dissected by branch streams, comes to be more rapid than the deepening of the main valleys, as shown by comparing the curves DFHK and CEGJ. The period 3–4 is the time of the most rapid consumption of the uplands, and thus stands in strong contrast with the period 1–2, when there was the most rapid deepening of the main valleys. In the earlier period, the relief was rapidly increasing in value, as steep-sided valleys were cut beneath the initial troughs. Through the period 2–3 the maximum value of relief is reached, and the variety of form is greatly increased by the headward growth of side valleys. During the period 3–4 relief is decreasing faster than at any other time, and the slope of the valley sides is becoming much gentler than before; but these changes advance much more slowly than those of the first period. From epoch 4 onward the remaining relief is gradually reduced to smaller and smaller measures, and the slopes become fainter and fainter, so that some time after the latest stage of the diagram the region is only a rolling lowland, whatever may have been its original height. So slowly do the later changes advance, that the reduction of the reduced relief JK to half of its value might well require as much time as all that which has already elapsed; and from the gentle slopes that would then remain, the further removal of waste must indeed be exceedingly slow. The frequency of torrential floods and of landslides in young and in mature mountains, in contrast to the quiescence of the sluggish streams and the slow movement of the soil on lowlands of denudation, suffices to show that rate of denudation is a matter of strictly geographical as well as of geological interest.

It follows from this brief analysis that a geographical cycle may be subdivided into parts of unequal duration, each one of which will be characterized by the strength and variety of relief, and by the rate of change, as well as by the amount of change that has been accomplished since the initiation of the cycle. There will be a brief youth of rapidly increasing relief, a maturity of strongest relief and greatest variety of form, a transition period of most rapidly yet slowly decreasing relief, and an indefinitely long old age of faint relief, on which further changes are exceedingly slow. There are, of course, no breaks between these subdivisions or stages; each one merges into its successor, yet each one is in the main distinctly characterized by features found at no other time.

The development of consequent streams

The preceding section gives only the barest outline of the systematic sequence of changes that run their course through a geographical cycle. The outline must be at once gone over, in order to fill in the more important details. In the first place, it should not be implied, as was done in Fig. 1, that the forces of uplift or deformation act so rapidly that no destructive changes occur during their operation. A more probable relation at the opening of a cycle of change places the beginning of uplift at O (Fig. 1), and its end at 1. The divergence of the curves OB and OA then implies that certain parts of the disturbed region were uplifted more than others, and that, from a surface of no relief at sea-level at epoch O, an upland having AB relief would be produced at epoch 1. But even during uplift, the streams that gather in the troughs as soon as they are defined do some work, and hence young valleys are already incised in the trough-bottoms when epoch 1 is reached, as shown by the curve OA'. The uplands also waste more or less during the period of disturbance, and hence no absolutely unchanged initial surface should be found, even for some time anterior to epoch 1. Instead of looking for initial divides separating initial slopes that descend to initial troughs followed by initial streams, such as were implied in Fig. 1 at the epoch of instantaneous uplift, we must always expect to find some greater or less advance in the sequence of developmental changes, even in the youngest known land-forms. "Initial" is therefore a term adapted to ideal rather than to actual cases, in treating which the term "sequential" and its derivatives will be found more appropriate. All the changes which directly follow the guidance of the ideal initial forms may be called consequent; thus a young form would possess consequent divides, separating consequent slopes which descend to consequent valleys; the initial troughs being changed to consequent valleys in so far as their form is modified by the action of the consequent drainage.

The grade of valley floors

The larger rivers soon—in terms of the cycle—deepen their main valleys, so that their channels are but little above the baselevel of the region; but the valley floor cannot be reduced to the absolute baselevel, because the river must slope down to its mouth at the sea-shore. The altitude of any point on a well-matured valley floor must therefore depend on river-slope and distance from mouth. Distance from mouth may here be treated as a constant, although a fuller statement would consider its increase in consequence of delta-growth. River-slope cannot be less, as engineers know very well, than a certain minimum that is determined by volume and by quantity and texture of detritus or load. Volume may be temporarily taken as a constant, although it may easily be shown to suffer important changes during the progress of a normal cycle. Load is small at the beginning, and rapidly increases in quantity and coarseness during youth, when the region is entrenched by steep-sided valleys; it continues to increase in quantity, but probably not in coarseness, during early maturity, when ramifying valleys are growing by headward erosion, and are thus increasing the area of wasting slopes; but after full maturity, load continually decreases in quantity and in coarseness of texture; and during old age, the small load that is carried must be of very fine texture or else must go off in solution. Let us now consider how the minimum slope of a main river will be determined.

In order to free the problem from unnecessary complications, let it be supposed that the young consequent rivers have at first slopes that are steep enough to make them all more than competent to carry the load that is washed into them from the wasting surface on either side, and hence competent to entrench themselves beneath the floor of the initial troughs, —this being the condition tacitly postulated in Fig. 1, although it evidently departs from those cases in which deformation produces basins where lakes must form and where deposition (negative denudation) must take place, and also from those cases in which a main-trough stream of moderate slope is, even in its youth, over-supplied with detritus by active side streams that descend steep and long wasting surfaces; but all these more involved cases may be set aside for the present.

If a young consequent river be followed from end to end, it may be imagined as everywhere deepening its valley, unless at the very mouth. Valley-deepening will go on most rapidly at some point, probably nearer head than mouth. Above this point the river will find its slope increased; below, decreased. Let the part up-stream from the point of most rapid deepening be called the headwaters; and the part down-stream, the lower course or trunk. In consequence of the changes thus systematically brought about, the lower course of the river will find its slope and velocity decreasing, and its load increasing; that is, its ability to do work is becoming less, while the work that it has to do is becoming greater. The original excess of ability over work will thus in time be corrected, and when an equality of these two quantities is brought about, the river is *graded*, this being a simple form of expression, suggested by Gilbert, to replace the more cumbersome phrases that are required by the use of "profile of equilibrium" of French engineers. When the graded condition is reached, alteration of slope can take place only as volume and load change their relation; and changes of this kind are very slow.

In a land-mass of homogeneous texture, the graded condition of a river would be (in such cases as are above considered) first attained at the mouth, and would then advance retrogressively up-stream. When the trunk streams are graded, early maturity is reached; when the smaller headwaters and side streams are also graded, maturity is far advanced; and when even the wet-weather rills are graded, old age is attained. In a land-mass of heterogeneous texture, the rivers will be divided into sections by the belts of weaker and stronger rocks that they traverse; each section of weaker rocks will in due time be graded with reference to the section of harder rock next downstream, and thus the river will come to consist of alternating quiet reaches and hurried falls or rapids. The less resistant of the harder rocks will be slowly worn down to grade with respect to the more resistant ones that are further down stream; thus the rapids will decrease in number, and only those on the very strongest rocks will long survive. Even these must vanish in time, and the graded condition will then be extended from mouth to head. The slope that is adopted when grade is assumed varies inversely with the volume; hence rivers retain steep headwaters long after their lower course is worn down almost level; but in old age, even the headwaters must have a gentle declivity and moderate velocity, free from all torrential features. The so-called "normal river," with torrential headwaters and well-graded middle and lower course, is therefore simply a maturely developed river. A young river may normally have falls even in its lower course, and an old river must be free from rapid movement even near its head.

If an initial consequent stream is for any reason incompetent to carry away the load that is washed into it, it cannot degrade its channel, but must aggrade instead (to use an excellent term suggested by Salisbury). Such a river then lays down the coarser part of the offered load, thus forming a broadening flood-land, building up its valley floor, and steepening its slope until it gains sufficient velocity to do the required work. In this case the graded condition is reached by filling up the initial trough instead of by cutting it down. Where basins occur, consequent lakes rise in them to the level of the outlet at the lowest point of the rim. As the outlet is cut down, it forms a sinking local baselevel with respect to which the basin is aggraded; and as the lake is thus destroyed, it forms a sinking baselevel with respect to which the tributary streams grade their valleys; but, as in the case of falls and rapids, the local baselevels of outlet and lake are temporary, and lose their control when the main drainage lines are graded with respect to absolute baselevel in early or late maturity.

The development of river branches

Several classes of side streams may be recognized. Some of them are defined by slight initial depressions in the side slopes of the main river-troughs: these form lateral or secondary consequents, branching from a main consequent; they generally run in the direction of the dip of the strata. Others are developed by headward erosion under the guidance of weak substructures that have been laid bare on the valley walls of the consequent streams: they follow the strike of the strata, and are entirely regardless of the form of the initial land surface; they may be called subsequent, this term having been used by Jukes in describing the development of such streams. Still others grow here and there, to all appearance by accident, seemingly independent of systematic guidance; they are common in horizontal or massive structures. While waiting to learn just what their control may be, their independence of apparent control may be indicated by calling them "insequent." Additional classes of streams are well known, but cannot be described here for lack of space.

Relation of river ability and load

As the dissection of a landmass proceeds with the fuller development of its consequent, subsequent, and insequent streams, the area of steep valley sides greatly increases from youth into early and full maturity. The waste that is delivered by the side branches to the main stream comes chiefly from the valley sides, and hence its quantity increases with the increase of strong dissection, reaching a maximum when the formation of new branch streams ceases, or when the decrease in the slope of the wasting valley sides comes to balance their increase of area. It is interesting to note in this connection the consequences that follow from two contrasted relations of the date for the maximum discharge of waste and of that for the grading of the trunk streams. If the first is not later than the second, the graded rivers will slowly assume gentler slopes as their load lessens; but as the change in the discharge of waste is almost infinitesimal compared to the amount discharged at any one time, the rivers will essentially preserve their graded condition in spite of the minute excess of ability over work. On the other hand, if the maximum of load is not reached until after the first attainment of the graded condition by the trunk rivers, then the valley floors will be aggraded by the deposition of a part of the increasing load, and thus a steeper slope and a greater velocity will be gained whereby the remainder of the increase can be borne along. The bottom of the V-shaped valley, previously carved, is thus slowly filled with a gravelly flood-plain, which continues to rise until the epoch of the maximum load is reached, after which the slow degradation above stated is entered upon. Early maturity may therefore witness a slight shallowing of the main valleys, instead of the slight deepening (indicated by the dotted line CE in Fig. 1); but late maturity and all old age will be normally occupied by the slow continuation of valley erosion that was so vigorously begun during youth.

The development of divides

There is no more beautiful process to be found in the systematic advance of a geographical cycle than the definition, subdivision, and rearrangement of the divides (water-partings) by which the major and minor drainage basins are separated. The forces of crustal upheaval and deformation act in a much broader way than the processes of land-sculpture; hence at the opening of a cycle one would expect to find a moderate number of large river-basins, somewhat indefinitely separated on the flat crests of broad swells or arches of land surface, or occasionally more sharply limited by the raised edge of faulted blocks. The action of the lateral consequent streams alone would, during youth and early maturity, sharpen all the vague initial divides into well-defined consequent divides, and the further action of insequent and subsequent streams would split up many consequent drainage slopes into subordinate drainage basins, separated by subdivides either insequent or subsequent. Just as the subsequent valleys are eroded by their gnawing streams along weak structural belts, so the subsequent divides or ridges stand up where maintained by strong structural belts. However imperfect the division of drainage areas and the discharge of rainfall may have been in early youth, both are well developed by the time full maturity is reached. Indeed, the more prompt discharge of rainfall that may be expected to result from the development of an elaborate system of subdivides and of slopes from divides to streams should cause an increased percentage of run-off; and it is possible that the increase of river-volume thus brought about from youth to maturity may more or less fully counteract the tendency of increase in river load to cause aggradation. But, on the other hand, as soon as the uplands begin to lose height, the rainfall must decrease; for it is well known that the obstruction to wind-movement caused by highlands is an effective cause of precipitation. While it is a gross exaggeration to maintain that the quaternary Alpine glaciers caused their own destruction by reducing the height of the mountains on which their snows were gathered, it is perfectly logical to deduce a decrease of precipitation as an accompaniment of loss of height from the youth to the old age of a landmass. Thus many factors must be considered before the life-history of a river can be fully analyzed.

The growth of subsequent streams and drainage areas must be at the expense of the original consequent streams and consequent drainage areas. All changes of this kind are promoted by the occurrence of inclined instead of horizontal rock-layers, and hence are of common occurrence in mountainous regions, but rare in strictly horizontal plains. The changes are also favoured by the occurrence of strong contrasts in the resistance of adjacent strata. In consequence of the migration of divides thus caused, many streams come to follow valleys that are worn down along belts of weak strata, while the divides come to occupy the ridges that stand up along the belts of stronger strata; in other words, the simple consequent drainage of youth is modified by the development of subsequent drainage lines, so as to bring about an *increasing adjustment of streams to structures*, than which nothing is more characteristic of the mature stage of the geographical cycle. Not only so: adjustments of this kind form one of the strongest, even if one of the latest, proofs of the erosion of valleys by the streams that occupy them, and of the long continued action in the past of the slow processes of weathering and washing that are in operation to-day.

There is nothing more significant of the advance in geographical development than the changes thus brought about. The processes here involved are too complicated to be now presented in detail, but they may be briefly illustrated by taking the drainage of a denuded arch, suggested by the Jura mountains, as a type example. AB, Fig. 2, is a main longitudinal consequent stream following a trough whose floor has been somewhat aggraded by the waste actively supplied by the lateral consequents, CD, LO, EF, etc. At an

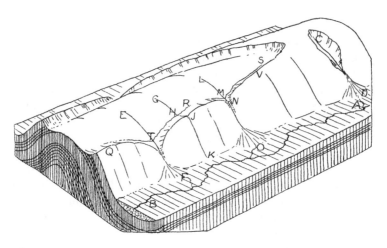

Figure 2

earlier stage of denudation, before the hard outer layer was worn away from the crown of the mountain arch, all the lateral consequents headed at the line of the mountain crest. But, guided by a weak under-stratum, subsequent streams, TR, MS, have been developed as the branches of certain lateral cousequents, EF, LO, and thus the hard outer layer has been undermined and partly removed, and many small lateral consequents have been beheaded. To-day, many of the laterals, like JK, have their source on the crest of the lateral ridge VJQ, and the headwaters, such as GH, that once belonged to them, are now diverted by the subsequent streams to swell the volume of the more successful laterals, like EF. Similar changes having taken place on the further slope of the mountain arch, we now find the original consequent divide of the arch-crest supplemented by the subsequent divides formed by the lateral ridges. A number of short streams, like JH, belonging to a class not mentioned above, run down the inner face of the lateral ridges to a subsequent stream, RT. These short streams have a direction opposite to that of the original consequents, and may therefore be called obsequents. As denudation progresses, the edge of the lateral ridge will be worn further from the arch-crest; in other words, the subsequent divide will migrate towards the main valley, and thus a greater length will be gained by the diverted consequent headwaters, GH, and a greater volume by the subsequents, SM and RT. During these changes the inequality that must naturally prevail between adjacent successful consequents, EF and LO, will eventually allow the subsequent branch, RT, of the larger consequent, EF, to capture the headwaters, LM and SM, of the smaller consequent, LO. In late maturity the headwaters of so many lateral consequents may be diverted to swell the volume of EF, that the main longitudinal consequent above the point F may be reduced to relatively small volume.

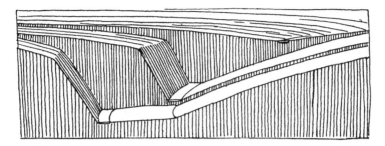

Figure 3

The development of river meanders

It has been thus far implied that rivers cut their channels vertically downward, but this is far from being the whole truth. Every turn in the course of a young consequent stream causes the stronger currents to press toward the outer bank, and each irregular, or, perhaps, subangular bend is thus rounded out to a comparatively smooth curve. The river therefore tends to depart from its irregular initial path (background block of Fig. 3) towards a serpentine course, in which it swings to right and left over a broader belt than at first. As the river cuts downwards and outwards at the same time, the valley-slopes become unsymmetrical (middle block of Fig. 3), being steeper on the side toward which the current is urged by centrifugal force. The steeper valley side thus gains the form of a half-amphitheatre, into which the gentler sloping side enters as a spur of the opposite uplands. When the graded condition is attained by the stream, downward cutting practically ceases, but outward cutting continues; a normal flood-plain is then formed as the channel is withdrawn from the gently sloping side of the valley (foreground block of Fig. 3). Flood-plains of this kind are easily distinguished in their early stages from those already mentioned (formed by aggrading the flat courses of incompetent young rivers, or by aggrading the graded valleys of overloaded rivers in early maturity); for these occur in detached lunate areas, first on one side, then on the other side of the stream, and always systematically placed at the foot of the gentler sloping spurs. But, as time passes, the river impinges on the up-stream side, and withdraws from the down-stream side of every spur, and thus the spurs are gradually consumed; they are first sharpened, so as better to observe their name; they are next reduced to short cusps; then they are worn back to blunt salients; and finally, they are entirely consumed, and the river wanders freely on its open flood-plain, occasionally swinging against the valley side, now here, now there. By this time the curves of youth are changed into systematic meanders, of radius appropriate to river volume; and, for all the rest of an undisturbed life, the river persists in the habit of serpentine flow. The less the slope of the flood-plain becomes in advancing old age, the larger the arc of each meander, and hence the longer the course of the river from any point to its mouth. Increase of length from this cause must tend to diminish fall, and thus to render the river less competent than it was before; and the result of this tendency will be to retard the already slow process by which a gently sloping flood-plain is degraded so as to approach coincidence with a level surface; but it is not likely that old rivers often remain undisturbed long enough for the full realization of these theoretical conditions.

The migration of divides must now and then result in a sudden increase in the volume of one river and in a correspondingly sudden decrease of another. After such changes, accommodation to the changed volume must be made in the meanders of each river affected. The one that is increased will call for enlarged dimensions; it will usually adopt a gentler slope, thus terracing its flood-plain, and demand a greater freedom of swinging, thus widening its valley. The one that is decreased will have to be satisfied with smaller dimensions; it will wander aimlessly in relatively minute meanders on its flood-plain, and from increase of length, as well as from loss of volume, it will become incompetent to transport the load brought in by the side streams, and thus its flood-plain must be aggraded. There are beautiful examples known of both these peculiar conditions.

The development of graded valley sides

When the migration of divides ceases in late maturity, and the valley floors of the adjusted streams are well graded, even far toward the headwaters, there is still to be completed another and perhaps even more remarkable sequence of systematic changes than any yet described: this is the development of graded waste slopes on the valley sides. It is briefly stated that valleys are eroded by their rivers; yet there is a vast amount of work performed in the erosion of valleys in which rivers have no part. It is true that rivers deepen the valleys in the youth, and widen the valley floors during the maturity and old age of a cycle, and that they carry to the sea the waste denuded from the land; it is this work of transportation to the sea that is peculiarly the function of rivers; but the material to be transported is supplied chiefly by the action of the weather on the steeper consequent slopes and on the valley sides. The transportation of the weathered material from its source to the stream in the valley bottom is the work of various slow-acting processes, such as the surface wash of rain, the action of ground water, changes of temperature, freezing and thawing, chemical disintegration and hydration, the growth of plant-roots, the activities of burrowing animals. All these cause the weathered rock waste to wash and creep slowly downhill, and in the motion thus ensuing there is much that is analogous to the flow of a river. Indeed, when considered in a very broad and general way, a river is seen to be a moving mixture of water and waste in variable proportions, but mostly water; while a creeping sheet of hillside waste is a moving mixture of waste and water in variable proportions, but mostly waste. Although the river and the hillside waste-sheet do not resemble each other at first sight, they are only the extreme members of a continuous series; and when this generalization is appreciated, one may fairly extend the "river" all over its basin, and up to its very divides. Ordinarily treated, the river is like the veins of a leaf; broadly viewed, it is like the entire leaf. The verity of this comparison may be more fully accepted when the analogy, indeed, the homology, of waste-sheets and water-streams is set forth.

In the first place, a waste-sheet moves fastest at the surface and slowest at the bottom, like a water-stream. A graded waste-sheet may be defined in the very terms applicable to a graded water-stream; it is one in which the ability of the transporting forces to do work is equal to the work that they have to do. This is the condition that obtains on those evenly slanting, waste-covered mountain-sides which have been reduced to a slope that engineers call "the angle of repose," because of the apparently stationary condition of the creeping waste, but that should be called, from the physiographic stand-point, "the angle of first-developed grade." The rocky cliffs and ledges that often surmount graded slopes are not yet graded; waste is removed from them faster than it is supplied by local weathering and by creeping from still higher slopes, and hence the cliffs and ledges are left almost bare; they correspond to falls and rapids in water-streams, where the current is so rapid that its cross-section is much reduced. A hollow on an initial slope will be filled to the angle of grade by waste from above; the waste will accumulate until it reaches the lowest point on the rim of the hollow, and then outflow of waste will balance inflow; and here is the evident homologue of a lake.

In the second place, it will be understood, from what has already been said, that rivers normally grade their valleys retrogressively from the mouth headwards, and that small side streams may not be graded till long after the trunk river is graded. So with waste-sheets; they normally begin to establish a graded condition at their base, and then extend it up the slope of the valley side whose waste they "drain." When rock-masses of various resistance are exposed on the valley side, each one of the weaker is graded with reference to the stronger one next downhill; and the less resistant of the stronger ones are graded with reference to the more resistant (or with reference to the base of the hill): this is perfectly comparable to the development of graded stretches and to the extinction of falls and rapids in rivers. Ledges remain ungraded on ridge-crests and on the convex front of hill spurs long after the graded condition is reached in the channels of wet-weather streams in the ravines between the spurs; this corresponds nicely with the slower attainment of grade in small side streams than in large trunk rivers. But as late maturity passes into old age, even the ledges on ridge-crests and spur-fronts disappear, all being concealed in a universal sheet of slowly creeping waste. From any point on such a surface a graded slope leads the waste down to the streams. At any point the agencies of removal are just able to cope with the waste that is there weathered *plus* that which comes from further uphill. This wonderful condition is reached in certain well-denuded mountains, now subdued from their mature vigour to the rounded profiles of incipient old age. When the full meaning of their graded form is apprehended, it constitutes one of the strongest possible arguments for the sculpture of the lands by the slow processes of weathering, long continued. To look upon a landscape of this kind without any recognition of the labour expended in producing it, or of the extraordinary adjustments of streams to structures, and of waste to weather, is like visiting Rome in the ignorant belief that the Romans of to-day have had no ancestors.

Just as graded rivers slowly degrade their courses after the period of maximum load is past, so graded waste-sheets adopt gentler and gentler slopes when the upper ledges are consumed and coarse waste is no longer plentifully shed to the valley sides below. A changing adjustment of a most delicate kind is here discovered. When the graded slopes are first developed, they are steep, and the waste that covers them is coarse and of moderate thickness; here the strong agencies of removal have all they can do to dispose of the plentiful supply of coarse waste from the strong ledges above, and the no less plentiful supply of waste that is weathered from the weaker rocks beneath the thin cover of detritus. In a more advanced stage of the cycle, the graded slopes are moderate, and the waste that covers them is of finer texture and greater depth than before; here the weakened agencies of removal are favoured by the slower weathering of the rocks beneath the thickened waste cover, and by the greater refinement (reduction to finer texture) of the loose waste during its slow journey. In old age, when all the slopes are very gentle, the agencies of waste-removal must everywhere be weak, and their equality with the processes of waste-supply can be maintained only by the reduction of the latter to very low values. The wastesheet then assumes a great thickness—even 50 or 100 feet—so that the progress of weathering is almost *nil*; at the same time, the surface waste is reduced to extremely fine texture, so that some of its particles may be moved even on faint slopes. Hence the occurrence of deep soils is an essential feature of old age, just as the occurrence of bare ledges is of youth. The relationships here obtaining are as significant as those which led Playfair to his famous statement concerning the origin of valleys by the rivers that drain them.

Old age

Maturity is past and old age is fully entered upon when the hilltops and the hillsides, as well as the valley floors, are graded. No new features are now developed, and those that have been earlier developed are weakened or even lost. The search for weak structures and the establishment of valleys along them has already been thoroughly carried out; now the larger streams meander freely in open valleys and begin to wander away from the adjustments of maturity. The active streams of the time of greatest relief now lose their headmost branches, for the rainfall is lessened by the destruction of the highlands, and the run-off of the rain water is retarded by the flat slopes and deep soils. The landscape is slowly tamed from its earlier strength, and presents only a succession of gently rolling swells alternating with shallow valleys, a surface everywhere open to occupation. As time passes, the relief becomes less and less; whatever the uplifts of youth, whatever the disorder and hardness of the rocks, an almost featureless plain (a peneplain) showing little sympathy with structure, and controlled only by a close approach to baselevel, must characterize the penultimate stage of the uninterrupted cycle; and the ultimate stage would be a plain without relief.

Some observers have doubted whether even the penultimate stage of a cycle is ever reached, so frequently do movements in the Earth's crust cause changes in its position with respect to baselevel. But, on the other hand, there are certain regions of greatly disordered structure, whose small relief and deep soils cannot be explained without supposing them to have, in effect, passed through all the stages above described—and doubtless many more, if the whole truth were told—before reaching the penultimate, whose features they verify. In spite of the great disturbances that such regions have suffered in past geological periods, they have afterwards stood still so long, so patiently, as to be worn down to pene-plains over large areas, only here and there showing residual reliefs where the most resistant rocks still stand up above the general level. Thus verification is found for the penultimate as well as for many earlier stages of the ideal cycle. Indeed, although the scheme of the cycle is here presented only in theoretical form, the progress of developmental changes through the cycle has been tested over and over again for many structures and for various stages; and on recognizing the numerous accordances that are discovered when the consequences of theory are confronted with the facts of observation, one must feel a growing belief in the verity and value of the theory that leads to results so satisfactory.

It is necessary to repeat what has already been said as to the practical application of the principles of the geographical cycle. Its value to the geographer is not simply in giving explanation to land-forms; its greater value is in enabling him to see what he looks at, and to say what he sees. His standards of comparison, by which the unknown are likened to the known, are greatly increased over the short list included in the terminology of his school-days. Significant features are consciously sought for; exploration becomes more systematic and less haphazard. "A hilly region" of the unprepared traveller becomes (if such it really be) "a maturely dissected upland" in the language of the better prepared traveller; and the reader of travels at home gains greatly by the change. "A hilly region" brings no definite picture before the mental eyes. "A maturely dissected upland" suggests a systematic association of well-defined features; all the streams at grade, except the small headwaters; the larger rivers already meandering on flood-plained valley floors; the upper branches ramifying among spurs and hills, whose flanks show a good beginning of graded slopes; the most resistant rocks still cropping out in ungraded ledges, whose arrangement suggests the structure of the region. The practical value of this kind of theoretical study seems to me so great that, among various lines of work that may be encouraged by the Councils of the great Geographical Societies, I believe there is none that would bring larger reward than the encouragement of some such method as is here outlined for the systematic investigation of land-forms.

Some geographers urge that it is dangerous to use the theoretical or explanatory terminology involved in the practical application of the principles of the geographical cycle; mistakes may be made, and harm would thus be done. There are various sufficient answers to this objection. A very practical answer is that suggested by Penck, to the effect that a threefold terminology should be devised—one set of terms being purely empirical, as "high," "low," "cliff," "gorge," "lake," "island;" another set being based on structural relations, as "monoclinal ridge," "transverse valley," "lava-capped mesa;" and the third being reserved for explanatory relations, as "mature dissection," "adjusted drainage," "graded slopes." Another answer is that the explanatory terminology is not really a novelty, but only an attempt to give a complete and systematic expansion to a rather timid beginning already made; a sand-dune is not simply a hillock of sand, but a hillock heaped by the wind; a delta is not simply a plain at a river mouth, but a plain formed by river action; a volcano is not simply a mountain of somewhat conical form, but a mountain formed by eruption. It is chiefly a matter of experience and temperament where a geographer ceases to apply terms of this kind. But little more than half a century ago, the erosion of valleys by rivers was either doubted or not thought of by the practical geographer; to-day, the mature adjustment of rivers to structures is in the same position; and here is the third, and to my mind the most important, answer to those conservatives who would maintain an empirical position for geography, instead of pressing forward toward the rational and explanatory geography of the future. It cannot be doubted, in view of what has already been learned to-day, that an essentially explanatory treatment must in the next century be generally adopted in all branches of geographical study; it is full time that an energetic beginning should be made towards so desirable an end.

Interruptions of the ideal cycle

One of the first objections that might be raised against a terminology based on the sequence of changes through the ideal uninterrupted cycle, is that such a terminology can have little practical application on an Earth whose crust has the habit of rising and sinking frequently during the passage of geological time. To this it may be answered, that if the scheme of the geographical cycle were so rigid as to be incapable of accommodating itself to the actual condition of the Earth's crust, it would certainly have to be abandoned as a theoretical abstraction; but such is by no means the case. Having traced the normal sequence of events through an ideal cycle, our next duty is to consider the effects of any and all kinds of movements of the land-mass with respect to its baselevel. Such movements must be imagined as small or great, simple or complex, rare or frequent, gradual or rapid, early or late. Whatever their character, they will be called "interruptions," because they determine a more or less complete break in processes previously in operation, by beginning a new series of processes with respect to the new baselevel. Whenever interruptions occur, the pre-existent conditions that they interrupt can be understood only after having analyzed them in accordance with the principles of the cycle, and herein lies one of the most practical applications of what at first seems remotely theoretical. A land-mass, uplifted to a greater altitude than it had before, is at once more intensely attacked by the denuding processes in the new cycle thus initiated; but the forms on which the new attack is made can only be understood by considering what had been accomplished in the preceding cycle previous to its interruption. It will be possible here to consider only one or two specific examples from among the multitude of interruptions that may be imagined.

Let it be supposed that a maturely dissected land-mass is evenly uplifted 500 feet above its former position. All the graded streams are hereby revived to new activities, and proceed to entrench their valley floors in order to develop graded courses with respect to the new baselevel. The larger streams first show the effect of the change; the smaller streams follow suit as rapidly as possible. Falls reappear for a time in the river-channels, and then are again worn away. Adjustments of streams to structures are carried further in the second effort of the new cycle than was possible in the single effort of the previous cycle. Graded hillsides are undercut; the waste washes and creeps down from them, leaving a long even slope of bare rock; the rocky slope is hacked into an uneven face by the weather, until at last a new graded slope is developed. Cliffs that had been extinguished on graded hillsides in the previous cycle are thus for a time brought to life again, like the falls in the rivers, only to disappear in the late maturity of the new cycle.

The combination of topographic features belonging to two cycles may be called "composite topography," and many examples could be cited in illustration of this interesting association. In every case, description is made concise and effective by employing a terminology derived from the scheme of the cycle. For example, Normandy is an uplifted peneplain, hardly yet in the mature stage of its new cycle; thus stated, explanation is concisely given to the meandering course of the rather narrow valley of the Seine, for this river has carried forward into the early stages of the new cycle the habit of swinging in strong meanders that it had learned in the later stages of the former cycle.

If the uplift of a dissected region be accompanied by a gentle tilting, then all the water-streams and waste-streams whose slope is increased will be revived to new activity; while all those whose slope is decreased will become less active. The divides will migrate into the basins of the less active streams, and the revived streams will gain length and drainage area. If the uplift be in the form of an arch, some of the weaker streams whose course is across the axis of the arch may be, as it were, "broken in half;" a reversed direction of flow may be thus given to one part of the broken stream; but the stronger rivers may still persevere across the rising arch in spite of its uplift, cutting down their channels fast enough to maintain their direction of flow unchanged; and such rivers are known as "antecedent."

The changes introduced by an interruption involving depression are easily deduced. Among their most interesting features is the invasion of the lower valley floors by the sea, thus "drowning" the valleys to a certain depth, and converting them into bays. Movements that tend to produce trough-like depressions across the course of a river usually give birth to a lake of water or waste in the depressed part of the river valley. In mountain ranges frequent and various interruptions occur during the long period of deformation; the Alps show so many recent interruptions that a student there would find little use for the ideal cycle; but in mountain regions of ancient deformation, the disturbing forces seem to have become almost extinct, and there the ideal cycle is almost realized. Central France gives good illustration of this principle. It is manifest that one might imagine an endless number of possible combinations among the several factors of structure, stage of development at time of interruption, character of interruption, and time since interruption; but space cannot be here given to their further consideration.

Accidental departures from the ideal cycle

Besides the interruptions that involve movements of a land-mass with respect to baselevel, there are two other classes of departure from the normal or ideal cycle that do not necessarily involve any such movements: these are changes of climate and volcanic eruptions, both of which occur so arbitrarily as to place and time that they may be called "accidents." Changes of climate may vary from the normal towards the frigid or the arid, each change causing significant departures from normal geographical development. If a reverse change of climate brings back more normal conditions, the effects of the abnormal "accident" may last for some small part of a cycle's duration before they are obliterated. It is here that features of glacial origin belong, so common in northwestern Europe and north-eastern America. Judging by the present analysis of glacial and interglacial epochs during quaternary time, or of humid and arid epochs in the Great Salt Lake region, it must be concluded that accidental changes may occur over and over again within a single cycle.

In brief illustration of the combined interruptions and accidents, it may be said that southern New England is an old mountain region, which had been reduced to a pretty good peneplain when further denudation was interrupted by a slanting uplift, with gentle descent to the south-east; that in the cycle thus introduced the tilted peneplain was denuded to a sub-mature or late mature stage (according to the strength or weakness of its rocks); and that the maturely dissected region was then glaciated and slightly depressed so recently that little change has happened since. An instructive picture of the region may be conceived from this brief description.

Many volcanic eruptions produce forms so large that they deserve to be treated as new structural regions; but when viewed in a more general way, a great number of eruptions, if not the greater number, produce forms of small dimensions compared to those of the structures on which they are superposed: the volcanoes of central France are good instances of this relation. Thus considered, volcanoes and lava-flows are so arbitrarily placed in time and space that their classification under the head of "accidents" is warranted. Still further ground for this classification is found when the effects of a volcanic eruption on the pre-existent processes of land-sculpture are examined. A valley may be blockaded by a growing cone and its lava-flows; lakes may form in the up-stream portion of such a valley, even if it be mature or old. If the blockade be low, the lake will overflow to one side of the barrier, and thus the river will be locally displaced from its former course, however well adjusted to a weak structure that course may have been. If the blockade be higher than some points on the headwater divides, the lake will overflow "backwards," and the upper part of the river system will become tributary to an adjacent system. The river must cut a gorge across the divide, however hard the rocks are there; thus systematic adjustments to structure are seriously interfered with, and accidental relations are introduced. The form of the volcanic cone and the sprawling flow of its lava-streams are quite out of accord with the forms that characterize the surrounding region. The cone arbitrarily forms a mountain, even though the subjacent rocks may be weak; the lava-flows aggrade valleys that should be degraded. During the dissection of the cone, a process that is systematic enough if considered for itself alone, a radial arrangement of spurs and ravines will be developed; in long future time the streams of such ravines may cut down through the volcanic structures, and thus superpose themselves most curiously on the underlying structures. The lava-flows, being usually more resistant than the rocks of the district that they invade, gain a local relief as the adjoining surface is lowered by denudation; thus an inversion of topography is brought about, and a "table-mountain" comes to stand where formerly there had been the valley that guided the original course of the lava-flow. The table-mountain may be quite isolated from its volcanic source, where the cone is by this time reduced to a knob or "butte." But although these various considerations seem to me to warrant the classification of volcanic forms as "accidental," in contrast to the systematic forms with which they are usually associated, great importance should not be attached to this method of arrangement; it should be given up as soon as a more truthful or more convenient classification is introduced.

The forms assumed by land waste

An extension of the subject treated in the section on Graded Valley Sides, would lead to a general discussion of the forms assumed by the waste of the land on the way to the sea; one of the most interesting and profitable topics for investigation that has come under my notice. Geographers are well accustomed to giving due consideration to the forms assumed by the water-drainage of the land on the way to the sea, and a good terminology is already in use for naming them; but much less consideration is given to the forms assumed by the waste that slowly moves from the land to the sea. They are seldom presented in their true relations; many of them have no generally accepted names—for example, the long slopes of waste that reach forward from the mountains into the desert basins of Persia; forms as common as alluvial fans are unmentioned in all but the most recent school-books; and such features as till plains, moraines, and drumlins are usually given over to the geologist, as if the geographer had nothing to do with them! There can be no question of the great importance of waste-forms to the geographer, but it is not possible here to enter into their consideration. Suffice it to say that waste-forms constitute a geographical group which, like water-forms, stand quite apart from such groups as mountains and plateaus. The latter are forms of structure, and should be classified according to the arrangement of their rocks, and to their age or stage of development. The former are forms of process, and should be classified according to the processes involved, and to the stage that they have reached. The application of this general principle gives much assistance in the description of actual landscapes.

Lack of space prevents due consideration here of the development of shore-lines, a subject not less interesting, suggestive, and helpful than the development of inland forms; but I shall hope to return on some later occasion to a discussion of shore features, when it may be found that much of the terminology already introduced is again applicable. In closing this article, I must revert, if even for a third time, to the practical side of the theoretical cycle, with its interruptions and accidents. It cannot be too carefully borne in mind that the explanation of the origin of land-forms is not for its own sake added to the study of geography, but for the sake of the aid that explanation gives to the observation and description of existing geographical features. The sequence of forms developed through the cycle is not an abstraction that one leaves at home when he goes abroad; it is literally a *vade-mecum* of the most serviceable kind. During the current year that I am spending in Europe, the scheme and the terminology of the cycle have been of the greatest assistance in my studies. Application of both scheme and terminology is found equally well in the minute and infantile coastal plains that border certain stretches of the Scotis shore-line in consequence of the slight post-glacial elevation of the land, and in the broad and aged central plateau of France, where the young valleys of to-day result from the uplift of the region, and the revival of its rivers after they had sub-maturely dissected a pre-existent peneplain. The adjustments of streams to structures brought about by the interaction of the waxing Severn and the waning Thames, prove to be even more striking than when I first noticed them in 1894.[1] The large ancient delta of the Var, between Nice and Cannes, now uplifted more than 200 metres, and maturely dissected, must come to be the type example of this class of forms. The Italian Riviera, west of Genoa, may be concisely described as a region of subdued mountains that has been partly submerged and that is now approaching maturity of shore-line features in the cycle thus initiated: one may picture, from this brief statement, the mountain spurs with wellgraded slopes, limited by a very irregular shore-line when first depressed, but now fronting in a comparatively simple shore-line of cliffed headlands and filled bays. The peninsula of Sorrento, on its northern side, once resembled the Riviera, but it has now been elevated 50 metres, and its uplifted bay-plains have cliffed fronts. The lower Tiber, whose mature valley floor is now somewhat wider than its meander belt, is consequent upon a volcanic accident, for it follows the trough between the slopes of the Bracciano volcanic centre on the north-west, and the Alban centre on the south-east; further up-stream, as far as Orvieto, the river, as a rule, follows a trough between the Apennines and the three volcanic centres of Bolsena, Vico, and Bracciano. The Lepini mountains, a maturely carved block of moderately deformed Cretaceous limestones south of the Alban volcanic group, has along a part of its north-eastern base a very young fault cliff, by which the graded slopes of the spurs and ravines are abruptly cut off; the fault cliff is easily recognized from the train on the line between Rome and Naples.

Botanists and zoologists know very well that a trained observer can easily recognize and describe many small items of form that pass without notice from the untrained observer. It is the same in geography, and the only question is—How can the desired training be secured? Of the many methods of geographical training, I believe that, as far as the forms of the land are concerned, no method can equal the value of one in which explanation is made an essential feature along with observation, for there is no other in which so many mental faculties are exercised.

Note

1 See *Geographical Journal*, 1895; and *Proceedings Geologists' Association*, 1899.

2

THE MID-TERTIARY CYCLE OF EROSION AND THE SUMMIT PENEPLAIN

S.W. Wooldridge and D.L. Linton

Source: S.W. Wooldridge and D.L. Linton, *Structure, Surface and Drainage in South-East England*, London: George Philip and Son, 1955, pp. 34–44.

In the physiographic evolution of most regions there comes a point in time when a history of predominant deposition is replaced by one of predominant erosion. The record provided by the stratigraphical succession is abruptly terminated, to be resumed only in an intermittent and fragmentary fashion, and the evidences available for physiographic reconstruction are thenceforward of a radically different character. In south-eastern England such a critical epoch was reached in Oligocene times when the long record of deposition was brought to an end by the onset of the Mid-Tertiary earth-movements. From that time down to the present day the phases of deposition have been so restricted in both space and time and their products so largely destroyed by later erosion that at best they can provide only a disjointed picture of a few episodes in the long period of later Tertiary and Quaternary time. Because of this inescapable deficiency in the stratigraphical evidence the main burden of elucidating these 'Dark Ages' must be borne by the evidences provided by the successive denudation stages. Instead of the clues available to the geologist in the lithological character and geographical distribution of sedimentary deposits, we have now to employ those afforded by the form, altitudinal relationships and geographical extent of *surfaces* which have resulted from former local or widespread base-levelling.

The essential basis of the technique for the morphological reconstruction of physiographic evolution thus comprises the recognition of former base-levelled surfaces now uplifted and partially dissected, or else buried and possibly in part exhumed. They represent periods of still-stand separated by relatively short periods of rising or falling base-level, and when their proper relationships are determined they offer a veritable chronology of denudation, which is independent of the doubtful witness of the superficial deposits, and is potentially more reliable and certainly more complete.

Little progress could, of course, be made toward the elucidation of such a chronology for south-eastern England until recent years, since the theoretical concepts upon which it is based have all been worked out within the last forty or fifty years. Nevertheless, the recognition of what we must now regard as the most important of all the base-levelled surfaces of the region, is nearly a century old. As long ago as 1846, Ramsay[16] recognized that the greater number of the hill tops and ridge crests of South and Central Wales, whatever their geological character, rise to much the same level and could, indeed, be regarded as having been derived, by the excavation of the present valleys, from a single continuous surface or plain of denudation, which in consonance with the views of his time, he attributed to the work of the sea. In the generation following Ramsay's classic observations, other workers had recourse to the idea of a plain of marine denudation represented by the present hill tops. Thus J.B. Jukes,[17] writing in 1862 of the rivers of southern Ireland, found it possible to explain their present courses only by assuming that the original streams had been formed upon 'a surface considerably above any part of the present surface', which had been previously planed down by the sea. Fourteen years later Topley adopted a similar point of view with regard to the Weald. In considering the views of Col. George Greenwood, he expresses doubt whether the result of the long-continued action of rain and rivers upon an upheaved dome of Chalk 'would be the present system of longitudinal and transverse valleys. To achieve this result it is necessary that the country should be *planed across* so that successive rocks should in turn crop up to the surface'.[18] His belief in such planation by the sea was, however, based also upon the evidence of accordant summit levels, both in the English Plain generally and in the Weald in particular. On this point he is again worthy of citation, and in regard to South-East England as a whole he writes: 'If we draw a line from the south-eastern end of the Eastern Moorlands of Yorkshire, past the southern end of the Pennine Chain, the Malvern Hills, and the Quantocks, to the south-eastern corner of Dartmoor, we shall find that all the summits to the south-east of that line, with very few exceptions, are below 1,000 feet. The only exceptions are both on the Cotteswolds (the Inferior Oolite escarpment) north-east of Cheltenham; they are Cleeve Cloud 1,080 feet and Broadway Hill 1,032 feet*. The summits of the various hills and escarpments rise to 700, 800 and in a very few cases to 900 feet. They evidently point to the former existence of a great plain of marine denudation over this area. This plain is far older than the glacial period; for the escarpments around which the Boulder Clay and

---

* Precise levelling by the Ordnance Survey later corrected these figures to 1,070 and 1,042 feet.

other deposits of that period have accumulated, have been carved out of it. It is later than the Eocene period; for outliers of that formation lie scattered over the plateaux, far up towards, and often upon the escarpments'.[18]

The Wealden region received more detailed discussion, and here Topley quoted in support of his thesis the altitudes of a fairly large number of summits, some of which have since been somewhat revised as the result of precise levelling. He recognized the importance of the close accordance of summit levels in a longitudinal sense, along each of the major uplands, the North and South Downs, and the Central Weald, and from his figures was led to conclude that the summit plane possessed a regional tilt toward the east. He also stressed the more significant though less perfect accord of summits in a transverse sense, along lines which passed across each of the geological formations in turn. Uniformity in this sense clearly argued that the planation occurred at some date after the rocks had been folded, and the age of the plain is thus broadly fixed as being later than the Mid-Tertiary earth-movements.

When W.M. Davis came to England in 1895, he brought with him clearer notions than those possessed by most of his contemporaries as to the power of fluvial erosion. He therefore felt himself under no compulsion to regard every planed-off surface as being the work of marine erosion; a plain of fluvial erosion was equally possible. On deductive grounds he argued that a scarpland region which had suffered such planation would show a number of distinctive features, viz. straightness of the escarpments in plan, a drainage system with a large proportion of subsequent streams well adjusted to structure, and bevelled summits to the uplands developed upon the more resistant formations. All these features, he observed, were to be found in the English Plain and to some extent found in combination.

He regarded their combined witness as sufficient evidence that 'the rivers of to-day ... appear to be the revived and matured successors of a well adjusted system of consequent and subsequent drainage inherited from an earlier and far advanced cycle of denudation'.[19]

Davis's views were so divergent from those prevalent in England at the time, that no attempt was made either to prove or to disprove them in their application to South-East England until H. Bury,[20] in 1910, made a critical analysis of the denudation of the western Weald. As a result of his examination he materially increased the weight of the evidence in favour of a planation of the region, but he regarded it as being due to marine agencies. He pointed out that, in the Weald, straightness of the escarpments appeared to be due chiefly, if not wholly, to to parallelism with longitudinal folds, and that, in some cases at least, Davis's argument concerning the subsequent streams could be shown to be invalid. Evidence on this point was chiefly drawn from the widespread occurrence of fragments of Lower Greensand chert upon the crest of the North Downs. Clearly such debris must have been carried to this position across the site of the present subsequent Vale of Holmesdale, and whether the emplacement of the chert-bearing gravels be referred to fluvial or marine action, it is clear that this particular subsequent valley must have arisen wholly in the present cycle. By similar arguments Bury was able to show that both the subsequent tributaries of the Wey are also of quite late development, having arisen well within the span of the present erosion cycle. This late out-growth of subsequents argued strongly against Davis's contention that the drainage system is in its second cycle. If the streams are survivors from a previous cycle, why did they not earlier develop subsequent tributaries? If, on the other hand, they are still in their first cycle how else could they have arisen but upon an emerged plain of marine denudation?

At this date, then, the evidence brought forward in favour of the idea of planation was drawn either from the accordance of summit levels and the bevelled form of the escarpments, or from features in the development of the drainage systems. On the whole, moreover, the evidence favoured a marine origin for the plain. But practically no attempt was made to relate this inferred marine plain, as regards its age and geographical extent, with the known Pliocene outliers of South-East England. Later it was shown that these high-level marine Pliocene deposits do rest upon a true plain of marine denudation which could be traced over most of the London Basin and in parts of the North Downs and Chiltern Hills.[21] The same work showed, however, that the highest portions of both uplands – roughly the areas above 700 feet – escaped submergence (p. 55). These observations have thus permitted a solution of the long-standing difficulty as to the marine or sub-aerial origin of the summit plain of South-East England and have reconciled the conflict of evidence upon which the older views were based. Over much of the area, including the region of South-West Surrey from which Bury drew his best evidence, true marine planation was proved. Other tracts were shown equally certainly not to have been submerged and here the planation, already inferred upon other grounds, must necessarily be attributed to sub-aerial agencies. Marine erosion, indeed, merely completed the work of rain and rivers by perfecting, over limited areas, a planation that was already far-advanced. Further work has brought to light other criteria which assist in the delimitation of the portions of the peneplain which were submerged, from those in which sub-aerial erosion was never interrupted. These points, however, may well be deferred until a later section where a full discussion of the Pliocene phase is attempted. Here it is important simply to emphasize the significance of the transgression as revealing definitively the extent of the unsubmerged areas, and as an invaluable index of the age of the planation. To whatever epoch in the Pliocene period the deposits of the marine phase are assigned by the palaeontologist, the preceding phase of widespread base-levelling must have been far advanced by early Pliocene times.

We are thus provided with a major landmark in our denudation chronology, the summit peneplain of south-eastern England being its visible expression. It marks a critical point in time, the end of one cycle and the beginning of another. Its production was the work of that long period of still-stand which we term the Mid-Tertiary erosion cycle. Its uplift to its present position and dissection in its present form have been accomplished during the cycle which is yet incomplete, the cycle of late Tertiary and Quaternary time. It is the oldest feature in the landscape of to-day, a surviving relic from a previous order of things, and its destruction and disappearance is now all but accomplished.

Here we may turn from the consideration of the evidence which has led to the understanding of the nature and age of the summit peneplain in South-East England to notice briefly the observations made by earlier writers upon its western extensions in Wessex and the south-west. Several portions of the upland surface of these regions have been recognized as plains of denudation, but their age has been largely a matter of rather vague inference and few objective criteria have been advanced to settle the question of their origin, which, following British tradition, has usually been attributed to marine agencies.

In the same year that Davis published his classic paper, 'On the Development of Certain English Rivers', Jukes-Browne[22] described the origin of the valley system of the Dorset Downs. He there recognized the existence of extensive flat interfluvial upland surfaces which bevelled indifferently the several stages of the Chalk including the hard bands of the Melbourne and Chalk Rocks. Associated with these upland flats he found deposits of 'reddish sandy loam or clay, full of flints', and he interpreted both the clay and the surface on which it rests as resulting from marine action. This surface of marine erosion he saw must be older than the present valleys, whose initiation he regarded as 'Pliocene or even Miocene'. Rather similar conclusions had been advocated some four years before by Rev. W.R. Andrews[23] in respect of the uplands on either side of the Vale of Wardour. He regarded their almost level ridge rops as relics of a former plain of erosion, tilted gently toward the east, which he tentatively attributed to the action of a Pliocene sea. As part of the evidence for this plain Andrews cited the drifts which cap the Great Ridge continuously for many miles, and the 'relics of old gravel with pieces of Upper Greensand chert' which occur on the crest of the southern escarpment, just as Jukes-Browne had attached much importance to the similar drifts in Dorset. In both cases, however, the drifts go far to negative the inference of marine planation which they were intended to support, for they are undoubtedly sub-aerial deposits and have, indeed, the characters of a land-wash rather than those of a fluviatile gravel. If they are, indeed, contemporary with the surface on which they lie, and the presence of chert fragments on the scarp crest does suggest they they are older than the present features of relief, then we should probably regard them as relics of the regolith or mantle of deeply-weathered rock-waste which would naturally be associated with the Mio-Pliocene peneplain.

Clement Reid was another worker who clearly believed that planation had been general in this region, but his writings on the subject are brief and probably do less than justice to his views. In describing the Ringwood district[24] in 1902, he suggested that the Mid-Tertiary folding episode was followed by 'a long period during which the (anticlinal) ridges were planed down by the sea'. He then speaks of 'the gently undulating plain thus formed', being trenched and cut into by rivers as the result of elevation in the Newer Pliocene period.

Ten years later H.J.O. White questioned whether this planation might not be sub-aerial rather than marine, and in a considerable series of survey memoirs from his pen this conception was much developed and extended. In particular we must notice the diagram with which he illustrated his account of the Shaftesbury district.[25] In this figure it is made clear that here at least the summit peneplain, represented by the bevelled upland surfaces of the Chalk downland, is to be clearly distinguished by the field evidence from the exhumed portions of the sub-Eocene surface which occur on the lower parts of the dip-slope. Both surfaces are inclined toward the south-east, but the earlier one much more strongly so that the two surfaces intersect and give rise to quite contrasted landscapes (see below, p. 56).

While the reality of the planation of South-East England is not now in question it is nevertheless true that no account of the character and geographical distribution of the summit plain has hitherto been attempted. Surviving remnants of the summit peneplain are nowhere common or extensive in the region. They are, in fact, preserved at all only by the favourable chance that the degradation of the permeable limestone uplands fails to keep pace with the reduction in the elevation of the region as a whole. These areas of calcareous sub-stratum have remained generally in the stage of late youth, while neighbouring outcrops of clay and sand have advanced to maturity, probably in more than one partial cycle. In particular, it is on the Chalk uplands of Wessex that the most considerable of these remnants of the summit plain survive, and a summary of the field relations in certain of these areas may be given here.

Standing anywhere in the south central part of Salisbury Plain one may look southward across the Wylye valley to the long even crest-line of the Great Ridge (Fig. 1). From such a viewpoint the ridge appears to have the character of a plateau, its summit surface being separated by a sharp topographic break from the valley slopes and spurs of lower levels. It suggests strongly by its appearance that its upper surface belongs to quite another morphological cycle than that represented by the smooth, well-modelled spurs and slopes that descend towards the modern valley. This upland comprises one of the largest and most instructive surviving remnants of the Mio-Pliocene peneplain. At its eastern end it pitches rather more definitely than elsewhere. If we except these easternmost two miles, the ridge rises gently from 600 feet near Grovely Lodge in a W.N.W. direction for some

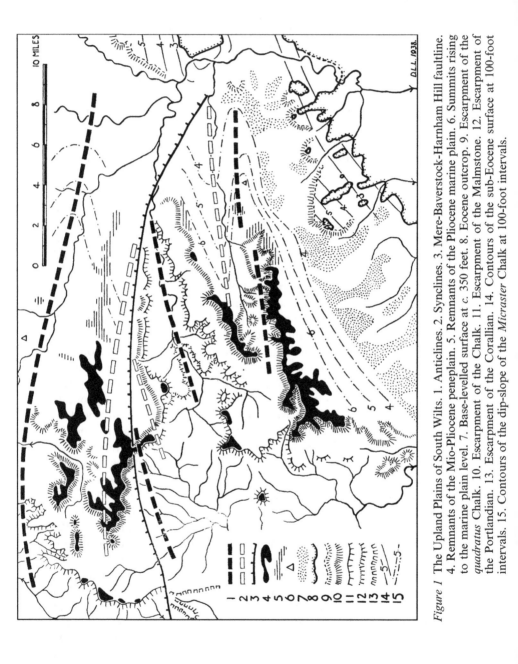

Figure 1 The Upland Plains of South Wilts. 1. Anticlines. 2. Synclines. 3. Mere-Baverstock-Harnham Hill faultline. 4. Remnants of the Mio-Pliocene peneplain. 5. Remnants of the Pliocene marine plain. 6. Summits rising to the marine plain level. 7. Base-levelled surface at *c.* 350 feet. 8. Eocene outcrop. 9. Escarpment of the *quadratus* Chalk. 10. Escarpment of the Chalk. 11. Escarpment of the Malmstone. 12. Escarpment of the Portlandian. 13. Escarpment of the Corallian. 14. Contours of the sub-Eocene surface at 100-foot intervals. 15. Contours of the dip-slope of the *Micraster* Chalk at 100-foot intervals.

11 miles to 784 feet near Brixton Deverill. Or passing due westwards it takes some 15 miles to rise to the same altitude at Whitesheet Hill north of Mere.* Concurrently the ridge broadens and becomes dissected by a branching dry-valley system, but ignoring the minor valleys incised below the plateau surface, the latter in its western part attains a breadth of about 5 miles at 700 feet. It thus ranks as a topographic feature of some magnitude with maximum dimensions of 15 by 5 miles and area of 35 to 40 square miles.

That this is not a structural surface is indicated by the fact that in the transverse direction it shows no relation to the well-marked underlying syncline, and that in the longitudinal direction it passes on to successively older horizons in the Chalk as one proceeds westwards. At its extreme eastern end it is not improbable that the zone of *Marsupites* is represented above Grovely Wood. At the western end on Whitesheet Hill it passes across the Chalk Rock on to the Middle Chalk. These relations to structure are rendered even clearer by consideration of the neighbouring uplands.

Immediately to the south lies the Vale of Wardour and on its opposite side rises what is perhaps the most stately of all the Wessex escarpments. From its crest the ground falls away fairly quickly on its farther side into the Ebble valley. This slope is very nearly a true dip-slope, for although the dip seems small enough when reckoned in degrees (of the order of 2°) it amounts to a fall of about 200 feet per mile. This marked gradient of the dip-slope serves admirably to reveal the narrow bevel of the crest, most noticeable at the salient points of the escarpment. The bevelled crests fall eastwards in evident parallelism with the eastward fall of the Great Ridge on the opposite side of the vale. From 766 feet at the western end the descent to 600 feet above Burcombe Ivers takes about $7^1/_2$ miles. This decline might actually be due to structural causes, since the scarp is slightly oblique to the strike. That this is not the case is shown by the fact that the Malmstone at the scarp foot falls from 500 feet to less than 250 feet between the same points, so that the thickness of the overlying Chalk is growing steadily from west to east, and the crest-line itself is passing on to successively higher horizons in the Chalk in this direction.

A third ridge of generally east-west trend, and almost as large as the Great Ridge itself, lies on the southern side of the Ebble valley. At its eastern end this ridge is actually developed upon the axis of the anticline of Bower Chalke, but for most of its length it is, like the last, of cuesta form with a north-facing escarpment. It, too, shows the same gentle fall from west to east and the same bevelled crest-line. The latter feature is well developed in the western part of the ridge where it forms the high ground at

* The highest points of the area are actually Long Knoll (945 feet) and Brimsdown Hill (933 feet), both a little further west. Consideration of these eminences is for the moment deferred.

700 feet to 850 feet north and west of Ashmore. The undulating plateau-like character of this facet of the cuesta contrasts strongly with the much more strongly inclined structural surface – actually the dip-slope of the *Micraster* Chalk – which underlies the immediately adjoining areas of Cranborne Chase and Vernditch Chase.

These three ridges of markedly contrasted structure – respectively, synclinal, homoclinal and anticlinal – thus exhibit the common feature of bevelled summit areas which decline gently eastwards. If points at corresponding altitudes on each ridge be joined, the resulting contour lines run smoothly across the map and may be taken as, at any rate, a first approximation to the contour lines of the Mio-Pliocene peneplain from which the ridges have been carved. Moreover, the drawing of these generalized contours upon a map brings to light other points belonging to the plain whose relation to it might otherwise have been overlooked.

A prolongation of the 600-feet contour which runs by way of Codford Circle across the Great Ridge near Grovely Lodge and so by Burcombe Ivers to Knighton Hill, passes in another 4 or 5 miles through the summit of Pentridge Hill. This bold and isolated hill lies well removed from the high ground so far considered and rises abruptly from the low downlands of North-East Dorset within a few miles of the main Eocene outcrop of the Hampshire Basin. Its steep northern face is actually the escarpment of the Belemnite Chalk; its gentler southern slopes carry a series of Eocene outliers and are obviously a dissected portion of the sub-Eocene surface. The most northerly of the outliers, some 3 miles from the main outcrop, caps the actual summit at a height of 600 feet. It is difficult to avoid the conclusion that this summit also is little below the surface of the Mio-Pliocene peneplain, and that the latter here passed on to the Tertiary formations. The relation between the two surfaces whose intersection was schematically represented by White more than thirty years ago can here be given precise expression. The sub-Eocene surface is declining in a southerly direction at an angle of 30°E. at approximately 120 feet per mile. In the opposite direction we rise from 600 feet at Pentridge to 691 feet at Marleycombe Hill above Bower Chalke in 3 miles, giving a gradient of 30 feet per mile. The older surface is thus tilted along this particular line of section four times as steeply as the newer one.

Pentridge, however, is not the only point on the plain brought to light by the generalized contour lines. In connecting points on the ridges on the two sides of the Vale of Wardour it was found that several points within the vale fall on or close to the lines so drawn and at the appropriate altitudes. They occur on the crests of the escarpments of the Malmstone and the Portlandian Beds, but, owing to the high dips prevailing, their extent is in all cases small. In the western part of the vale, Cleave Hill (700 feet) and Haddon Hill (737 feet) developed on Malmstone near West Knoyle, continue the summit plain at the 700-foot level from the Chalk right across the major fault which forms the northern boundary of the vale. Farther south-east, two other points on the Malmstone rise to 700 feet and Stop Beacon culminates at 778 feet. On the southern side of the vale the Malmstone scarp again rises to the appropriate level, here about 800 feet, in conformity with the greater altitude of the Chalk downs north of Cranborne Chase, and on the road between Shaftesbury and Semley station it shows some signs of bevelling. No point in the Portlandian outcrop actually attains the presumed full altitude of the summit plain, although in rising to 700 feet (2 miles west of Tisbury) it is perhaps only 70 feet short of it. Farther east a similar relation to the plain is shown by the culminating points of the Malmstone outcrop, notably at Castle Ditches and Wick Balls, but in spite of their present defect of altitude it seems just to regard these summits as having conformed to the general level in the fairly recent past. It is to be remembered that the extent of hard rock upon which they depend for their preservation is strictly limited, and unlike the case of some of the plateau-like Chalk ridges the smallest recession of the escarpment is attended by a notable loss in the altitude of the crest.

From this discussion it would thus appear that we may recognize surviving remnants of the Mio-Pliocene peneplain over a considerable area in South Wilts. and Dorset. Their total area is admittedly small, but the area in which the former existence of the plain seems satisfactorily demonstrated is not less than 200 square miles. Throughout this region the plain shows complete independence of structure, bevelling all the Mid-Tertiary flexures indifferently. It truncates the powerful anticlinal axis of the Vale of Wardour, the subsidiary axis of Bower Chalke and the synclines of Hindon, the Ebble valley and the Great Ridge. It is quite unaffected by the presence of the powerful strike fault of Mere. It must originally have bevelled all horizons from the Kimmeridge Clay to the Eocene, and it is still preserved on the chief hard or permeable rocks of this sequence – the Portlandian, the Malmstone and the Upper Chalk. Finally, it is intersected on Pentridge Hill by the sub-Eocene surface dipping toward the Hampshire Basin with a gradient at least four times that of the summit peneplain itself. In no other part of South-East England can the Mio-Pliocene surface be satisfactorily recognized over so large and continuous a tract of country and its relations to the Mid-Tertiary flexures so convincingly demonstrated. We may perhaps even go so far as to regard it as in some sense a type locality, the more so since it not only furnishes evidence regarding the present form and structural relations of the summit plain, but also as to its original character. We have already noticed that the drifts which cap some of the summits may possibly be regarded as portions of the regolith of a sub-aerial peneplain, and a little evidence pointing also to the sub-aerial character of the surface is furnished by the detail of its present form.

At the western end of the Great Ridge the plain attains a height of 784 feet, on Whitesheet Hill, where it is interrupted by the main Chalk escarpment. Rising from the Malmstone dip-slope beyond, are the Lower Chalk outliers of Little Knoll and Long Knoll. The former, and the greater part of the ridge of Long Knoll rise just over 800 feet and appear to be detached remnants of the summit plain, but at the western end of Long Knoll the ground rises in about 200 yards, smoothly but quickly, to 945 feet. This final eminence can hardly be part of the plain, being marked off from it by slopes of 10°–20°. If the plain be of sub-aerial origin this rise is best explained as a residual hill some 120 feet high; but if the plain be marine it must be regarded as an unconsumed stack or island. In the latter case it might be expected to show signs of marine cliffing, but, in fact, the thinly covered Chalk slopes pass down smoothly into the remainder of the ridge. A mile or so away the larger masses of Brimsdown Hill (933 feet) and Cold Kitchen Hill (844 feet) appear to be similar residual eminences rising slightly but definitely above the peneplain. Similar relations are again found in the area north of Cranborne Chase where Melbury Hill (863 feet), Breeze Hill (859 feet), Win Green (911 feet) and Winkelbury Hill (852 feet), appear as the culminating points of subdued but definite rises above the plateau. Individually they might perhaps be passed over as 'accidental'; but taken together they are too numerous and too systematically disposed upon hard rocks or watersheds to favour this hypothesis; they are residual eminences rising from a sub-aerial peneplain.

From our type locality on the southern border of Wiltshire the Mio-Pliocene peneplain may be traced along the Chalk uplands, north-eastwards to the Chilterns, eastwards to the Weald and south and west into Dorset and Devon. In the first-named direction we note that the surface fringes the whole north-western margin of Salisbury Plain. The latter is a synclinal area whose upturned rim has been bevelled at altitudes between 650 and 750 feet, but mature dissection in the present cycle has made the bevel very difficult to appreciate except in the large view. Farther east, in the Downs of northern Hampshire the summit peneplain forms fairly large upland areas, readily distinguished from the marine Pliocene bench at about 650 feet, and rising with very gentle slopes to the broad swelling culmination of Walbury Hill (975 feet), the most elevated point on the English Chalk. It is difficult to say whether this high ground represents a residual hill rising above the general peneplain level or whether it should be regarded as a gentle convexity induced by the warping which let in the Pliocene sea. The gentleness of the slope perhaps favours the second suggestion. It is quite otherwise, however, with the culminating points of the Chalk which overlook the Vale of Pewsey from the north. Here without doubt we have a series of residual hills of subdued but well-defined form which rise 100 feet or so above the general level of the plain, here at 750–800 feet. It is reasonable to suppose that they mark the line of the old watershed between the drainage to the North Sea, via the Kennet, and that to the Channel, via the Avon.

In the Marlborough and Berkshire Downs the summit plain is readily inferred although little of it remains. It is represented to-day only by the long convex and often undulating crests of the finger-like downland ridges. These are generally united only at their north-western extremities where they attain altitudes of 900 feet or more. Generalized contours permit us to infer that the plain declined gently south-eastwards to the 700-feet level, at which altitude a visible feature on most spurs separates the summit surface from the Pliocene marine bench. Similar relations exist in the Chilterns, where in spite of the even more restricted development it is possible to mark off with some precision the area occupied by the Mio-Pliocene peneplain. To-day it comprises merely the narrow convex crests of a series of branching ridges at 650–850 feet, in a zone only two or three miles broad lying immediately behind the scarp crest between Watlington and Luton. The contrasts of form, drainage–pattern and superficial deposits between this zone and the adjoining zone of the Pliocene abrasion platform are here excellently displayed (p. 56).

In and around the Weald all the higher ground, say above 700 feet, may be referred to the Mio-Pliocene peneplain. The North Downs from the Medway gap westwards to Dorking rise well above the level of the Pliocene marine bench, and just as in the Chilterns this higher ground may be regarded with some certainty as part of the sub-aerial peneplain. Behind this portion of the Chalk cuesta, the Greensand outcrop just rises to the peneplain level in Ide Hill near Sevenoaks, while in the western Weald, Leith Hill (965 feet), Hindhead (895 feet) and Blackdown (918 feet) indicate the elevations to which the peneplain rises over the Central Wealden axis. There is good reason for supposing that along the western and southern margins of the Weald the higher parts of the Chalk upland from Holybourne Down near Alton southwards by Wheatham Hill to the South Downs and so east to the Arun also represent the peneplain surface. Butser Hill which rises to 889 feet is probably a low residual rising slightly above the old plain. The peneplain is again represented farther east by the crest of the Downs between Washington and Plumpton, and in the Isle of Wight by the crests of the Southern Downs and of the upland south of Calbourne.

We have now traced the Mio-Pliocene peneplain continuously across southern England from the margins of the London Basin, where it is dated by the fossiliferous deposits left upon its down-warped surface by the transgressing Pliocene sea, to the western limits of the Cretaceous uplands. Considering its development over South-East England as a whole it is seen that it shows no broad regional variations in altitude such as would suggest tilting as a whole in one direction or another. Rarely, if ever, does it descend below 700 feet, or rise to 900 feet, though residual eminences may stand 100 feet or so higher than the plain itself. In contrast to this general uniformity of altitude, however, we find perceptible gradients ranging from 10–40 feet per mile in the plain within each of the regions studied. The gradients vary considerably in direction and very commonly decline toward the main lines of existing drainage. These facts taken in conjunction with the not infrequent occurrence of the low residuals rising above the general level convey a strong suggestion as to the origin of the surface with which we are dealing. These are clearly the characters of a sub-aerial peneplain in which base-levelling, though far advanced, was not complete even upon rocks as soft as the Chalk. It is in entire conformity with this conception that, when we try to trace the peneplain westwards into the regions of harder rocks, we find it to be more limited in extent and increasingly interrupted by unreduced residuals. Thus, in the Mendip region, the inlying areas of resistant Old Red Sandstone rise above remnants of an upland plain which may well be the representative of the Mio-Pliocene peneplain. In the South-West Peninsula the peneplain must be represented by one of the high-level platforms fringing the uplands of Dartmoor and Bodmin Moor. It is probable that if we were able to trace the peneplain into the Welsh massif we should find it reduced to the status of a high-level valley bench, representing merely an important stage in the history of valley excavation.[26]

Bibliographic references

16 A. Ramsay, 'On the denudation of South Wales and the adjacent counties of England', in vol. 1 'Essays', *Mem. Geol. Survey*, 1846, pp. 326–8.

17 J.B. Jukes, 'On the mode of formation of some of the river valleys in the south of Ireland', *Quart. Journ. Geol. Soc.*, vol. 18, 1862, pp. 378–400.

18 W. Topley, 'The Geology of the Weald', *Mem. Geol. Survey*, 1875, pp. 27, 285, 286.

19 W.M. Davis, 'On the origin of certain English rivers', *Geogr. Journ.*, vol. 5, 1895, pp. 128–46.

20 H. Bury, 'On the denudation of the western end of the Weald', *Quart. Journ. Geol. Soc.*, vol. 66, 1910, pp. 640–92.

21 S.W. Wooldridge, 'The Pliocene Period in the London Basin', *Proc. Geol. Assoc.*, vol. 38, 1927, pp. 49–132.

22 A.J. Jukes-Browne, 'The origin of valleys of the chalk Downs of North Dorset', *Proc. Dorset Nat. Hist. and Antiq. Field Club*, vol. 16, 1895, p. 8.

23 W.R. Andrews, 'The origin and mode of formation of the Vale of Wardour', *Wilts. Archaeol. and Nat. Hist. Mag.*, vol. 26, 1891, p. 258.

24 C. Reid, 'The Geology of the country around Ringwood', *Mem. Geol. Survey*, 1902, p. 29.

25 H.J.O. White, 'The Geology of the country south and west of Shaftesbury', *Mem. Geol. Survey*, 1923.

26 Other opinions have since been expressed on the age and origin of the high-level erosion surfaces in Wales and South-West England. See especially O.T. Jones, *Quart. Journ. Geol. Soc.* 1952, and W.G.V. Balchin, *Geogr. Journ.*, vol. 118, 1952, pp. 453–76, and a summary relating these areas to South-East England, S.W. Wooldridge, *Geogr. Journ.*, vol. 118, 1952, pp. 297–308.

3

THE PROBLEM OF TORS

D.L. Linton

Source: *Geographical Journal* 121 (1955): 470–87.

Everyone knows the tors of Dartmoor. They rise as conspicuous and often fantastic features from the long swelling skylines of the moor, and dominate its lonely spaces to an extent that seems out of proportion to their size. Approach one of them more closely and the shape that seemed large and sinister when silhouetted against the sunset sky is revealed as a bare rock mass, surmounted and surrounded by blocks and boulders; rarely will the whole thing be more than a score or so feet high. But if on closer examination the tor loses something of grandeur it loses nothing of its strangeness. Usually its main mass will be a solid rock outcrop as big as a house rising abruptly from the smooth and gentle slopes of a rounded summit or broadly convex ridge. The contrast between the gentle turf- or heather-covered slopes that lead up to the tor and the near-vertical bare rock walls of the tor itself is striking. The rock is usually traversed by bold and widely spaced near-vertical fissures, and these as well as the bounding walls are clearly the expression of joint planes in the granite. Powerful divisions also traverse the rock more or less horizontally—the so-called pseudo-bedding—and give the whole mass the rudely architectural aspect that is well described by the term "cyclopean masonry." The observer who permits himself to wonder by what giant hands these blocks were piled, however, will find his wonder increased when he observes how many of them have rounded and pillowy forms. Some tors closely simulate a great heap of piled woolsacks. Some comprise nothing more than a few rude ellipsoids of naked rock balanced one upon another; and in the extreme cases such stones have so small an area of contact with their supports that they will move under the hand. These are the "rocking-stones" or "logan stones" for which supernatural origin has often been invoked.

The origin of tors

Even the most casual observer confronted with these strange assemblages of stones must ask himself by what means they have been shaped, and further, why they should occur where they do and not elsewhere. Geologists have also asked these questions, but being perhaps a little disinclined to pay overmuch attention to objects which are regarded with wonder and curiosity by the layman, have been too ready to assume that they know the answers. Tors are mentioned or illustrated in almost every British text-book of physical geology[1] as examples of the "weathering of granite," and it is explicitly or implicitly clear that the weathering in question is atmospheric. Holmes,[2] for example, in the caption to an illustration figuring Bowerman's Nose, speaks of "a Dartmoor tor . . . carved from well-jointed granite by rain and wind." But in no case known to me has the *modus operandi* of the weathering process been clearly described and shown to be adequate to produce the forms actually seen. For my own part, indeed, I find it very hard to see how atmospheric weathering—that is, the action of the elements on bare rock under the atmosphere—can produce tors or tor-like forms. In fact if we examine a few tors closely we shall find that present-day atmospheric weathering is acting upon them destructively and not constructively. Many tors are conspicuously in a state of collapse. The large rounded masses that in some cases are delicately poised are more commonly found dislodged and tumbled, wedged in fissures or lying at the foot of the tor. There are cases in which a whole rock-stack has fallen outwards away from the main mass of the tor and lies, as it were, face down upon the turf. In yet others a joint-bounded slice of rock, standing at the head of a fairly steep slope, has apparently lost its foothold and collapsed backwards so that it now leans drunkenly against the rest of the tor. One gains a strong impression from such instances that however the tors were produced, exposure to the elements is highly destructive to their architecture. Moreover when we look at the individual pillows and woolsacks which are so characteristic of tors it is common to find them broken by frost along an almost plane surface. Sometimes the severed halves are only inches apart; sometimes they lie on opposite sides of the pedestal that bore the unbroken rock. The fracture plane appears as unrelated to the rounded form as the cut surface of a half potato to the rest of its shape: and just as knives were not in question when the potato was forming in the ground it seems evident that the component masses of tors were not exposed to frost action when they were receiving their rounded forms. Finally, when we look more minutely at the surface we find that since the rounded masses received their forms their more exposed portions have been notably modified by atmospheric weathering. They display, on their upper and windward portions, surface roughening and pitting with granular disintegration of the rock constituents leaving large crystals of quartz or felspar projecting as much as half an inch from the general surface. More striking still there may be groovings and flutings several inches in depth, as well as round basin-like hollows variously referred to as "pot-holes" or as "rock basins" (though neither term can be acceptable to geomorphologists since both have long been pre-empted for larger and more important features). The disposition of these grooves and hollows makes it clear that they are of later origin than the acquisition of the general rounded and pillowy forms. Some of the "pillows" which remain in place have been pitted and grooved on their exposed upper parts while their "underbellies" which have been sheltered from the elements remain smooth. Other blocks which are clearly fallen masses have developed flutings or "rock-basins" while in their new positions. There is in all this no evidence which suggests that atmospheric weathering acting on an exposed rock surface will produce from it a tor, but much to show that the weathering of the last few thousand years is in various ways modifying and destroying tors that were produced in some earlier time and by some other means.

These other means, however, must operate in a fashion similar to that in which geologists have supposed atmospheric weathering to act. They must act *in situ* and they must act selectively. The piled and delicately poised logans and rocking-stones cannot possibly have been transported to their present situations: they have been produced where they now stand by some form of wastage and removal of the surrounding solid rock. And since the shape and architectural character of the Dartmoor tors is clearly related to the arrangement of joints and pseudo-bedding planes in the granite, it follows that the wasting process has been guided and controlled by these internal division planes in the rock. The process most likely to act in this way is the sub-surface rotting of rock by acidulated ground water percolating along joints and from them into the body of the rock.

It is well known that in the humid tropics the downward percolation of ground water, acidified by its passage through the soil, leads to extensive rock decomposition, and further that embedded in the layer of decomposed material there may be detached masses of relatively sound rock. These have a rounded or ellipsoidal form and their production from quadrangular joint blocks has been indicated in standard texts[3] as the natural result of more effective penetration by the decomposing solutions where they work into the block from two sides (i.e. at an edge) rather than from one (in the middle of a plane face), and *a fortiori* at a corner as compared with an edge. Such residual masses have been well described, for example, by Scrivenor in his "Geology of Malaya"[4] and called by him "core boulders." This term, especially in the form *core-stones* (which carries no implication of any limit of size either upward or downward) seems admirable for the purpose and will be so used in what follows.

Core-stones are not, however, confined to the tropics. Philippson mentions the production of round boulders (*rundliche Blöcke*) from cubical rock-masses and indicates how by the removal of the surrounding

Verwitterungsmulm such boulders may be left strewn over the landscape as a *Blockmeer* or *Felsenmeer*, of which the granite areas of the Harz and Odenwald offer striking examples. De Martonne[5] speaks similarly of the granite country of the Sidobre, near Castres in the southernmost part of the Central Plateau; "les boules sur les sommets" he says "sont les parties qui ont résisté à la décomposition, isolées par entraînement de l'arène." His description is illustrated by photographs of slopes littered with great rounded corestones laid bare "par le ruissellement," including a fine example of a perched boulder. Comparable forms are cited also from the granite mass of Huelgoat in Brittany and we are thus brought within a metaphorical stone's-throw of the tors of Devon and Cornwall. Moreover, the production of tor-like forms by the same processes that can produce core-stones had also been envisaged by Philippson who mentions "die bizarrsten, von Schluchten durchsetzten Felsformen" which he calls *Felsburgen*, and by Cotton[6] in New Zealand. Indeed, there appears to be an idea here which is of rather general application but has never been explicitly formulated—namely that tors, core-stones, and possibly other residual rock forms, *are the result of a two-stage process*, the earlier stage being a period of extensive sub-surface rock rotting whose pattern is controlled by structural considerations, and the later being a period of exhumation by removal of the fine-grained products of rock decay. It would seem desirable to formulate this hypothesis more precisely with the aid of diagrams (Figs. 1a and 1b).

Figure 1a represents a vertical section of some 20 or 30 feet of a "uniform" but massively jointed rock such as granite. Layers numbered from ground level downwards from 1 to 6 are separated by unevenly spaced pseudo-bedding planes. Columns lettered A to H are separated by unevenly spaced near-vertical joints. The rock is thus divided, or potentially divided, into a number of parallelopipedal blocks of very varying sizes. Some of these blocks, like those which would occupy the (un-numbered) spaces A_2, B_2, F_3, F_5 are relatively small, while others like C_5 and D_6 are large. Chemical weathering is assumed to result from the inter-granular penetration of the rock by groundwater moving down from the surface, inwards from joint-planes, and downwards, and to a small extent upwards, from pseudo-bedding planes. After a considerable lapse of time decomposition will have affected all the shaded portion. Some original joint-bounded blocks near the surface, or between closely-spaced joints, will be rotted throughout; others further from the surface or bounded by widely separated joints will be affected only by rounding at the edges and corners and some reduction in size.

In Figure 1b a radical change is presumed to have occurred. The finer-grained products of rock rotting have been removed—by a phase of active surface wasting following local rejuvenation of the rivers or by some other agency. Ground level has effectively been brought down to the base of the diagram. If the removal has been gradual, core-stones such as D_2, and D_3

74

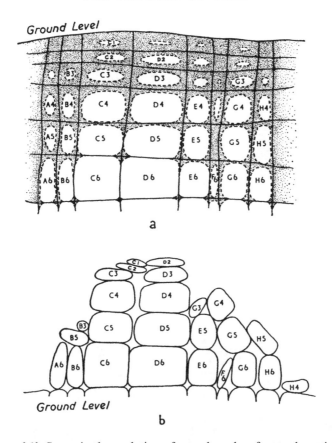

Figure 1a and 1b Stages in the evolution of a tor by subsurface rock rotting.

will settle down on the supporting composite pedestal D_4, D_5, D_6. Some stones such as C_1 and C_2 may come to rest in very delicately poised or precarious positions; others, like G_3, will topple and become chock-stones, or like H_4 will fall to the ground. The whole will have an architectural aspect deriving from the original joint pattern of the rock. The upper and outer parts will be rounded and well separated; the lower parts will approximate to joint-bounded blocks firmly rooted in bed rock.

Tors which conform in every essential to this model are very numerous—Hound Tor on Dartmoor may be cited as a typical example (Photos A and B). This is strong circumstantial evidence that they have been formed in the way that has just been described; but direct evidence is also available. At Two Bridges on Dartmoor a tor has been partly exhumed by quarrying (Photo C). The exposed portion is of the woolsack type, its form having been determined by pseudo-bedding planes at fairly close intervals and rather

A. *General view of Hound Tor, Dartmoor, showing the abrupt rise of the tor from the gently convex surface of the moor and the influence of vertical joints and pseudo-bedding on its architecture.*

B. *Part of Hound Tor, Dartmoor, to be compared with Fig. 1b. Note the rounding of the upper masses, some of which are perched like D_2 and D_3, or wedged like G_3 and G_4, or fallen like H_4. The lower blocks are conspicuously joint-bounded like C_6 and D_6, while wide joint-bounded fissures traverse the tor which correspond to vanished masses such as might have been numbered F_1 to F_5.*

C. *"Woolsack type" tor partly exhumed from the surrounding and overlaying growan by quarrying operations at Two Bridges, Dartmoor. In the architecture of this tor the pseudo-bedding dominates.*

widely separated joints. The remainder of the tor is still surrounded and sur-mounted by the relatively fine-grained products of rock decay, the material known on Dartmoor as the "growan." The growan, though incoherent, preserves the structure of the granite with its lines of veining and has clearly been formed *in situ* with the residual mass of relatively unaltered rock within it. Instances of the same kind occur elsewhere. In a granite quarry at Cintra in Portugal a tor and core-stones have similarly been artificially exhumed by removal of the *saibro* (Portuguese equivalent of growan: Photo D), and I have noted numerous whaleback masses and core-stones partially exhumed by natural agencies in the Serra da Estrella, e.g. near S. Comba de Dao and north-east of Castello Branco. Again in a hollow at 3,600 feet near the summit of Cairngorm in Scotland core-stones and rounded masses were exposed in 1954 by excavation and in one instance a vein in the solid rock could be traced continuously into its altered representative in the surrounding and overlying decomposed material. In the face of such evidence I think it must be conceded that tors resembling the model represented in Figure 1b may all be assumed to have had the two-stage history postulated for that case. Even when we concede this, however, we have only answered part of our question. It remains to be explained why the tors are located where they are.

D. *Well rounded core-stones of granite in place in decomposed rock* (saibro) *exposed by quarrying at Cintra, Portugal. The strong contrast between the sound rock and the rotted* saibro, *in which pick marks are clearly visible, should be noted. It is possible that if the exhumation were completed by careful hosing down a tor of perched rounded masses would be disclosed.*

The location of tors

The tors of Dartmoor sometimes crown actual summits; just as commonly they rise from gentle slopes that lead to a summit; sometimes they are found at the break of slope above a valley-side; sometimes they occur at spur ends. In almost all cases there is an abrupt change from the flat turf platform or the gentle surface slopes to the bold rock walls of the tor itself. The latter are usually determined by master joints and where tors occur in groups a community of joint direction can be recognized. There have clearly been both topographic and structural factors at work in selecting the location of tors. Summits, ridge tops, valley brows and spur ends all provide situations

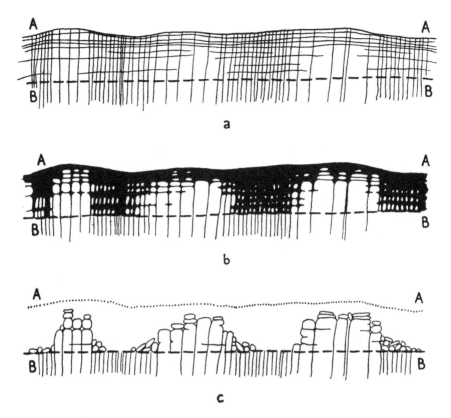

Figures 2a, 2b and 2c Stages in the evolution of a group of tors, illustrating the importance of joint spacing.

where ground water is relatively deficient since it is supplied only by precipitation and rather freely removed by underdrainage: there can be no accession of groundwater, already well charged with acids, by down-slope percolation from higher levels. Rock rotting will therefore be minimized in such locations. But tors are more narrowly localized than this and the selection of a particular mass of rock for survival while neighbouring masses are decomposed is clearly the result of structural guidance. There is every indication that the main factor at work on Dartmoor and other granite outcrops is the spacing of joints.

Figures 2a, 2b and 2c will serve to illustrate the point. In Figure 2a a stretch of granite a few score feet in depth and 200 or 300 yards long is shown in vertical section with the ground surface at AA. Joints and planes of pseudo-bedding are shown by vertical and horizontal lines and considerable variations in their frequency are assumed. In nature it is unlikely that such large variations would be found in so short a section.

Figure 2b represents the state of affairs after a considerable period of rock rotting by percolating groundwater, the decomposed rock being shown in black. It will be seen that in the areas of closely spaced jointing rotting has in places been complete so that no solid rock survives, and is commonly so thorough that it survives only in the form of core-stones embedded in and isolated from each other by growan. This is probably the state of affairs over the greater part of the landscape, but locally where joints or pseudo-bedding planes are scanty rock decay has been much less effective and tors of varied architecture have been outlined. In Figure 2c the fine-grained products of decay have been removed and three striking tors have been exposed. Between them the ground may be, and commonly is, littered with core-stones. Further it will be noted that the three tors are shown rising abruptly from a solid rock platform. As has already been emphasized this is a common characteristic of tors—one of the most striking examples being the well-known Haytor (Photo E). This *basal platform* may be locally flat, but in a wider view is seen to undulate. Over an area of an acre or so it may be determined by pseudo-bedding planes as is suggested in Figures 2a, 2b and 2c, though it must be remembered that pseudo-bedding planes on Dartmoor and elsewhere are not necessarily flat as in the diagram but are commonly broadly convex to the sky in upland situations. It is evident that the surface BB in the diagrams has acted in some way as a downward limit to rock decay. It is not clear that any pseudo-bedding plane could act as a

E. *The north-eastern portion of Haytor Rocks, Dartmoor. This large and massive tor is evidently bounded towards the camera by a major joint-face, and rises very abruptly from the* basal platform *on which the sight-seers are standing. This platform is of rock which frequently shows through the thin turf.*

general limit in this way, and as suggested long ago by Cotton[7] the surface most likely to do so is the water-table. In areas of very tight joints the water-table may be irregular, but where joints are numerous and well opened it will be a regular surface undulating in conformity with the topography of the time. Below its level the restricted circulation will inhibit chemical reactions since equilibrium concentrations of the soluble reaction products will soon be built up in the nearly stagnant ground water. This view seems both inherently probable and in conformity with the facts of observation. Tors on ridges and at spur ends commonly rise from inclined bases, the inclination being downslope and often quite marked. And the possible height of any tor is evidently predetermined by the available interval between the original surface AA and the water-table BB. It is in conformity with this that some of our highest tors—e.g. Vixen Tor which is 50 feet high on its uphill and 90 feet on its downhill side—are in spur end positions.

Thus far our discussion has been concerned with the questions of origin and location posed by the rock features of Dartmoor and the other granite masses of the south-west of England that are by common consent—both in geological writing and common usage—regarded as tors. Such features, however, are by no means confined to one type of rock or one part of the country, and as we move away from the type locality it may be necessary to consider forms and instances which vary from the norm. There is obviously a danger here of including under the name "tor" features superficially similar, but in fact differing progressively more and more until at last we admit features which are not tors at all. Our definition must be genetic to be effective as a basis for classification. Using the hypothesis developed in our previous discussion we may therefore offer the following.

A *tor* is a residual mass of bedrock produced below the surface level by a phase of profound rock rotting effected by groundwater and guided by joint systems, followed by a phase of mechanical stripping of the incoherent products of chemical action. Ellipsoidal rock masses produced in the same way but entirely separated from bed-rock are designated *core-stones*. The upper parts of a tor approximate in form to core-stones, and like them may be completely detached, though still perched; the lower portions of a tor approximate to massive joint-bounded blocks. Great variations in form and architectural style thus result from variations in the original joint disposition. Tors may rise sharply from a *basal platform* of bed-rock which may be flat or inclined and is interpreted as representing the position of the water-table during the period of rock-decomposition, so that tor height cannot exceed the depth of the vadose zone of that period.

Since our definition is genetic we must also make use of empirical terms to designate features whose origin is undetermined or in question. An upstanding rock mass rising from a slope or a hill-top surface, clear of its surroundings on all sides, and with near vertical sides that may approximate to joint planes is best termed a *stack*. It would probably be so designated by rock-climbers and similar features on the sea shore are usually called sea-stacks or marine stacks. A similar feature rising from a valley side, and separated from the valley slope only on two or three sides should, I suggest, be called a *buttress*. Again, many of them would be so called by climbers. R.A. Pullan[8] in an unpublished account of some geomorphological features of the Pennines above Wharfedale has used the term "bastion" for such features: this would seem inadmissable since it has been used by von Engeln[9] for a glacial feature described by Flückiger[10] and has been accepted in this sense in standard texts.[11] As far as I know, no such objection applies to the term buttress, which in any case better describes most features of this type. Other descriptive terms are naturally used of rock masses of distinctive form and such terms are valuable. *Pedestal rocks* are well known in Britain and elsewhere; some of them I believe to be tors, others doubtless are not. *Whaleback* is a term often used to describe granite masses, especially in the tropics; some British whalebacks are undoubtedly *roches moutonnées*, some others, and probably many tropical examples, are genetically related to tors; others may be shown to be of yet other origins. Association of a rock-stack, buttress, pedestal rock or whaleback with an incompletely removed mantle of decomposed rock will of course demonstrate it to be a tor but the evidence of form, topographic relations or indirect association with core-stones or sapprolite may be adequate to justify ascription to the genetic category "tor."

Upstanding rock features that may be tors occur in Britain in fair abundance. In the south-west in addition to the undoubted tors of the granite outcrops there are tor-like masses of schorl rock at, for example, Roche in Cornwall[12] and Ausewell Rocks in Devon. Tor-like features occur on some of the doleritic outcrops of the Prescelly Hills and on flinty rhyolitic rocks at lower levels in Pembrokeshire. The Stiper Stones in Shropshire appear to be tors of Ordovician quartzite. Stacks and buttresses in association with core-stones and pedestal rocks are abundant in the gritstone moors of the Pennines in several areas and most that I have seen appear, like the Prescelly masses, to be genetically tors. J. Palmer[13] has described stacks and buttresses in Corallian limestone in the Tabular Hills west of Scarborough for which he would suggest a different mode of origin, and R.A. Pullan[8] would prefer to interpret some of the Wharfedale features in a way that would put them outside the category of tors as I have defined it. In the Cheviot, R. Common[14] has noted granitic tors on Standrop Rigg and elsewhere and attributed them to multigelation. In the Ochils, Miss J.M. Soons has shown me outcrops of basic andesite on the summit of Bencleuch which, though small, have some of the characteristics of tors. Further north in Scotland tors are widespread and numerous. The Kincardine granite carries the striking tor of Clachnaben and the granite mass of Bennachie in Aberdeenshire carries a fine tor group. The summit of the Buck, just inside the Banffshire boundary, has a rough and broken tor-like mass developed in "schists" which have possibly been rendered tough and massive by thermal metamorphism. On the outcrop of the adjoining Cabrach intrusion of gabbroic rocks are numerous minor tor-like features, rather like the summit rocks of Bencleuch, and a good deal of evidence of former rock rotting in depth. The granite of Ben Rinnes carries a conspicuous tor group while on the granite of the Cairngorms tor development reaches impressive proportions. Several of the mountains carry large and numerous tors at altitudes well above 3,000 feet, those of Ben Avon being the most numerous and the Barns of Bynack surely the most fantastic. Nor is this all. The sandstones and arkoses of the Old Red Sandstone give rise to striking tors on the summits of Morven, Smean and the Maiden Pap in Caithness, and in the extreme north of Scotland the syenitic summit of Ben Loyal is crowned with conspicuous tors. Although there are a few of these that I have not personally visited and a few more about which I have reservations, I am satisfied that the majority are genetically tors. They differ a great deal among themselves in form and characteristics and their full description must await a more comprehensive publication, but it seems to me already certain that tors are not a local peculiarity of the granite outcrops of the south-west of England, but a type of land form widely and significantly distributed through the country from extreme north to extreme south. Tors are developed from a considerable range of rock types, but it is clear that the Hercynian granites of south-west England, the Caledonian granites of Scotland, the felspathic gritstones of the Carboniferous series in the Pennines, and the arkoses of the Caithness Old Red Sandstone are the most noteworthy parent rocks. One may see in this a combination of apparently opposed characters that predisposes these rocks to tor formation. On the one hand the presence of felspar renders them all susceptible to chemical attack, while on the other the presence of a good deal of unweatherable quartz means that that attack is slow, and wherever joint-spacing is wide is so retarded that even after long periods of rock decay core-stones and tors remain unreduced. Tors are also formed in other siliceous rocks, the Stiper Stones quartzite, the Pembrokeshire rhyolites, and the Ben Loyal syenite, and on basic rocks, the gabbroic and doleritic intrusions of Banffshire and Prescelly, and appear to assume distinctive characteristics. They seem not to be formed in Britain, however, from any fine-grained or foliated rocks, an observation entirely consonant with the hypothesis of origin here advanced, though we may note that tors of schist certainly occur elsewhere.[15]

The age of the British tors

Finally, the question arises—when were the British tors produced? This is a question which cannot be adequately answered without drawing on more detailed evidence than can be assembled here, but the general indications of that evidence can be outlined. The tors of glaciated regions such as Scotland

and the north of England pose the question of age in a particular form—what is the relation of the period of tor formation to the glacial episodes? Moreover we should speak, not of "the period of tor formation" but of the two phases of the process. We need to date first the lengthy period of rock rotting in relation to a stable water-table and base-level, and second, the nature and date of the change which led to the tors being exposed to view. Of these two problems the first is in some ways the easier to approach.

Profound rock rotting to depths equivalent to the heights of our larger tors would appear to require not only a great deal of time but probably also a warmer climate than we now enjoy anywhere in these islands. However that may be, it is certain that the decomposition of the rock that once surrounded the large tors of the Cairngorms could not have occurred under the virtually tundra climate of the Cairngorm summits today—a climate in which solifluction terraces and patterned ground are currently developed. Even during the climatic optimum these summits were 1,000 feet or more above the tree line, and we may rule out the whole of post-glacial time, and *a fortiori* the whole period of the last glaciation, as the time when such deep decay occurred. The period of rock rotting cannot in fact be younger than the last inter-glacial. The same conclusion may be reached in a different way in other localities. Ben Loyal in Sutherlandshire bears on its higher parts between 2,200 and 2,500 feet several indubitable tors, but its ridges and spurs up to 1,800 feet have been overrun by the last ice and bear plain *roche moutonnée* surfaces. It is clear that if post-glacial weathering has not destroyed the characteristic features of the *roche moutonnée* surfaces it cannot have converted any pre-existing forms above 2,200 feet into tors. The deep rotting responsible for the latter must therefore be at latest inter-glacial.

In places where the mantle of rotted rock has not been swept away by glacial erosion it may be covered either by solifluction deposits, as appears to be the case in places on Dartmoor, in the Pennines and below Bennachie, or by indubitable glacial deposits as in the city of Aberdeen where T.C. Phemister and S. Simpson[16] have described up to 30 feet of thoroughly decomposed granitic rock with undisturbed veins overlain by several feet of morainic gravels or boulder clay with erratics of undecayed rock. Clearly the period of rock decay here antedates the local glacial deposits. Indeed, I conclude from the evidence at present available that everywhere in Britain the latest date that can be assigned to the period of profound rock decay is the last inter-glacial.

The possibility is not thereby excluded that the period of rock rotting was in fact much earlier, possibly late Tertiary. Workers in other countries[17] have associated tors with relatively old erosion surfaces: even Walther Penck is included here though he did not admit base-levelling and, so I think, misinterpreted his tors. And it is noteworthy that many of our most striking tors stand on upland surfaces that probably all British geomorphologists regard as having been shaped in pre-Pleistocene times. And if the present hypothesis of tor formation is accepted it is hard to imagine that the deep decomposition implied by the tors, and the long period of still-stand implied by the deep decomposition were not the products of the long period of still-stand implied by the past-mature form of the surfaces themselves. This possibility that the weathering implied by the upland tors is late Tertiary is attractive because of the known warmth of our Pliocene climate.

To admit such a possibility, however, is to emphasize the distinction between the two phases of tor formation; exhumation may be much later than the decomposition phase. The date of stripping is indeed very difficult to assess. Mention has already been made of tors and core-stones that were still buried until they were artificially laid bare in quarries, while many examples could be cited where the process of exhumation appears to be currently in progress, but all would be on valley sides or below convex breaks of slope. Most upland tors on the other hand have been exposed to the elements long enough to be considerably affected by them, in some cases long enough for virtual destruction or collapse. Moreover the degree of weathering of individual stones of upland tors greatly exceeds that seen on stones of similar kind (in some cases almost certainly core-stones) used in megalithic monuments. The three to four millenia since the erection of those monuments is thus much too short a period to have affected this degree of weathering and it becomes likely that such tors have been exposed throughout post-glacial time. It may well be that the melt water from the ice and, more important, the snows, of the last glacial episode were the active agents in washing away the finer material and laying many tors bare. There is reason to think however that others have been exposed by solifluction, by the downhill flow of the whole layer of rotted material, embedded core-stones and all, when saturated with thaw water. In either case the last glacial episode appears to be the probable time of exhumation of many of our tors.

It will be clear that these views have implications regarding the glacial history of the areas in which tors are found. It is, for instance, hard to interpret the evidence of the tors and *roches moutonnées* on Ben Loyal except on the view that the summit of this mountain was a nunatak projecting above the surface of the ice of the last glaciation, though whether the nunatak area was still mantled by rotted rock or not is, to say the least of it, difficult to decide. But there is nothing inherently improbable in such an interpretation. Eilif Dahl's[18] suggestion that where mountains lie fairly close to the shore line of a warm-water sea the gradient of the ice surface will be high enough to permit some of the higher seaward mountain peaks to project as nunataks, may well be applicable here.

The evidence of the tors on the Cairngorms and in Aberdeenshire is more difficult to interpret. The generally held supposition that such an area as the Cairngorms must, because of its height and the rigour of its present climate, have been a centre of ice radiation in Pleistocene times, gives one pause before offering the suggestion made for Ben Loyal, that the area may have been a nunatak. Yet it is clear that since the deep rock rotting implied by the massive tors is at latest inter-glacial, the erosion accomplished by the last ice on these upland surfaces was negligible. The sections cited by Phemister and Simpson in Aberdeen remind us that the soft mantle of rotted rock is not necessarily removed by the passage of ice across it, though to be sure, the predominant results of ice action in the Aberdeen area were depositional rather than erosional. This, however, is not usually true of the ice over the uplands and it becomes doubtful if they could have witnessed any really active streaming of ice at this period. Indeed, as I have suggested elsewhere,[19] it seems necessary to regard parts of these uplands, with their tors and other topographic features which may be regarded as pre-glacial or inter-glacial survivals, as essentially unglaciated during the last Ice Age, though doubtless deeply covered by névé. Such a hypothesis obviously calls for detailed discussion of the evidence area by area. This is not appropriate here and we may perhaps conclude by noting the more general point conveyed in Figure 3. On that map the positions of more than a hundred known British tors are represented by about half that number of heavy black dots. On the same map the main centres of intense local glaciation are shown in solid black—North Wales, Cumbria, Galloway, the highlands of Argyll, western Inverness and Wester Ross and the outlying centres of Arran, Mull and Skye. It will be seen that uplands with tors and uplands of intense glaciation are distributed in two complementary zones. The latter lie close to the western seaboard of our island from the Irish Sea to the Minch: the former are found in a broad zone to the east and south of this. Within this zone the distribution of tors is limited by the distribution of the outcrops of suitably constituted, massively jointed rocks. Particularly is this apparent in southern Scotland where the morphological features of the uplands strongly resemble those of the eastern Grampians, but tors are, as far as I know, unknown except on the Cheviot granite. Further, no tors are shown in lowland Britain although some may well exist,* so that the eastern limitation of the zone of tors may be more apparent than real. But the absence of tors from the western uplands is undoubted and it is hard to escape the conclusion that this is so because tors are part and parcel of an order of things which has been swept away from these areas by the heavy drubbing they received from the ice. Where tors are found they rise from hilltops and uplands of smooth and rounded aspect with forms and features that we associate with mature and past-mature development in the cycle of normal erosion. These landscapes can only be regarded as survivals from, in

---

* Mention has already been made of the Bridestones in north-east Yorkshire[13] which may prove to be valley-side tors; and it may be that some of the fantastic rock forms of the Tonbridge Wells sandstone outcrop in the Weald are also genetically tors, but they have yet to be examined.

Figure 3 The general distribution of the principal groups of British tors (see pp. 85–87).

the local sense, pre-glacial times, some of them possibly from a period which is pre-glacial in any sense. Such landscapes, but for the Ice Age, would have been normal on suitably constituted rocks throughout upland Britain; today they are to be found only to the east and south of the heavy line drawn on Figure 3. That line marks the real western limit of the zone of tors and closely envelops the areas of intense glacial sculpture. It probably represents a climatic limit of Pleistocene times and it is striking how closely it corresponds with the generalized modern annual isohyet of about eighty inches.

DISCUSSION

Afternoon meeting, 9 May 1955

Before the paper the CHAIRMAN (Professor S.W. WOOLDRIDGE) said: I rise to introduce my old friend and co-worker, Professor Linton, who in the case of many of you needs no introduction. I will permit myself to say, having heard the substance of the paper delivered to the International Geographical Congress at Washington in 1952, that it was easily the outstanding paper out of many read at that meeting. I am glad that Professor Linton has had the opportunity to present a rather fuller exposition of the exceedingly interesting and significant views he has formed on this question of tors.

Professor Linton then read his paper.

The Chairman

After that careful piece of reasoning, based upon shrewd and critical observation, as Professor Hollingworth is here I will ask him to open the discussion.

Professor S.E. Hollingworth (University College)

It was indicated in the introductory remarks to Professor Linton's paper that the problem of "tors" has largely been ignored by geomorphologists. The field of "process" in earth sculpture is common to geologists and geographers and much remains to be achieved, but I doubt the justice of the implied reproach of the former for past neglects. In this country geologists have, for example, largely pioneered the study of Pleistocene periglacial solifluction as a landscape moulding process. They would all welcome the striking growth of the study of process among British geographers both in the field, including excavation, and by laboratory study. Professor Linton's lucid and fascinating presentation of the tor problem is an outstanding example of careful analysis of field evidence.

Few students of land form would disagree with Professor Linton in his recognition of the two phases in the history of tor formation, especially in his application of it to granitic areas where deep chemical sub-soil weathering is so clearly demonstrable. As in the glaciated terrain such as North Wales and the Lake District the recognition of rounded pre-glacial upland, maturing in late Tertiary time, is the basis from which so much of our geomorphological reasoning springs. On one point in Professor Linton's thesis, I feel some reserve: I was not convinced that the downward limit to the second phase of tor formation, with its mechanical removal of the weathered material, was determined by the former water table, and so I am hesitant about the desirability of including this point in the formulation of a genetic definition. The water table in granite hills is subject to wide seasonal fluctuation and there appears to be inadequate evidence that the level of the base of the upstanding tors is related to the lower limit of advanced sub surface weathering. It would seem preferable to envisage that the lowering or mechanical stripping was a gradual process operating in a favourable environment, slowing down in its rate as a less weathered rock was approached, but was nevertheless a process which could be halted at any stage by a change of external conditions and so was a factor in producing a high or low tor rising from a smooth, gently inclined surface. In some cases the theoretical limit of "available relief" coinciding with the base of the weathered zone, might be reached.

I would like to ask Professor Linton whether in his studies of the tor problem he has found any evidence bearing on the relative rate of weathering of the granite under sub-surface attack as compared with the direct attack by surface agencies—rain, wind, etc.? In limestone terrain the development of deeply griked surfaces could largely be attributed to the more active solution of rock under vegetation than that which was bare and exposed to the direct action of rain only. Does Professor Linton think there is any significance in this analogy? Incidentally one of the finest examples of tor topography in the British Isles is the great piles of tabular masses of Old Red Sandstone conglomerate which form one summit of the Galtee Mountains in Southern Eire. Their Pleistocene history makes it reasonable to consider that this summit was not overriden by ice and so the occurrence would fall into line with others described by the author. In the case of such upstanding masses as the Prescelly Hills, the limited extent of the weathering along the joints in the exposed rock masses would appear to weaken the case for regarding these crags as being formed by a process closely similar to that of tors of granitic and other rocks in which chemical and mechanical disintegration go hand in hand. I warmly welcome the paper and look forward to its publication as an important contribution to the scientific study of landscape.

Professor D.L. Linton

I would like first to express my thanks to Professor Hollingworth for his appreciative and sympathetic reception of my paper. It is with considerable satisfaction that I learn that he feels there is good ground for accepting the two-stage hypothesis of tor formation that I chiefly wish to urge. It is by no means surprising that he should be less inclined to accept all the details of my interpretation. Regarding the feature which I have termed the *basal platform* I would point out that this is not an integral or essential part of my genetic definition of tors. In the latter I said only that "tors may rise sharply from a basal platform"—and the word I would like to emphasize now is "may." By no means all tors do. Only if the phase of rock rotting has locally proceeded to its downward limit, and only if the rotted material has been fairly comprehensively removed is a smooth basal platform likely to appear. I hope that more decisive field evidence bearing on this question may be found. In the case of the tors of the Prescelly region of Pembrokeshire— by no means all of which are on the Prescelly hills—it is fair to say that in some cases, e.g. the Maiden Castle built of Treffgarne rhyolites, rotting guided by joint planes has been so deep, even in a highly uniform and siliceous rock, as to produce masses of quite fantastic outline, and that there are instances where the products of rock rotting still remain *in situ*. As to the question of the relative rates of sub-surface rock rotting and "atmospheric weathering" my own observations are insufficient to lead me to any conclusion in the matter. The rock rotting which I believe to be responsible for our British tors may well have taken place under a climate warmer than we now enjoy, and there may have been available for it quite substantial periods of geological time. Finally I must thank Professor Hollingworth for drawing my attention to the tor topography of the Galtee Mountains: the relation of the tors on these summits to the corries that hold Lough Muskry and Lough Carra would certainly seem parallel to the relations observed between tors and corries on Lochnagar and in the Cairngorm Mountains.

Mr. W.V. Lewis (Cambridge)

I would like first to say how greatly I have enjoyed this lecture and even more being shown some of these features in the field by Professor Linton. I was inclined to think that he was overdoing the deep weathering process until he showed me deep weathering occurring in basalts in Tideswell Dale. Even if Professor Linton were wrong, and he certainly is not in the main, our gratitude is due to him for bringing before us so clearly important matters which we have neglected.

If I may try to be constructive and mildly critical, I do not like the lines drawn on the diagrams (Figs. 2a and 2b); the sequence of changes is unlikely to be represented as one line (BB), then a halt for deep weathering, a further halt for removal of the soil, and a revealing of the tors; yet another halt with the tors in the state in which we see them today, and so on. I think the process is more continuous. Whilst agreeing that the deep weathering occurs in warmer climates, as in Portugal, I think that in the colder areas, such as the Cairngorms, there is a wearing away of material around the larger blocks, which may form something like a tor, whether or not there is soil round the block. The large blocks on the tops of such hills are probably still being attacked today, when for half the year or more there is snow wrapped round the granite. Professor Linton showed an admirable instance of a circular hole which water had eroded in Millstone Grit, which is tough. If unacidulated, relatively innocent rainwater can corrode a hole in Millstone Grit, I really think that snow resting against granite or some such massive rock, keeping the side of the tor soggy, will help surely though slowly to eat into it and steepen the sides.

Concerning the fact that tors occur on the summits of hills, this is simply due to the tendency of soil to creep down hill. True, tors will tend to occur where the rocks are most massive, but statistically we can say that hilltops, when not too flat, are favourite places, and I suspect that the down hill creep of material by solifluction or any process away from the top will tend to leave the summit area of the rock naked because no more soil is supplied to the summit area.

Professor Linton

I agree that the diagrams perhaps suggest something more categorical than is actually the case. Nevertheless, I think that more than soil creep is involved in producing the difference between the blocks that remain as a tor and material that creeps away. Once a rock mass has become exposed it is not likely to be destroyed very rapidly by atmospheric weathering. Whereas the surrounding material is relatively rapidly attacked by sub-surface rotting, the tors once they are exposed have a good chance of a long life. Rainwater, as Mr. Lewis has suggested, can have some effect—in fact, may have; having recently found a hole 21 inches deep corroded in Pennine gritstone, I have a healthy respect for what acidulated water can do.

Mr. Lewis said he did not think I was wrong; but even if I am not right I hope I have rendered a valuable service. I cannot believe that all I have suggested is right; but I hope I have stimulated interest in the tor problem. It needs close examination and by as many workers as possible. I hope some of those present will work on it.

Mr. John Palmer (University of Leeds)

Professor Linton has kindly referred to my recent study on tor formation in N.E. Yorkshire, read to the Institute of British Geographers in January 1955, and has suggested that the features there discussed might be a special class of valley-side tor, and, if they originated in a manner other than that envisaged by Professor Linton for tors in general, that they should have a name other than tor. Some of the tors described by me occur *on* the watershed and have all the other appearances associated with tors. When my study was made I did not have the advantage of Professor Linton's evidence, but was acquainted with the work of Crickmay (*J. Geol.*, **43** (1935) 745–58) which has some similarities. I am still unable to find that the evidence in N.E. Yorkshire allows the tors there to be explained by Professor Linton's theory.

At the same time, but with great diffidence since I do not know Dartmoor intimately, it has occurred to me that the tors there might be explained in a manner similar to that which I have favoured elsewhere. Many workers on Dartmoor have found that the rock in the tors is more resistant than the rock around them. The maps indicate that most of the tors occur around the upper edge of steep slopes, be they valley sides or steps between plateaux levels. Tors are absent over much of the plateaux surfaces, and I am led to wonder at this distribution. If at some time the rate of removal at the top of steep slopes was rapid, differential resistance to weathering might have allowed portions of the ground to emerge above the general level. Once emerged, the pseudo-bedding and joint structure would allow the detailed tor form to develop. The reason for the emergence might be related to changes consequent upon river rejuvenation, or perhaps climate change.

I should like to offer thanks to Professor Linton for this the first attempt to grapple seriously with the intriguing problem of tors on a national and rational scale, but would add my opinion that in view of the revolutionary morphological implications, a series of detailed studies to test the theory locally are clearly called for.

Mr. Ronald Waters (University College of the South-West)

I would like to refer to two characteristics of the Dartmoor tors to which Professor Linton has already directed our attention, namely, their structure and their alignment.

On Dartmoor, tor structure varies from massive to what the late Mr. Hansford Worth described as lamellar. Lamellar tors are tors in which vertical joints are relatively rare, but quasi-horizontal divisional planes are frequent; they are tors in which the restriction of vertical jointing has led to the development of numerous closely spaced pseudo-bedding planes. Massive tors, on the other hand, exhibit better developed vertical jointing and fewer horizontal joints. The massive castellated tors are the more striking, often rising abruptly from the edge of a fairly level surface, Professor Linton's "basal platform"; moreover, their vertical extent is commonly greater than that of the lamellar variety. But the lamellar tor is found most frequently crowning a hillock or mound which rises above the general level of the surrounding surface; the surface slopes of the mound often conform almost exactly to the lie of the pseudo-bedding in the exposed rock mass.

These observations suggest that so far as tor formation is concerned there is an optimum spacing for the vertical joints. On the porphyritic granite of Dartmoor the horizontal dimension of the joint-bounded blocks building the biggest tors is approximately 300 square feet. This is the size of the largest joint blocks on Vixen Tor, the biggest Dartmoor tor which stands 54 feet high. The vertical extent of tors decreases with a decrease in the horizontal extent of their constituent blocks until the tor becomes no more than a group of low pinnacles barely breaking the surface of growan. An increase in the horizontal dimension of constituent blocks beyond the optimum is also accompanied by a decrease in the vertical extent of the tors as they become progressively more lamellar until once again they scarcely rise above the ground surface. In this extreme case the surface is, of course, developed on sound granite. In the double tor of Haytor Rocks these relations are well shown. The eastern, more massive and higher portion rises some 50 feet; its constituent blocks are comparable in size to those building Vixen Tor. The western portion exhibits fewer vertical joints, is consequently more lamellar, extends over a greater surface area and is lower.

There are on Dartmoor distinct rectilinear alignments of the tors and many other associated features owing their origin to differential etching. These alignments are, of course, valuable indicators of the structural grain of the country. They may or may not be parallel with valleys and divides; in fact, they can be used as a yardstick against which may be measured the amount of adjustment to structure attained by different elements of the drainage pattern. A field survey of the relations existing between the streams and structural grain of Dartmoor is now being made, and the promising results already obtained would seem to indicate that the significance of tors and related features for geomorphology may be very great when their space relations are considered in some detail.

Professor Linton

I thank Mr. Palmer and Mr. Waters not only for what they have said but for their assurances that they are both working on the problem in the field. Mr. Palmer will perhaps seek more evidence for his view that the details of tor structure are fashioned after the tors have emerged. The suggestion I am putting forward tonight is that the tors are rock masses which were pre-selected before they were exhumed and is based on local studies in different parts of the country. Naturally, the process that leads to the pre-selection will be governed and controlled by structure and notably by the spacing of joints.

Dr. G. T. Warwick (University of Birmingham)

The many granite blocks littering the slopes below the Dartmoor tors are often concentrated into distinct lines along minor gullies and spread out on reaching flatter ground. This indicates a considerable reduction of the tors after exposure, and the present apparent lack of movement of the blocks leads me to suggest solifluction as the mechanism responsible for their transport to their present positions. On the Millstone Grit moorlands of the Middle Wharfe Valley, Mr. R.A. Pullan has demonstrated (in an unpublished M.SC. thesis) that tors are only common on those grits with a high felspar content.

In the southern part of the granitic section of the Odenwald, the hills are gently rounded like chalk downland, and there are no tors. Roadside exposures reveal at least 10 feet of extremely weathered rock, in which original banding, and in one case a small reversed fault, were clearly to be seen, indicating that there had been little subsurface creep here. This area was unglaciated, and may have preserved its vegetation cover during the cold periods of the Pleistocene so that the unweathered blocks have not been exposed. In strong contrast, in the Trifels district on the western side of the Rift Valley, gigantic, bare "stacks" of siliceous Middle Bunter Sandstone rise some 150 feet abruptly from their gentler, forested basal slopes. This type of feature appears to be more closely related to the hypothesis of Walther Penck.

Professor J.A. Steers

I know nothing directly about this subject, but Professor Linton's arguments seem very sound. While listening to his paper and looking at his slides, my thoughts took me to Queensland. At Cape Melville there are hills some hundreds of feet high and a few miles long composed almost entirely of great boulders. I do not know how they were formed, but if Professor Linton's arguments apply also to them, a great deal of rock has been removed in some way. Professor Linton has referred to Portugal: I hope he will extend his activities because I think he will find some admirable material for his purpose in the Tropics. I congratulate him on his paper.

Professor Dudley Stamp

I do not often venture into this field, but I am tempted to do so because only a few weeks ago I was observing with Professor Davis, Professor of Geography at Hong Kong, what is happening in that colony. In these days of land shortage there it is nothing for it to be decided to remove a hill of 300 feet high. They are granite hills, nominally, and the removal is done by Chinese contractors. If the contractor is lucky and the hill is of completely decomposed granite he makes a fortune; if he is unlucky he goes bankrupt. We found by measurement that the rotting of the granite goes down to at least 250 feet. It seemed that to use the word "weathering" when speaking of that sort of depth was wrong. We said we would refer to it as the "rotting" of granite, which does not presuppose whether it is atmospheric in origin or not.

I would ask Professor Linton if he would link up his studies in the southwest with another phenomenon of granite masses, china clay, and particularly with its origin, because there are those extraordinarily deep pits from which the china clay is extracted not very far away from the area of the tors. Professor Davis and I were wondering whether it is pneumatolysis at the lower levels which gives rise to the areas of rotting rather than atmospheric weathering from above. Does Professor Linton think there is any possibility of the same being true in the south-west?

Professor Linton

That last suggestion has been considered by other geologists who found that there is no connection between pneumatolysis and the growan. Some of the china clay beds are 100 feet or more deep, but they are very sharply defined and clearly distinct from the areas in which rotting from above has given rise to the growan.

I am greatly indebted to Professor Steers and Dr. Stamp for their observations on occurrences of deep weathering and core-stones in the Tropics. And since I asked for a better definition, I accept Dr. Stamp's suggestion that rotting is undoubtedly the right word. I shall be glad to use it.

The Chairman

I know it will be the wish of all present that I should express on their behalf and my own also our cordial thanks to Professor Linton. I will not attempt to add further to the discussion but simply record my own satisfaction and conviction that Professor Linton is throwing a great deal of continuous light on this phenomenon. He has not only described in some particularity the origin of the tors but has thrown a great deal of light on their significance.

References

1 Geikie, A., "Text-book of geology," 4th ed. London, 1903, vol. 1, p. 456 and Figure 98. Lake, P., and Rastall, R.H., "A text book of geology." London, 1910, Plate VII.
Smith, Bernard, "Physical geography for schools." London, 1917, p. 63, Figure 72.
2 Holmes, A., "Principles of physical geology." London, 1944, p. 76 and Figure 62.
3 Philippson, A., "Grundzüge der allgemeinen Geographie," 2nd ed. Leipzig, 1931, II Band 2 Hälfte, pp. 20–22.
Strahler, A.N., "Physical Geography." New York, 1951, p. 129.
4 Scrivenor, J.B., "The geology of Malaya." London, 1931, pp. 136–8.
5 Martonne, E. de., "Traité de géographie physique," 4th ed. Paris, 1926, Tome II, pp. 631–4.
6 Cotton, C.A., "The Geomorphology of New Zealand." Wellington, 1926, pp. 26–9, with Figures 23, 24 and 25.
7 Cotton, C.A., op. cit., p. 29.
8 Pullan, R.A., Unpublished M.Sc. thesis of the University of Birmingham, 1954.
9 Engeln, O.D. von, "Rock sculpture by glaciers: a review." *Geogr. Rev.* **27** (1937) 482.

10 Flückiger, O., "Glaziale Felsformen." *Petermanns geogr. Mitt.*, Erganzungsheft, 218, 1934.

11 Cotton, C.A., "Climatic accidents in landscape making." Christchurch N.Z., 1942, pp. 232–3.

Engeln, O.D. von, "Geomorphology." New York, 1942, p. 461.

12 Well illustrated in Plate VII of "British regional geology: South-West England," 2nd ed. H.M.S.O., 1948.

13 Palmer, J., "Tor formation at the Bridestones in North East Yorkshire." *Trans. Inst Brit. Geogr.* for 1955 (in the press).

14 Common, R., "The geomorphology of the East Cheviot Area." *Scot. Geogr. Mag.* **70** (1954) p. 131.

15 Cotton, C.A., *op. cit.*, 1926, Figures 24 and 25.

Raeside, J.D., "The origin of Schist Tors in Central Otago." *N.Z. Geogr.* **5** (1949) pp. 72–6.

16 Phemister, T.C., and Simpson, S., "Pleistocene deep weathering in North-East Scotland." *Nature, Lond.* **164** (1949) p. 318.

17 Raeside, J.D., *op. cit.*, p. 74.

Handley, J.R.F., "The geomorphology of the Nzega area of Tanganyika with special reference to the formation of granite tors." *Comptes rendus de la XIXe congrès géologique international.* Algiers, 1954, fascicule XXI, pp. 201–210.

Penck, W., "Morphological analysis of land forms." London, 1953, pp. 202–206 and Fig. 14.

18 Dahl, E., "On different types of unglaciated areas during the Ice Ages and their significance to phytogeography." *New Phytol.* **45** (1946) p. 225–42.

19 Linton, D.L., "Unglaciated areas in Scandinavia and Great Britain." *Irish Geogr.* **2** (1949) pp. 25–33 and pp. 77–9.

"Unglaciated enclaves in glaciated regions.!" *J. Glaciol.* **1** (1950) pp. 451–3.

A paper entitled "The significance of tors in glaciated lands" was read to the XVIIth International Geographical Congress at Washington on 12 August 1952, but no *comptes rendus* of this congress have yet appeared.

4

INTERPRETATION OF EROSIONAL TOPOGRAPHY IN HUMID TEMPERATE REGIONS

J.T. Hack

Source: *American Journal of Science* 258-A (1960): 80–97.

Abstract

Since the period 1890 to 1900 the theory of the geographic cycle of erosion has dominated the science of geomorphology and strongly influenced the theoretical skeleton of geology as a whole. Some of the principal assumptions in the theory are unrealistic. The concepts of the graded stream and of lateral planation, although based on reality, are misapplied in an evolutionary development, and it is unlikely that a landscape could evolve as indicated by the theory of the geographic cycle.

The concept of dynamic equilibrium provides a more reasonable basis for the interpretation of topographic forms in an erosionally graded landscape. According to this concept every slope and every channel in an erosional system is adjusted to every other. When the topography is in equilibrium and erosional energy remains the same all elements of the topography are downwasting at the same rate. Differences in relief and form may be explained in terms of spatial relations rather than in terms of an evolutionary development through time. It is recognized however that erosional energy changes in space as well as time, and that topographic forms evolve as energy changes.

Large areas of erosionally graded topography in humid regions have been considered to be "maturely dissected peneplains." According to the equilibrium theory, this topography is what we should expect as the result of long continued erosion. Its explanation does not necessarily involve changes in base level. Pediments in humid regions and some terraces are also equilibrium forms and commonly occur on a lowland area at the border of an adjacent highland.

Introduction

The part of geologic theory that deals with the interpretation of landforms and the history of landscape development has been dominated for several generations by the ideas of William Morris Davis and his followers. Davis' theory of landscape evolution was first fully presented in his essay, "The Rivers and Valleys of Pennsylvania" (Davis, 1889).[1] The important concepts that he introduced include the geographic cycle, the peneplain, and the formation of mountains by a succession of interrupted erosion cycles. Davis' theories became immensely popular among geologists in Europe as well as in America, though there were dissenters, including, for example, Tarr (1898) and Shaler (1899). His theory of the evolution of mountains as topographic features through the mechanism of multiple erosion cycles was especially influential and came to have a great influence on the theoretical skeleton of the whole science of geology. Its impact is still felt. Many of our ideas relating to the history of mountains, the internal constitution of the earth and the origin of some ore deposits are closely related to this theory. The idea that mountain ranges are vertically uplifted after they have been folded was conceived in order to explain the widespread existence of dissected peneplains (Daly, 1926). Another example is the theory of origin of bauxite and of manganese ores and other residual concentrates in the Appalachian Highlands, that are thought by some to have formed on a Tertiary peneplain surface (Hewett, 1916; Stose and Miser, 1922, p. 52–55; Bridge, 1950, p. 196.)

In the last 20 years, however, Davis' ideas have become less popular and the small but ever-present number of geologists who were skeptical of his theories has increased. Though many geologists have been dissatisfied with it, the theory of the geographic cycle and its application to the study of landforms has not generally been replaced by any other concept. Several alternative theories have been proposed, including the theory of Penck (1924, 1953) which relates the form of slopes to changes in the rate of uplift relative to the rate of erosion, and the "pediplain" theory of L.C. King (1953), an elaboration and expansion of Penck's concept of slope retreat. Both of these theories, however, are also cyclic concepts and hold that the landscape develops in stages that are closely dependent on the rate of change of position of baselevel.

During the course of my work in the Central Appalachians which began in 1952, seeking a different approach to geomorphic problems, a conscious effort was made to abandon the cyclic theory as an explanation for landforms. Instead, the assumption was made that the landforms observed and mapped in the region could be explained on the basis of processes that are acting today through the study of the relations between phenomena as they are distributed in space. The concept of dynamic equilibrium forms a philosophical basis for this kind of analysis. The landscape and the processes molding it are considered a part of an open system in a steady state of balance in which every slope and every form is adjusted to every other. Changes in topographic form take place as equilibrium conditions change, but it is not necessary to assume that the kind of evolutionary changes envisaged by Davis ever occur. The consequences and results of this kind of analysis in most cases differ from conclusions arrived at through the use of the cyclic concepts of Davis.

On rereading some of the classic American literature in geomorphology I realized that G.K. Gilbert used essentially this approach and that I have followed a way of thinking inherited either directly from him or from some of his colleagues. Even though Davis and Gilbert were contemporaries and friends, Gilbert makes little use of and few references to the theory of the geographic cycle or any of its collateral ideas. This omission is so conspicuous that it is difficult to believe Gilbert ever wholeheartedly accepted the idea. It seems to me that Gilbert's famous paper, "Geology of the Henry Mountains" (Gilbert, 1877, p. 99–150) outlines a wholly satisfactory basis for the study of landscape that does not foreshadow the developments in geomorphology that followed in the next 50 years.

In the pages that follow some concepts inherent in the theory of the geographic cycle that seem to me unsound are briefly discussed. The alternative approach to landscape studies based on spatial relations in a system in equilibrium is briefly presented. Very few of the ideas are original and most of them have been published in one form or another in the works of other geologists. In addition I wish to acknowledge the considerable assistance obtained in stimulating discussions with my friends and colleagues, especially C.S. Denny, J.C. Goodlett, C.B. Hunt, L.B. Leopold, C.C. Nikiforoff, and M.G. Wolman with whom I have been associated at various times during the formulation of these ideas. The manuscript has been read and criticized by R.P. Sharp of the California Institute of Technology, Sheldon Judson of Princeton University, C.C. Nikiforoff (formerly of the Department of Agriculture) as well as by some generous colleagues in the U.S. Geological Survey.

The geographic cycle and the peneplain concept

The theory of the geographic cycle rests on the assumption that there is a base level toward which every area erodes and to which the streams become graded. After an initial uplift or rise of a part of the earth's crust, erosion proceeds through successive stages of youth, maturity, and old age; from a time in which stream grades are irregular, through a time when they become smooth, to a stage of low relief when the entire landscape is reduced close to base level. An important stage in the cycle is reached when the slopes of the larger streams are so reduced that they are able to transport just the amount of debris supplied from upstream and no more. At this point the stream is said to be graded and the stage of maturity begins. The trunk streams, unable to erode their beds, shift laterally, forming floodplains and meanders.

The debris-covered area on the valley floor expands laterally as the inter-stream divides are lowered. At the final stage of old age the landscape is one in which the streams meander across broad plains covered by a sheet of waste. The divide areas are graded to the streams and are covered by a waste mantle transported by creep. They rise only slightly above the shallow valleys (Davis 1909, p. 254–272; 1899, p. 485–499). Davis envisaged that there must be interruptions of the ideal cycle of erosion and in fact that an ideal cycle is rarely completed. Alternate periods of uplift and stability of base level result in successive incomplete cycles during which the uplifted peneplains are dissected and new ones form along the streams. Hilly areas in which the hilltops rise to roughly the same height above the streams, like the Piedmont region of the central Atlantic States, are peneplains that have been uplifted and dissected by stream erosion until a stage of maturity has been reached (Davis, 1909, p. 272–274; 1899, p. 499–501).

The concept of planation

The concept of *planation* or *lateral planation* is lucidly presented in the Henry Mountain report (Gilbert, 1877, p. 127). Gilbert described the process of planation in connection with the formation of smoothly graded, gravel-covered surfaces, cut on soft Mesozoic rocks at the foot of the Henry Mountains. Surfaces like these have come to be known as *pediments* and have been restudied in the Henry Mountains by Hunt, Averitt, and Miller (1953). It was recognized by Gilbert that the planation in this region occurs on soft rocks, such as weak sandstones and shales where slopes or declivities are small in comparison with the trachyte mountains in which the streams originate. Gilbert thought that lateral shifting of the streams is dependent on the fact that the bed load transported by the stream is more resistant than the rock through which it flows so that the stream cuts laterally against the soft bank. Where one of these streams cuts again through hard rock lateral planation ceases (Gilbert, 1877, p. 130) and canyons are formed. The process is dependent on the geology of the drainage basin of the laterally shifting stream, and on a contrast in rock resistance and slope between the upper and lower parts of the basin.

Davis applied the erosion cycle concept to the idea of lateral planation. In his theory a stream always has a tendency to erode laterally against its banks. As the landscape passes through the evolutionary stages of the erosion cycle, first the larger streams and later the tributaries approach the base level of erosion. As they do so their ability to cut downward diminishes. They migrate laterally eroding the valley walls, producing a floodplain or surface of planation.

It is interesting to note the contrast between the planation observed and described by Gilbert and the planation envisaged by Davis. Gilbert's explanation of lateral planation involves a dynamic equilibrium of forces existing at the present time in actual drainage basins and the relation of these forces to the rocks. Davis' theory on the other hand assumes that lateral planation occurs in any drainage basin with the passage of time, regardless of its geology. In fitting the concept of planation into the framework of the geographical cycle Davis attempted to rationalize relations between things that change through time and hence cannot be observed or measured. In the transfer from a scheme of ideas that involves space to a scheme that involves time Davis ignored the spatial relations cited by Gilbert that make the concept valid.

Surfaces of planation are produced by streams under certain circumstances, but there is no reason to believe that such surfaces enlarge through time as relief is lowered, merely as a consequence of a reduction in slope. On the contrary it is likely that as gradation proceeds, the efficiency of the stream system in removing the waste of its drainage basin may increase.

The graded stream

One of the key ideas in the theory of the geographic cycle is the concept of the graded stream. The word "grade" was used by Gilbert (1877, p. 112) in discussing the stable slope of the stream channel in the same sense that an engineer uses it to describe the slope of a railroad or highway. Davis borrowed the term at Gilbert's suggestion and used it in a more special sense to designate a certain stage in the evolution of stream profiles when the stream's ability to transport the load supplied to it from above is just balanced by the load that it has to carry (Davis, 1909, p. 392; 1902, p. 89).

This concept of grade has probably received more discussion among geomorphologists than any other aspect of the geographic cycle (for example Kesseli, 1941, Mackin, 1948, Woodford, 1951, Rubey, 1952, Leopold and Maddock, 1953, Wolman, 1955). As suggested by Kesseli, the concept as outlined by Davis seems rather elusive, so that it is difficult to identify a graded stream in nature. Mackin, however, in his study of the graded stream, clarifies some of the ideas and suggestions of Davis. The examples of graded streams cited by him are migrating laterally, depositing on the floodplain an amount of material equal to what they erode by lateral cutting. The graded stream is not actively cutting vertically downward and its longitudinal profile is being changed only very slowly as the relief or other conditions in the drainage basin change. Since it is cutting laterally, the channel of the graded stream is bordered by a floodplain underlain by thin river deposits and by terraces whose composition is entirely material carried from upstream and different from the underlying rock (Mackin, 1948, p. 472).

Leopold and Maddock (1953) considered the graded stream in relation to the hydraulic geometry of the channel. Their study of stream gaging and cross section data indicates just as consistent a pattern in the relationships between the variables width, depth, velocity, and sediment load in ungraded as in graded streams. They conclude that Mackin's concept of grade cannot be demonstrated by consideration of stream gaging data and they use the term "quasi-equilibrium" in reference to the equilibrium in stream channels observed by them in all the streams studied. They recognize that this equilibrium is distinct from the equilibrium implied by Davis and Mackin in the concept of the graded stream.

In Davis' concept of grade, high velocity of flow and a high capacity are associated with a steep channel slope. As slope diminishes during the evolution of the landscape through the erosion cycle, velocity diminishes as well as the capacity to transport debris (Davis, 1909, p. 397–398; 1902, p. 95–96). This idea may have seemed reasonable to Davis because he shared with many others the belief that mountain streams with steep channels have higher velocities of flow and therefore greater capacity than do large streams with lower slopes in lowland areas. This observation is not necessarily true. Actual measurements in many natural streams demonstrate that for equivalent frequencies of discharge average velocities tend to increase downstream rather than decrease (Leopold, 1953). Studies of some Appalachian streams, furthermore, indicate that the size of material a stream has on its bed and banks is not related directly to slope, but is related also to discharge and other variables in such a way that in many streams the competence (or size of material that a stream can transport) increases downstream as slope diminishes (Hack, 1958). These facts make it appear doubtful that streams reach a balanced condition through any evolutionary sequence involving a gradual reduction in slope. Probably the balance that exists in most streams is the quasi-equilibrium described by Leopold and Maddock (1953, p. 51). This is a balance among at least seven variables. It is so complex and there are so many alternative adjustments possible that equilibrium can be achieved under many conditions and is arrived at very quickly, almost immediately, in the development of a valley. The uniform, or regular concave-upward longitudinal profile that is characteristic of many streams and has been called "the profile of equilibrium" results not from the attainment of a certain stage in the evolution of a valley, but merely from the regular change downstream in some of the many variables involved in channel equilibrium. Most important of these is probably discharge that increases downstream as a consequence of a regular enlargement of the drainage area.

The streams cited by Mackin (1948) as examples of graded streams represent special cases that are exceptional rather than general. Like Gilbert's streams in the Henry Mountains, such streams head in hard rock areas of high slope and altitude. Their lower courses are in soft rocks and as a consequence have a low slope relative to the increased discharge. They migrate laterally and have a diminishing competence only because of the geologic pattern of the terrain they traverse. They represent a class of streams intermediate between those whose competence with respect to the load derived from upstream is increasing in a downstream direction or remains the same, and those whose competence decreases downstream so abruptly that they aggrade actively enough to build fans. They are no more in a state of equilibrium or disequilibrium than a mountain torrent that is engaged in cutting a gorge. The torrent also has a bed load and lag deposit or floodplain along the bank composed of material too coarse for the stream to move in the ordinary flood, but in this case the lag deposit is locally derived by washing or sliding down the adjacent slope or by plucking from the bed.

The stage of old age and the maturely dissected peneplain

In the concept of the geographic cycle the appearance of the land surface in the stage of old age is dependent on the process of lateral planation. The ideal surface is a plain partly graded by planation and covered by a veneer of waste. Divide areas with convex upward slopes exist, but are relatively smaller in area than in earlier stages. Such graded surfaces, as stated above, do not now exist in nature. The extensive plainlands of the earth are either depositional surfaces like alluvial plains, deltas, drift plains, and coastal plains, or if they are erosion surfaces in humid areas, they are hilly with rounded divides and steep-walled valleys that have generally come to be described as "maturely dissected peneplains." Exceptions are pediments and terraces, that in humid regions occupy relatively small areas. Excellent examples of maturely dissected landscapes in America are found in the Piedmont region of eastern North America, or in the Central United States where the so-called Ozark peneplain has been "dissected and uplifted." Large areas of the Canadian shield have been said to be a dissected peneplain whose drainage has been disrupted by glaciation. The great limestone valley of the Appalachians, similarly, has been said to be a dissected peneplain as is the plateau area to the west of the Appalachians. Thus land surfaces that are worn down to the stage of old age, as conceived by Davis, are virtually nonexistent; on the other hand former old age surfaces that have been dissected to Davis' stage of maturity are ubiquitous in the older terrains of the earth, especially in humid regions. Indeed this kind of topography is so universal it suggests that the end product or end surface toward which erosion proceeds resembles the "maturely dissected" surface rather than the "old age" surface or peneplain. Such an end surface is one whose forms are graded for the efficient removal of waste rather than one on which the waste products accumulate and stagnate.

The principle of dynamic equilibrium in landscape interpretation

An alternative approach to landscape interpretation is through the application of the principle of dynamic equilibrium to spatial relations within the drainage system. It is assumed that within a single erosional system all elements of the topography are mutually adjusted so that they are downwasting at the same rate. The forms and processes are in a steady state of balance and may be considered as time independent. Differences and characteristics of form are therefore explainable in terms of spatial relations in which geologic patterns are the primary consideration rather than in terms of a particular theoretical evolutionary development such as Davis envisaged.

The principle of dynamic equilibrium was applied to the study of landforms both by Gilbert (1877, p. 123) and by Davis (1909, p. 257–261, 389–400; 1899, p. 488–491; 1902, p. 86–98). Recently Strahler has outlined the principle in more modern terms as it might be applied to landscapes (Strahler, 1950, p. 676). The concept requires a state of balance between opposing forces such that they operate at equal rates and their effects cancel each other to produce a steady state, in which energy is continually entering and leaving the system. The opposing forces might be of various kinds. For example, an alluvial fan would be in dynamic equilibrium if the debris shed from the mountain behind it were deposited on the fan at exactly the same rate as it was removed by erosion from the surface of the fan itself. Similarly a slope would be in equilibrium if the material washed down the face and removed from its summit were exactly balanced by erosion at the foot.

In the erosion cycle concept of Davis, equilibrium is achieved in some part of the drainage system when there is a balance between the waste supplied to a stream from the headwaters and the ability of the stream to move it, or in other words, when the slope of the channel is reduced just enough so that the stream can transport the material from above with the available discharge. As argued on page 9 this kind of equilibrium probably is achieved in a stream almost immediately and is not related to a particular stage in its evolution. Davis' concept would imply that some parts of a drainage system would be in equilibrium whereas at the same time other parts would not, and that the condition of equilibrium is in time gradually extended from the downstream portion to the entire drainage system.

Rather than a concept of balance between the load of a stream and the ability of the stream to move it, it is more useful in the analysis of topographic forms to consider the equilibrium of a particular landscape to involve a balance between the processes of erosion and the resistance of the rocks as they are uplifted or tilted by diastrophism. This concept is similar to Penck's concept of exogenous and endogenous forces (1924, 1953). Suppose that an area is undergoing uplift at a constant rate. If the rate of uplift is relatively rapid, the relief must be high because a greater potential energy is required in order to provide enough erosional energy to balance the uplift. The topography is in a steady state and will remain unchanged in form as long as the rates of uplift and erosion are unchanged and as long as similar rocks are exposed at the surface. If the relative rates of erosion and uplift change, however, then the state of balance or equilibrium constant must change. The topography then undergoes an evolution from one form to another. Such an evolution might occur if diastrophic forces ceased to exert their influence, in which case the relief would gradually lower; it might occur if diastrophic forces became more active, in which case the relief would increase; or it might occur if rocks of different resistance became exposed to erosion. Nevertheless as long as diastrophic forces operate gradually enough so that a balance can be maintained by erosive processes, then the topography will remain in a state of balance even though it may be evolving from one form to another. If, however, sudden diastrophic movements occur, relict landforms may be preserved in the topography until a new steady state is achieved.

The area in which a given state of balance exists and that may be considered a single dynamic system may be conceived as very small or very large. In the Appalachian region, it may be that large areas are essentially in the same state of balance. In the West, however, in an active diastrophic belt, a single dynamic system may constitute only a small area such as a single mountain range or a small part of a mountain range. Furthermore, because of sudden dislocations of the crust relict forms may be preserved in the landscape that reflect equilibrium conditions that no longer exist.

The crust of the earth is of course not isotropic and within a single erosional system, no matter how small, there is a considerable variation in the composition and structure of the crust. These variations are reflected by variations in the topography. Consider, for example, an area composed partly of quartzite and partly of shale. To comminute and transport quartzite at the same rate as shale, greater energy is required; and since the rates of removal of the two must be the same in order to preserve the balance of energy, greater relief and steeper slopes are required in the quartzite area. Similarly geometric forms differ on different rock types. An area that is underlain by mica schist or other igneous or metamorphic rock subject to rapid chemical decay, has more rounded divides than an area underlain by qaurtzite, if both are in equilibrium in the same dynamic system, for the schist is comminuted by weathering to silt and clay particles that are rapidly removed from hill tops on low slopes. On the other hand to remove quartzite from a divide at the same rate, steeper slopes and sharper ridges are required because the rock must be moved in the form of larger fragments.

The analysis of topography in terms of spatial or time-independent relations provides a workable basis for the interpretation of landscape. This kind of analysis is uniformitarian in its approach, for it attempts to explain landscapes in terms of processes and rates that are in existence today and therefore observable. It recognizes that processes and rates change both in space and time, and, by clarifying the relation between forms and processes, it provides a means by which the changes can be analyzed.

The relation of soil to topography

Cyclic concepts of soil evolution have developed in a manner parallel to the erosion cycle concept of geomorphology. The idea of a cyclic evolution of soil through a stage of maturity to senility in which the profile becomes intensified and thickened through time is dependent on the concept of a topography that is stable, such as might exist on a peneplain or on a remnant of a dissected peneplain. Naturally enough, this idea lends support to the cyclic concept of landscape evolution.

An alternative theory of soil evolution based on dynamic equilibrium has been forcefully presented by C.C. Nikiforoff (1942, 1949, 1955, and 1959, p. 188) and parallels the concept of equilibrium in landscape evolution. As explained by Nikiforoff, nearly all soils achieve a state of dynamic equilibrium if they are exposed to the surface for a sufficient time. Factors in the equilibrium include climate, slope, rate of erosion, composition of the parent material, vegetation, and others. The horizons of the soil become diversified and owe their existence to an equilibrium among processes that tend to accumulate certain substances at certain depths, and those that tend to remove them. Take the clayey "B" horizon as an example:

The cyclic viewpoint holds that the clay in the "B" horizon accumulates through leaching of the "A" horizon above it. The concentration of clay increases through time, though at a slower and slower rate until further accumulation is impossible. At this point the soil is mature and remains in this state until removed by erosion.

In terms of dynamic equilibrium, on the other hand, the amount of clay present in any horizon of the soil is the result of a balance at that place between the rate of clay accumulation and clay removal. These rates differ in different horizons and subhorizons and the balance between them determines the amount of clay present. Similar balance between rates of removal and of accumulation and the interactions between them determine the composition of all the horizons.

From the point of view of landscape interpretation this view of the soil has many advantages. It permits the lowering of the hilltop by erosion at a more or less constant rate, at the same time maintaining the equilibrium of the soil. In fact the soil profile is dependent on erosion as one of the factors in the equilibrium.

Compare a hilltop underlain by pure limestone or by clayey or silty lime-stone with a hilltop nearby underlain by cherty limestone containing massive beds of chert (fig. 1). The hilltop underlain by silty limestone does not accumulate weathering products because they are removed by erosion as soon as they are freed from the rock by solution of the carbonate that binds them. On the other hill, underlain by cherty rocks, though solution goes on at the same rate, a debris of chert fragments is produced in addition to fine silt. These are too coarse to be removed by sheet erosion on slopes as gentle

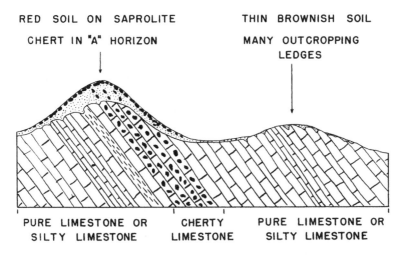

RED SOIL ON SAPROLITE

CHERT IN "A" HORIZON

THIN BROWNISH SOIL

MANY OUTCROPPING
LEDGES

PURE LIMESTONE OR
SILTY LIMESTONE

CHERTY
LIMESTONE

PURE LIMESTONE OR
SILTY LIMESTONE

Figure 1 Hypothetical cross section through two hills in a limestone region showing the relation of the bedrock and topography to the residual soils.

as those of the first hill. The chert accumulates forming a protective armor that prevents finer insoluble residues, also present, from being eroded. Accumulation of this deposit continues until the slopes are steep enough for the chert to be removed. The topography is now in equilibrium. One hill is covered with saprolite or residuum and the other is not, but the saprolite-covered hill has steeper slopes and rises higher above the streams than the other (Hack, 1958b). It might be said that the residual material is "stored" on top of the hill as its covering armor is comminuted to sizes that can be removed by creep and wash. During the period of "storage" the material becomes oxidized and reddened and a profile develops. The time of storage may be very long, even thousands of years, and may be long enough for the soil to have survived major changes in climate and to owe some of its characteristics to irreversible reactions that took place in the past (Nikiforoff, 1955, p. 48).

Examples of erosionally graded or equilibrium topography

As an area is graded by erosional processes the differences in the bedrock from one place to another cause a differentiation of the forms on them. Landscapes that develop on intricate and actively rising fault blocks may bear a closer relation to major structural features than to the underlying rock, but in a landscape like that of the Appalachian region in which large areas are mutually adjusted, the diversity of form is largely the result of differential erosion of rocks that yield to weathering in different ways. Such topography may be referred to as erosionally graded.

Ridge and ravine topography

Many of the erosionally graded landscapes in humid temperate regions belong to the almost ubiquitous type that is commonly known as the "maturely dissected peneplain". Preferring a term that has no genetic connotation a more descriptive one such as *ridge and ravine* topography is suggested. This term refers to the monotonous network of branching valleys and intervening low ridges that make up the landscape of large areas. This topography may be concisely explained in terms of dynamic equilibrium in the words used by Gilbert (1909) in his discussion of rounded hilltops. He conceived that the important elements of the topography could be divided into two domains. The first, a domain of stream sculpture represented by channels in which the slopes are concave upward, because, as he says (Gilbert, 1909, p. 344), the transporting power of a stream per unit of volume increases with the volume; the transporting power also increases with the slope; and a stream automatically adjusts slope to volume in such a way as to equalize its work of transportation in different parts. The other domain is that of creep, represented by the slopes between the channels. In this domain slopes are mostly convex. Gilbert states (1909, p. 345) that:

> This is because the force impelling movement of material is gravity which depends for its effectiveness on slope. On a mature or adjusted profile the slope is everywhere just sufficient to produce the proper velocity. It is greatest where the velocity is greatest and therefore increases progressively with distance from the summit.

The forms of well-graded ridge and ravine landscapes vary within wide limits. Typical examples, both in areas of high relief, have been described by Strahler (1950) and by Hack and Goodlett (in press, 1960). Somewhat gentler topography of the same type is widespread in the Piedmont Province. An example of such an area is shown in figure 2. This is in a drainage basin tributary to the Patapsco River in Carroll County, Maryland. The bedrock is phyllite that is cut by veins of quartz. The interstream divides are convex upward and if measured on a coordinate system in which the origin, or zero point, is the top of the hill or ridge, they have the form of a parabolic curve like the one shown in fig. 3. They intersect the stream bottoms in steep slopes and sharp angles, though in places the foot of the slope is concave upward, especially in slopes that intersect the floodplain of a stream at a point opposite the channel. The regularity of the landscape and the rather uniform height of the hills owe their origin to the regularity of the drainage pattern that has developed over long periods, by the erosion of rocks of uniform texture and structure.

Differences in form from one area to another, including the relief, form of the stream profile, valley cross sections, width of floodplain, shape of hill

0   1   2

SCALE IN MILES
CONTOUR INTERVAL 20 FEET

Figure 2 Topographic map of area in the Piedmont Province of Maryland, showing typical ridge and ravine landscape (1950, U.S. Geol. Survey Winfield Quadrangle, Maryland).

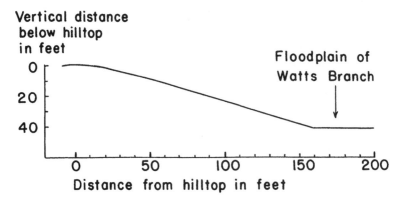

Figure 3 Profile of hill slope near Rockville, Maryland, on fine-grained mica schist.

tops and other form elements are explainable in terms of differences in the bedrock and the manner in which it breaks up into different components as it is handled on the slopes and in the streams.

Pediments

Where differences in rock resistance in graded landscapes are slight or confined to narrow or small areas, the differences in topography are small. Where, however, two large areas, one of resistant rock and the other of much softer rock, are juxtaposed, the differences are not only pronounced but there is a zone of transitional forms on the less resistant rock. In both areas ridge and ravine landscapes are developed, but as the more resistant area has greater relief, steeper slopes, and sharper divides, debris is shed from the higher to the lower area. This kind of situation is a common one in the Appalachian Highlands where many valley or lowland areas are underlain by limestone and shale and are bordered by ridges or series of ridges underlain by sandstone and quartzite. The transitional forms are broadly fan-shaped gravel-covered and dissected surfaces cut by streams on bedrock that closely resemble typical pediments in many western areas. They are called pediments because of this resemblance, but it is recognized that similar surfaces may be produced by different processes.

An example of a pediment area in the headwaters of the South Fork, Shenandoah River, Augusta and Rockingham Counties, Va., may clarify the equilibrium relations involved. As shown in figure 4, Dry River, Briery Branch, and the North River have their headwaters in an area of resistant rock of Silurian. Devonian and Mississippian age consisting mostly of sandstone, quartzite, and shale. The relief in this area averages 1,500 to 2,000 feet and ridges rise to altitudes as high as 4,400 feet. The soft rock area into

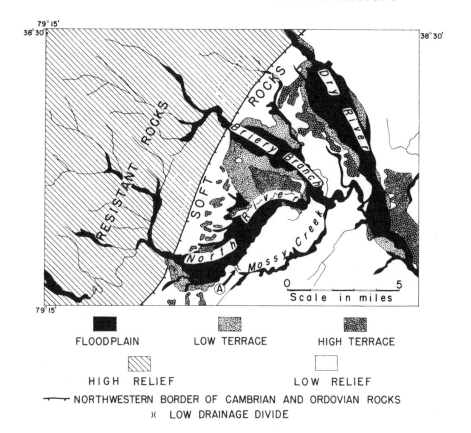

Figure 4 Simplified map showing bottomlands and terraces on the northwest side of the Shenandoah Valley, Va., to illustrate the formation of pediments by lateral planation and piracy.

which the three streams flow is underlain by Cambrian and Ordovician carbonate rocks and shale and has an average relief of only a few hundred feet. Most of it is less than 1,800 feet in altitude. On entering the carbonate rocks each of the three streams is bordered by a broad floodplain and terraces composed of cobbles and sand derived from the mountains upstream, The low hilly plain at the foot of the mountains is an extensive complex of dissected terrace remnants that resembles, and in fact is similar to the dissected pediments common at the foot of mountain ranges in the western United States.

Mossy Creek, a tributary of the North River, does not head in the sandstone mountains, but in the limestone area. It has a flatter gradient than the North River even though a smaller stream and at locality A (fig. 4) where the two streams are separated by a low divide, Mossy Creek is about 60 feet below the elevation of the North River floodplain. In 1949 during a severe flood, water actually spilled over the divide into Mossy Creek. This is therefore an example of a stream piracy in progress.

In the resistant rock area the relief is high because the bedrock is removed largely through mechanical processes. Slopes are steep, divides sharp and there are many rock slides. Stream slopes throughout are adjusted for any given drainage area and discharge to transport rock fragments of large size. In the soft rock, or carbonate area, on the other hand, chemical weathering is more important and although the surface is being lowered at the same rate as in the resistant rock area the graded slopes are much gentler and the relief lower. Where the North River leaves the resistant rock area it moves a large load of sandstone cobbles on its bed. They are carried out onto the soft rock area where the channels are adjusted for the transportation of much finer debris. As a consequence the stream shifts laterally and deposits cobbles on the banks, forming a floodplain. Being more resistant to weathering than the carbonate rocks on which they have been deposited, the cobbles persist in the landscape for long periods and as the river continues to erode they form terraces and eventually cap divides.

Mossy Creek will continue to erode its channel and since it is not required to move a load of cobbles, momentarily cuts faster than the North River. Eventually piracy will occur and Mossy Creek Valley will be aggraded by cobbles brought in by the North River. The North River floodplain below locality A will then be abandoned and will become a dissected terrace. Captures similar to this one have already occurred at other places. Note for example in figure 4 the low terraces that connect Briery Branch with the North River.

In this explanation it has been assumed that the resistant and soft rock areas are in the same erosional system and that the average rate of erosion in the two parts is the same. The pediment area exists because cobbles are shed from one part to the other, thus introducing and maintaining a belt of resistant cobbles at the margin of the soft rock area. The pediment area is of course itself in equilibrium in the same system. Its size and the amount of relief are determined by the rate at which cobbles are carried into the soft rock area (a function of size of drainage basin) and by the rate at which they are weathered and broken into pebbles that can be moved out in the streams draining the soft rock area. Hunt, Averitt, and Miller (1953, p. 189) applied the same kind of explanation to the pediments in the Henry Mountains. In that area, however, because the climate is semi-arid the processes are not quite the same, for there is a loss of discharge involved as the streams leave the more humid mountain area.

In the Valley and Ridge Province of the Appalachian Highlands there are many extensive surfaces produced by the processes just described. Many valleys are floored by shale and bordered by relatively high sandstone ridges. The gravels shed by the ridges are too coarse to be carried off by the master stream flowing in shale, and so are stored in floodplains, terraces, and dissected terraces, that because of their resistance to erosion form high benches on either side of the master streams. Eventually the cobbles and gravels in these benches are reduced by weathering and reworking in the laterally shifting streams to sizes that can be carried off down the main valley. Such extensive pediment-like landscapes have long been mistaken for a former gravel-covered broad valley stage or peneplain that is now dissected. Actually such gravelly surfaces testify to the contrast in resistance between the rocks of the mountains and the rocks of the valley and they are part of the equilibrium between the two.

Terraces

Some stream terraces may have their origin as equilibrium forms. Mapping of surficial deposits in the Shenandoah Valley, Va., indicated that terraces are most common in soft rock areas along streams that originate in hard rock areas. This coincidence suggests that the terraces are preserved essentially because they contain components more resistant than the underlying rock. Terraces composed of chert cobbles are common in areas of cherty limestone. They are not common, however, in areas of homogeneous rocks of any kind that do not provide a possibility for a contrast in resistance between the stream load and the rock through which it moves.

The retreating escarpment and parallel retreat of slopes

According to cyclic theories elaborated by Penck (1953) and King (1953) the evolution of topographic forms involves parallel retreat of slopes. As the hill or mountain slope retreats away from the main stream a pediment or network of interconnected pediments forms at the foot of the retreating slope and is extended as erosion continues. The Badlands area of South Dakota provides an example (Smith, 1958). Such forms are, however, far from universal. The retreating escarpment appears to be characteristic of gently dipping stratified rocks in which some beds are more resistant to erosion than others. Pediments and foot slopes may or may not be associated with them. The Highland Rim of Tennessee is an example of an escarpment that may be retreating but is without pediments at its foot. The Badlands of South Dakota, on the other hand, have in front of them a broad network of miniature pediments.

Retreating escarpments and associated pediments appear to be especially characteristic of dry climates, a relationship pointed out by Frye (1959). They are certainly less common features, however, in the Appalachian Highlands and in the Piedmont of the eastern United States where not only is the climate humid but horizontally bedded rocks are rare. In these areas the escarpments are fixed in space by geologic contacts. There may, however, be exceptions. The Blue Ridge escarpment of Virginia and the Carolinas appears to divide rock areas that are in some places identical. This scarp may indicate a condition of disequilibrium between two areas (Davis, 1903; White, 1950; Dietrich, 1958).

Evolution of topographic forms through time

The theory of dynamic equilibrium explains topographic forms and the differences between them in a manner that may be said to be independent of time. The theory is concerned with the relations between rocks and processes as they exist in space. The forms can change only as the energy applied to the system changes. It is obvious, however, that erosional energy changes through time and hence forms must change. It is of interest, therefore, to speculate on the effect of a gradual reduction in relief of a well-graded landscape such as we assume occurs through long periods of geologic time as diastrophic forces cease to exert their influence and an isostatic balance is approached.

In a typical ridge and ravine landscape the general character of the topography is probably maintained as the relief is lowered. There is no reason to believe that the efficiency of the forms for the shedding of waste becomes any less. The forms in which the waste is removed may change, however, and the rate of removal may diminish. In an area of high relief the waste may be largely in the form of boulders and cobbles that are removed mechanically. As relief is lowered in the same area, perhaps chemical weathering becomes relatively more important. In a high relief area, the divides are sharp and slopes steep. As relief is lowered in the same area the slopes in interstream areas become more rounded, and the divides more blunt.

Speculating on the evolution of pediment landscapes that occur in soft-rock areas adjacent to hard-rock areas it is evident that if relief becomes lower the difference in the energy potential between the two areas will become less marked. It can be expected therefore that the pediment areas will diminish in size and may eventually disappear.

The application of geomorphic concepts to general geologic problems

The theory of the geographic cycle has been widely used by geologists. Its abandonment must result eventually in changes in many of our concepts. Though it is not my purpose here to elaborate fully such changes, some examples are cited in order to illustrate the extent to which a change in theory may affect geologic problems.

Theories of ore genesis, particularly of deposits classed as supergene, are affected by abandonment of the cyclic concept of landscape evolution. The manganese deposits of the Appalachians associated with lower Cambrian carbonate rocks are a good example. These deposits generally occur in thick residuum preserved beneath quartzite gravels shed from adjacent high-land areas. They have been interpreted as of Tertiary age, formed in the Harrisburg cycle of erosion (Hewett, 1916, p. 43–47). By application of the theory of the equilibrium landscape they may be interpreted not as relics of a Tertiary weathered mantle preserved beneath younger gravels, but as deposits that are forming at the present time, or under conditions like the present: They form beneath the gravelly mantle, and are preserved because the gravel covers them, and protects both the ore minerals and the residuum around them from erosion (Hack, 1959).

Some of the greatest changes in concept required relate to the concept of the dissected peneplain. Because we have accepted for many years the idea that a ridge and ravine landscape is formed by the dissection of a peneplain we have also accepted the idea that many highland areas like the Appalachians eroded in steps or cycles and that the orogenies that deformed the rocks of such highland belts were followed by long periods of vertical uplift of a cyclic nature involving repeated changes in the rates of deformation. Having abandoned the peneplain we must reexamine the history of such areas and apply areal studies of erosional process and form to the interpretation of their past history. In the Appalachian Highlands, for example, the general outlines of the present drainage may be inherited in part from conditions that existed as early as Permian or Triassic time. The present landscape may have formed through one continuous period of dying orogeny or isostatic adjustment. Differences in relief and form in different areas are explainable partly by the reaction of various erosive processes on a complex bedrock, and partly by what is probably a long history of complicated diastrophic movements.

Cyclic theories of landscape origin are close relatives of the theory of periodic diastrophism which holds that orogenies have generally occurred in geologic time in brief episodes of world wide extent. This theory is questioned and critically discussed by Gilluly (1949) who shows that the evidence of the sedimentary rock column supports the idea that diastrophism has not been periodic but was almost continuous through time, though the form and location of diastrophic movements has continually changed. This concept of continuity of diastrophic processes is, of course, discordant with cyclic geomorphic theories, but is in harmony with the equilibrium concept outlined here.

Note

1 Davis' major papers dealing with the sculpture of landscapes by streams (Davis, 1889, 1890, 1896a, 1896b, 1899a, 1899b, 1902a, 1902b, 1903, 1905a, 1905b) as well as others of his papers were collected in one volume published as "Geographical Essays" (Davis, 1909) and reprinted in 1954 (Davis, 1954). The 1954 edition of "Geographical Essays" has the same page numbers as the 1909 edition.

References cited

Bridge, Josiah, 1950, Bauxite deposits in the southeastern United States, *in* Snyder, F.G., ed., Symposium on mineral resources of the southeastern United States: Knoxville, Tenn., Tennessee Univ. Press, p. 170–201.

Daly, R.A., 1926, Our mobile earth: New York, Charles Scribner's and Sons, 342 p.

Davis, W.M., 1889, The rivers and valleys of Pennsylvania: Natl. Geog. Mag., v. 1, p. 183–253.

——, 1890, The rivers of northern New Jersey, with notes on the classification of rivers in general: Natl. Geog. Mag., v. 2, p. 81–110.

——, 1896a, Plains of marine and subaerial denudation: Geol. Soc. America Bull., v. 7, p. 377–398.

——, 1896b, The Seine, the Meuse, and the Moselle: Natl. Geog. Mag., v. 7, p. 189–202, 228–238.

——, 1899, The geographic cycle: Geog. Jour., v. 14, p. 481–504.

——, 1899b, The peneplain: Am. Geologist, v. 23, p. 207–239.

——, 1902a, River terraces in New England: Harvard College, Mus. Comp. Zoology Bull., v. 38, p. 281–346.

——, 1902b, Base level, grade, and peneplain: Jour. Geology, v. 10, p. 77–111.

——, 1903, The mountain ranges of the Great Basin: Harvard College, Mus. Comp-Zoology Bull., v. 42, p. 129–177.

——, 1903, The stream contest along the Blue Ridge: Geog. Soc. Philadelphia, Bull., v. 3, p. 213–244.

——, 1905a, The geographical cycle in an arid climate: Jour. Geology, v. 13, p. 381–407.

Davis, W.M., 1905b, Complications of the geographical cycle: Internat. Geog. Cong., 8th Rept., p. 150–163.

——, 1909, Geographical assays: Boston, Ginn and Co., 777 p.

——, 1926, Biographical memoir of Grove Karl Gilbert, 1843–1918: Natl. Acad. Sci., 5th Mem., v. 21, 303 p.

——, 1954, Geographical Essays: Dover Publications, Inc., 777 p.

Dietrich, R.V., 1958, Origin of the Blue Ridge escarpment directly southwest of Roanoke, Virginia: Virginia Acad. Sci. Jour., v. 9, New Series, p. 233–246.

Frye, John C., 1959, Climate and Lester King's "Uniformitarian Nature of Hillslopes": Jour. Geology, v. 67, p. 111–113.

Gilbert, G.K., 1877, Geology of the Henry Mountains (Utah): Washington, D.C., U.S. Geog. and Geol. Survey of the Rocky Mts. Region, U.S. Govt. Printing Office, 160 p.

——, 1909, The convexity of hill tops: Jour. Geology, v. 17, p. 344–350.

Gilluly, James, 1949, Distribution of mountain building in geologic time: Geol. Soc. America Bull., v. 60, no. 4, p. 561–590.

Hack, John T., 1958a, Studies of longitudinal stream profiles in Virginia and Maryland: U.S. Geol. Survey Prof. Paper 294B, p. 45–97.

——, 1958b, Geomorphic significance of residual and alluvial deposits in the Shenandoah Valley, Virginia (abs.): Virginia Jour. Sci., v. 9, p. 425.

——, 1959, The relation of manganese to surficial deposits in the Shenandoah Valley, Virginia (abs.): Washington Acad. Sci. Jour. Proc., v. 49 p. 93.

Hack, J.T., and Goodlett, J.C., 1960, Geomorphology and forest ecology of a mountain region in the Central Appalachians: U.S. Geol. Survey Prof. Paper 347 in press.

Hewett, D.F., 1916, Some manganese mines in Virginia and Maryland: U.S. Geol. Survey Bull. 640-C, p. 37–71.

Hunt, C.B., Averitt, Paul, and Miller, R.L., 1953, Geology and Geography of the Henry Mountains region, Utah: U.S. Geol. Survey Prof. Paper 228, 234 p.

Kesseli, J.E., 1941, The concept of the graded river: Jour. Geology, v. 49, no. 6, p. 561–588.

King, L.C., 1953, Canons of landscape evolution: Geol. Soc. America, Bull., v. 64, no. 7, p. 721–752.

Leopold, L.B., 1953, Downstream change of velocity in rivers: AM. JOUR. SCI., v. 251, no. 8, p. 606–624.

Leopold, L.B., and Maddock, Thos., Jr., 1953, The hydraulic geometry of stream channels and some physiographic implications: U.S. Geol. Survey Prof. Paper 252, 57 p.

Mackin, J.H., 1948, Concept of the graded river: Geol. Soc. America Bull., v. 59, no. 5, p. 463–511.

Nikiforoff, C.C., 1942, Fundamental formula of soil formation: AM. JOUR. SCI., v. 240, no. 12, p. 847–866.

——, 1949, Weathering and soil evolution: Soil Sci., v. 67, p. 219–230.

——, 1955, Harpan soils of the Coastal Plain of southern Maryland: U.S. Geol. Survey Prof. Paper 267-B, p. 45–63.

——, 1959, Reappraisal of the soil: Science, v. 129, no. 3343, p. 186–196.

Penck, Walther, 1953, Morphological analysis of landforms (translation by Hella Czeck and K.C. Boswell): New York, St. Martin's Press, 429 p.

——, 1924, Die morphologische Analyse. Ein Kapitel der physikalischen Geologie: Stuttgart, Geog. Abh. 2 Reihe, heft 2, 283 p.

Rubey, W.W., 1952, Geology and mineral resources of the Hardin and Brussels quadrangles (in Illinois): U.S. Geol. Survey Prof. Paper 218, 179 p.

Shaler, N.S., 1899, Spacing of rivers with reference to hypothesis of base-leveling: Geol. Soc. America Bull., v. 10, p. 263–276.

Smith, K.G., 1958, Erosional processes and landforms of Badlands National Monument, South Dakota: Geol. Soc. America Bull., v. 69, no. 8, p. 975–1008.

Stose, G.W., and Miser, H.D., 1922, Manganese deposits of western Virginia: Virginia Geol. Survey Bull. 23, 206 p.

Strahler, A.N., 1950, Equilibrium theory of erosional slopes approached by frequency distribution analysis: AM. JOUR. SCI., v. 248, no. 10, p. 673–696; no. 11, p. 800–814.

Tarr, R.S., 1898, The peneplain: Am. Geologist, v. 21, p. 341–370.

White, W.A., 1950, The Blue Ridge front—a fault scarp: Geol. Soc. America Bull., v. 61, no. 12, pt. 1, p. 1309–1346.

Wolman, M.G., 1955, The natural channel of Brandywine Creek, Pennsylvania: U.S. Geol. Survey Prof. Paper 271, 56 p.

Woodford, A.O., 1951, Stream gradients and the Monterey sea valley: Geol. Soc. America Bull., v. 62, no. 7, p. 799–851.

5

EVOLUTIONARY GEOMORPHOLOGY OF AUSTRALIA AND PAPUA–NEW GUINEA

C.D. Ollier

Source: *Transactions of the Institute of British Geographers* 4 (1979): 516–39.

Abstract

By means of examples it is demonstrated that much of the geomorphology of Australia dates back to early Cenozoic, Mesozoic and even Palaeozoic times. Papua–New Guinea is built of several fragments that had different histories before they collided with the Australian plate, and it is meaningless to construct a geomorphic history on the present map of the country. Despite signs of geomorphic youth, such as Pleistocene granite and gneiss domes, even Papua–New Guinea has many relic landforms. Conventional models of geomorphic evolution are evaluated and it is considered that process studies, dynamic equilibrium, and even cyclic theory and uniformitarianism, are inadequate for the time-scale involved and the many unique events recorded in landscape history. Evolutionary geomorphology, a branch of evolutionary earth science, is suggested as a more appropriate concept.

It is well known that Australia is an old continent, worn down over a long period of continental erosion to become the flattest continent. Over the past few years the development and application of absolute dating techniques has put the geomorphic history on a firm basis with a large number of accurate dates. It has been found that the development of the Australian landscape that we see today started before the break-up of Gondwanaland, which created the present continents of the southern hemisphere, and that there are significant relationships between landforms and continental drift. But even more important than the decipherment of this long geomorphic history is the light it throws on geomorphic concepts. Many of the ruling theories of today seem to be inadequate on the long time-scale, and a new view of the evolution of ancient landscapes is required that can accommodate the many unique events that occur in such long spans of time.

Palaeogeography

The palaeogeographic maps (Fig. 1) reveal that much of Australia has been land since pre-Cretaceous times, and therefore has a terrestrial geomorphic history going back that far. There is no assurance that we shall find individual landforms of such age, but this knowledge does give us some idea of the time-scale that might be involved in landscape evolution, and it is a proper part of geomorphology to try to find the earliest evidence of earth history as revealed in existing landforms. In older geology books 'earth history' was almost synonymous with stratigraphy, and most stratigraphy was concerned with the history of shallow seas. Geomorphology studies the landward equivalent of the marine story revealed by stratigraphy, and of course the two stories should tally—periods of rapid erosion should correspond with periods of rapid sedimentation; periods of little or no erosion should match unconformities or periods of little deposition. Also, the type of material deposited reveals aspects of geomorphic history in the catchment areas. Has a granite been exposed? Have volcanoes erupted? Is the hinterland deeply weathered? Examples of landform/sediment relationships will be presented later from Lake George and the Broughton Delta.

In Australia, to find a fresh start comparable to that created by Quaternary glaciation of large parts of Europe we have to go back to the glaciation of Permian times about 250,000,000 years ago. There are many places in Australia where Permian glacial striae are exposed at the gound surface, and in some places Permian glacial landforms such as roches moutonnées have been exhumed. In Western Australia some sandstone-capped hills are formed by inversion of relief along a Permian glacial outwash deposit which has probably been at the ground surface since its formation. With these possible exceptions the oldest traces of land surfaces in Australia are probably Triassic. In Cretaceous times there was a great spread of shallow seas that could only have been achieved if much of the continent had already been worn down to a plain before Cretaceous times.

In northern Australia much of the country has been tectonically stable for long periods, and multiple erosion surfaces relate partly to tectonic uplift, but also to distinct periods of weathering and stripping of regolith.

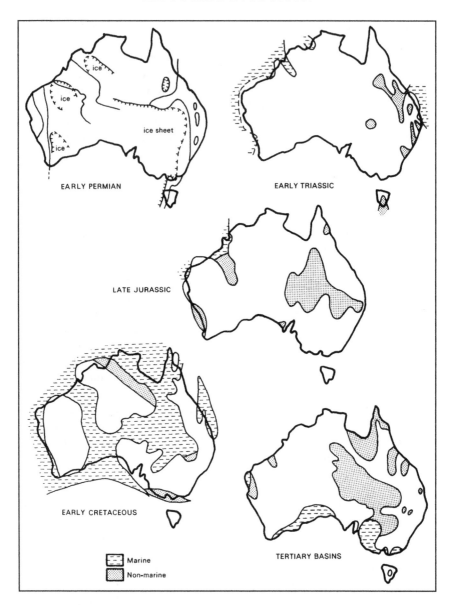

Figure 1 Palaeogeographic maps of Australia.

An account of erosion surfaces in the Daly Lowland of the Northern Territory provides a detailed example. Wright (1963) recognized three surfaces:

1. Bradshaw surface. This forms main divides, is of considerable perfection, and has a deep lateritic profile with a strongly silicified horizon in the lower part of the pallid zone. This is equivalent to the Tennant Creek surface found elsewhere in the Territory, where it is seen to be topographically lower than the Ashburton surface which is preserved only on very resistant rocks such as Precambrian quartzites.
2. Maranboy surface. This forms secondary divides, with related rock-cut terraces. It was produced by stripping of the upper, less silicified parts of the Bradshaw pallid zone (therefore silicification was pre-Maranboy). The Maranboy surface is associated with a lateritic weathering profile less deep than the Bradshaw profile and mainly developed in the Bradshaw weathering mantle.
3. Tipperary surface. This advanced by removal of the Maranboy pre-weathered layer, exposing the resistant Bradshaw silicified rock which commonly provides a structural base level for denudation. The broad plains of the Tipperary surface are relatively unweathered.

At Camfield a Miocene limestone (probably lacustrine) is found on a surface probably equivalent to the Tipperary, so the sequence is Miocene or older. The oldest surface is probably older than Cretaceous.

In the New England region of northern New South Wales, an area of tectonism in the Eastern Highlands, a summary of geomorphic history is as follows:

1   Formation of a planation surface of Cretaceous or older age.
2   Uplift of at least 500 m.
3   Formation of an Eocene surface on which Eocene fluvial deposits accumulated, straddling the present divide and thickening westwards.
4   Deposition of Palaeocene-Eocene and Oligocene volcanics.
5   Formation of the first post-basaltic planation surface.
6   Eruption of Miocene volcanics.
7   Formation of second post-basaltic planation surface.
8   Later uplift and dissection.

These examples show that uplift, and also climatic and weathering history, affect the sequence of planation surfaces. Nevertheless it is clear that in Australia planation surfaces are real, multiple, and old (Figs 15 and 16).

River patterns and tectonics

The study of river patterns, like that of planation surfaces, is sometimes regarded as old-fashioned, but in Australia the relationship between drainage pattern, tectonic activity, and datable events provides many interesting data as indicated by the following examples:

Western Australia

Much of Western Australia consists of a plateau which is generally less than 600 m high and is known as the Great Plateau. This is deeply weathered, partially stripped of regolith, and traversed by valleys. Some ancient broad valleys, partly filled with sediments and now containing chains of salt lakes, are remnants of a drainage system which apparently had its headwaters to the south of the present coastline (Fig. 2). The break-up of Australia and Antarctica therefore post-dates the drainage system. The separation really became effective about 55,000,000 years ago, but rifting as a precursor to drift probably started earlier, perhaps 70,000,000 years ago, so there seems little doubt that the Great Plateau, the period of great weathering and the early drainage pattern belong to the Mesozoic (Ollier, 1977; Van de Graaff et al., 1977).

Eastern Australia

Rivers like the O'Hare, Cordeaux, Cataract and Kangaroo rise within a few kilometres of the coast and flow to the west, with broad valleys (several are

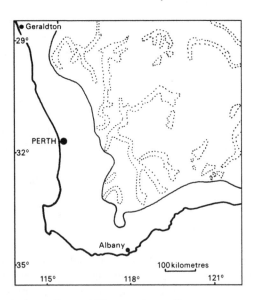

Figure 2 Ancient drainage lines of Western Australia.

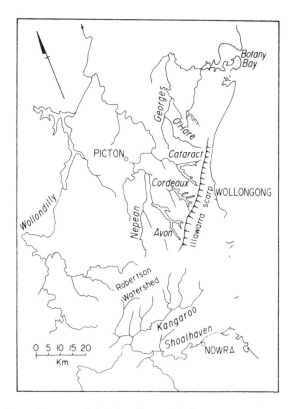

Figure 3 Drainage features of the Shoalhaven–Sydney area.

used for reservoirs) and with perfectly simple dendritic patterns (Fig. 3). It would take a long time to erode these valleys, and as Taylor pointed out in 1911

> when these valleys were produced there could not have been coastal cliffs a thousand feet high immediately at their origins as at present. We are led to believe, therefore, that not far back in geological time there was an extensive area to the east of this divide, perhaps one or two hundred miles wide, which has subsided lately beneath the waves.

The situation is similar to that in Western Australia, but if the lost headwaters of the Western Australian rivers are in Antarctica, where can those of eastern Australia be? It seems quite probable that the headwaters of the eastern Australian rivers have not subsided beneath the waves, but have drifted off to form the Lord Howe Rise and other continental fragments to the east of the Australian mainland.

123

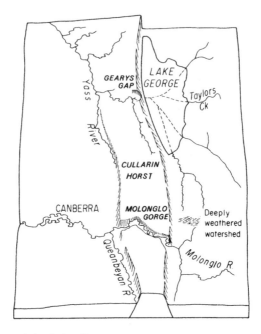

Figure 4 Diagram of the Lake George area.

Lake George

In New South Wales, Lake George and its associated features provide a simple and elegant example of drainage modified by faulting (Fig. 4). Before faulting, Taylors Creek was continuous with Shingle House Creek, a tributary of the Yass River, and to the south the Molonglo flowed along the line of its present route. Faulting later produced a horst bounded by the Queanbeyan Fault on the west and the Lake George Fault on the east, the horst being known as the Cullarin Horst. The Molonglo River eroded its bed to keep pace with uplift, forming an antecedent course across the horst, but Taylors Creek was defeated by the faulting. Lake George was formed on the downthrown side, and Gearys Gap, marked by many patches of ancient river gravels as well as a topographic depression marks the former continuation of the valley. Studies of the gravels show that they came from the east. They are very mature, well-rounded quartz gravels, clearly derived from a thoroughly weathered and low-lying source area. If the basin were to fill, water would overflow through Gearys Gap and not over the watershed to the north or south. It is easy to build a relative chronology here: the lake is younger than the drainage pattern, which is older than the fault but younger than the surrounding plain. Can we put absolute ages to the sequence?

Ever since the lake was formed it has accumulated lacustrine sediments, so a core through the sediments might reveal the history of the lake. A maximum of 71 m of sediment has been found. Palaeomagnetic examination of a core has shown the Brunhes-Matuyama reversal (dated at 700,000 years) at 18 m, and the Matuyama-Gauss reversal (of 2,400,000 years) at 31 m. At 50 m the varied upper sediments give way to barren laminated clays which do not yield any dates, but clearly the lake has been in existence for several million years, and probably has a Miocene origin (Jennings, 1978).

The deep weathered lateritic profile found on the watershed between the Taylors Creek and Molonglo River (preserved by its location below the fault scarp) suggests that the oldest planation surface of which we have evidence in the area was of middle or lower Tertiary age at least. The drainage system was initiated on this surface and the rivers carried a load of quartz pebbles like those preserved in Gearys Gap. The fault created the Lake George basin in Miocene time, since when it has accumulated sediments. To the south the Molonglo has maintained an antecedent course, and is currently carrying a load of river gravels of varied petrology.

The River Murray

Williams and Goode (1978) have documented another fine example of drainage/tectonics inter-relationships. The River Murray turns abruptly south at the town of Morgan (Fig. 5). A former straight continuation of the Murray can be traced from this bend along Burra Creek and Broughton River. Nowadays the Broughton fails to reach the sea even in flood times and has long been thought to be 'underfit'. But there is a large delta at the mouth of the Broughton, which could only have been built when the Murray flowed this way to the sea. The oldest dated deposits in the delta sediments are Middle to Upper Eocene at a depth of about 150 m. We therefore conclude that the Murray flowed to Spencer Gulf in Upper Eocene times, noting also that Spencer Gulf was in existence at that time, not long after the separation of Australia and Antarctica. At some time later the uplift of the Mt Lofty and Flinders Ranges defeated the Murray, which was deflected to the south. This may have happened as recently as the close of the Pliocene.

Volcanic geomorphology and tectonics

Volcanic activity affected eastern Australia throughout Tertiary times and into historical times. Three different styles of volcanic activity can be recognized, each of which has a clear time relationship. Wellman and McDougall (1974) recognized two kinds—the lava field provinces and central volcano provinces, and Ollier (1978) distinguished a third, the areal volcano province.

The lava field provinces were produced from eruptive areas of dyke swarms up to 100 m across, with widespread flows building shield volcanoes and lava piles up to 1,000 m thick, consisting exclusively of rocks of basaltic

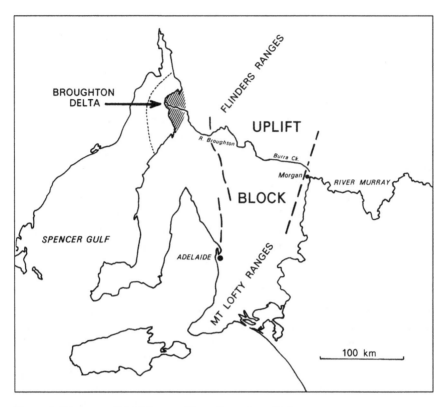

Figure 5 Diagram of the Murray River–Broughton Delta area.

composition. The main divide appears to be closely associated with lava field provinces (Fig. 6), and in Queensland the divide follows lava field provinces rather than the sometimes higher coastal ranges. If the association is genetic we might expect the age of the formation of the divide to be roughly the same as the age of the associated lavas. This reasoning suggests that the divide was formed between about 65,000,000 and 20,000,000 years ago—that is in Palaeocene, Eocene and Oligocene times. The lava field provinces were most active between 55,000,000 and 34,000,000 years ago, so the period of uplift of the divide may similarly be more limited.

The central volcano provinces are related to large vents, now sometimes represented only by central plugs, and were originally large volcanoes up to several thousand metres high. They were predominantly basaltic but also have some felsic flows and intrusions and have a higher and more variable potassium content than the lava field volcanics.

The central provinces reveal a remarkable correlation of age with latitude, ranging in age from 33,000,000 to 6,000,000 years with volcanoes

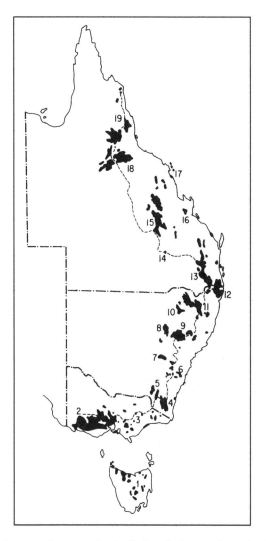

Figure 6 Volcanic area of eastern Australia in relation to the main divide.

getting younger to the south (Fig. 7). An intriguing explanation suggested by Wellman and McDougall is that the Australian continent has moved in a northerly direction over two magma sources fixed in the asthenosphere beneath the crust (hot spots). The apparent southward migration rate of the centres of vulcanism in eastern Australia, 66.5 m/1,000 years, then becomes the rate of northerly movement of the Australian plate over the asthenosphere. This rate corresponds with the rate of drift derived from the study of sea-floor spreading between Australia and Antarctica, which is

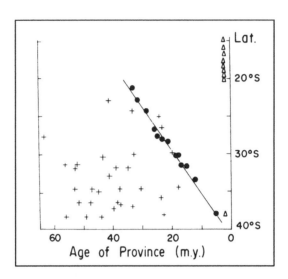

Figure 7 Relation between the latitude and age of Cainozoic volcanism in Australia for central volcano (dots), lava field (crosses) and areal provinces (triangles). *Source*: Wellman and McDougall, 1974.

somewhere between 50 and 74 m/1,000 years and suggests that Antarctica has been stationary while Australia, and the spreading ridge, moved north.

The third province type, the areal province, is characterized by the absence of any tendency for eruption centres to be localized at a point for any length of time, while individual volcanoes are small and short-lived, with scoria cones and maars dominant. In Australia the provinces of this type are the youngest (all Pleistocene) and the westernmost, but the further tectonic significance of this is not yet clear. Wellman and McDougall suggest that when all volcanic activity is considered together the rate of activity is found to be nearly constant from about 60,000,000 years to the present day at approximately 300 km³/1,000,000 years which may result from crustal extension being also constrained in some way to occur at a constant rate.

Dated lava flows can also be used to determine, or at least set limits on, ages of uplift and erosion. Dated flows on plains give a minimum age for the plain, but do not tell us what the elevation of the plain was at the time of eruption. Dated flows in valleys give us the age of the valley, the minimum available relief at the time of eruption, and from that the minimum amount of uplift in pre-eruption times.

In the Gelantipy district of eastern Victoria, for example, late Eocene flows (38–42,000,000 years) lie on the Nuniong Plateau at about 1,200 m (Fig. 8). The plateau was eroded by a valley 600 m deep, and along this a younger lava flow poured, which is also of late Eocene age. This shows that the Nuniong Plateau existed as a plateau in late Eocene times and was

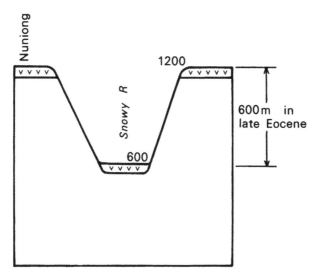

Figure 8 Diagrammatic cross section of the Snowy River near Gelantipy.

dissected even then by a valley 600 m deep. That valley must have been above sea level so uplift in this area has been no more than 600 m since late Eocene times.

In the Aberfeldy area there is a plateau at about 1,350 m, which was dissected by a valley that cut down to 650 m, and along which a lava flow poured (Fig. 9). This lava is 27,000,000 years old (Oligocene) so the old

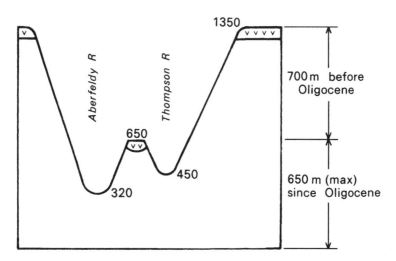

Figure 9 Diagrammatic cross section of the Aberfeldy and Thompson Rivers near Aberfeldy.

plain was uplifted at least 700 m before Oligocene times. There has been a further uplift of about 600 m since the Oligocene flow.

In contrast there is a lava flow at Seven Creeks, 7,000,000 years old, which is at the same level as the nearby river, indicating no downcutting over the past 7,000,000 years.

The Eastern Highlands

The Great Dividing Range of eastern Australia is a cartographic myth, but the Eastern Highlands are real enough. There is no range in the usual sense of the word and the divide is commonly a region of undulating country or broad plains. It is generally agreed that the Eastern Highlands were created by earth movements warping up a gentle arch along the eastern edge of the continent. The uplift was far from uniform, and the highlands can be regarded as consisting of a series of culminations or domes. Uplift was accompanied by faulting and volcanic activity, and the short steep rivers flowing to the Pacific rapidly eroded the eastern slopes producing the steep and rugged escarpments which, seen from the coast, look like mountain ranges.

Many examples of drainage modification are associated with the divide, including river capture, drainage reversal, and antecedent and superimposed rivers. On the very gentle slopes of the divide some rivers were so back-tilted that they gave rise to lakes, and a line of lakes can be traced all along the divide (Fig. 10). Many of these are very small, but some such as Lake Buchanan and Lake Galilee are tens of kilometres across.

The traditional view of the geomorphology of the Eastern Highlands was expressed most clearly by Andrews (1910). He saw a great unity in the Australian landscape, with a single great peneplain of Miocene age and a single great uplift—the Kosciusko Uplift—at the very close of the Tertiary period. Modern work (e.g. Ollier, 1978) suggests that there was no single great peneplain, and that tectonic uplift took place over a longer period, starting in the early Tertiary. The new interpretations are very much dependent on the potassium/argon dates obtained on volcanic rocks of eastern Australia, mainly resulting from the work of Wellman and McDougall (1974).

We have seen that the Eastern Highlands are associated with lava field volcanics dated 55–34,000,000 years so uplift was about that time too. The central volcanics record the northward drift of Australia between 33,000,000 and 6,000,000 years so these are not associated with the uplift. Ocean floor evidence suggests that Australia has been drifting since 55,000,000 years ago, but for much of this time the divide was already in existence on the drifting plate.

The direction of drifting, roughly NNE, has no relation to the sinuous but more northerly direction of the divide. Certainly the idea of a range of mountains pushed up at the leading edge of a drifting continent can be

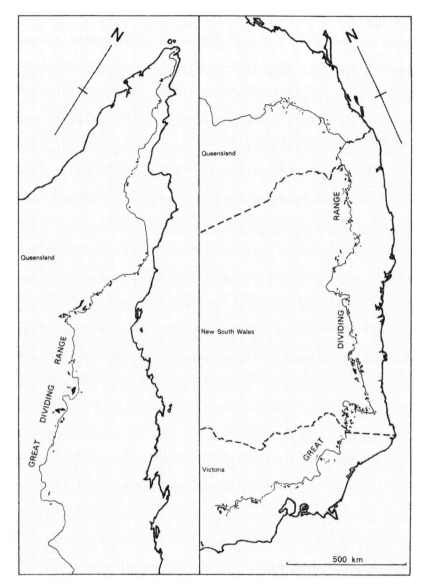

Figure 10 Lakes and swamps along the divide. Lakes in other areas are not shown, so this figure is not meant to show there are more lakes on the divide than elsewhere, but simply that many lakes have this tectonic position.

discounted. On the contrary the area shows signs of tension in the style of faulting and in the dyke-emplaced volcanics, a tension possibly associated with doming.

Most modern ideas on mountain formation are associated with plate tectonics, and most ranges are thought to be somehow related to the underthrusting (subduction) of one plate beneath another, which rises. There is no evidence whatever for subduction associated with the Eastern Highlands, and the seas off eastern Australia appear to be spreading sites. The combination of tension, vulcanicity and uplift, with occasional downwarped areas, marks the Eastern Highlands as a region of vertical tectonics, but the more fundamental cause of the vertical movement is not yet known.

Australia appears to be surrounded by spreading sites on all sides, except the north. In this it is very like Africa, and the similarity may go further. Along the eastern side of Africa there is a chain of swells, commonly marked by rift valleys. Perhaps some of the higher parts of the Eastern Highlands are equivalent structurally to the swells of Africa, with the geocols equivalent to gaps between swells. The African swells commonly have a rise towards the rift valleys with their associated vulcanism. The rift valley-swell landscape is attributed to vertical tectonics (Le Bas, 1971). Petrologically the rift valley volcanics differ from the eastern Australian basalts so we cannot pursue the analogy too far, but the map of swells in Africa is certainly reminiscent of the Eastern Highland situation (Fig. 11). The African rift valleys are tectonically continuous with the Red Sea, which is in turn continuous with sub-oceanic ridges: the rift valleys may be the initiation of spreading sites. If the Australia/Africa analogy has any validity, the Eastern Highlands may reflect a line of future splitting of the continent. Perhaps the present eastern coastline is along an earlier split.

We have already seen some suggestion that there was a former land mass to the east of Australia that has disappeared, either by downfaulting or in some other way. One possibility is that suggested by Carey (1976), who envisages the various fragments of land from eastern Australia being incorporated in the Indonesia–New Guinea arc, Lord Howe Rise, New Zealand, and other islands (Fig. 12). On this basis it may be reasonable to assume that another sliver of the Australian continent will follow suit eventually, and the line of fracture is marked by the Eastern Highlands.

Papua–New Guinea

In studying the tectonic geomorphology of Australia it has become clear that it is necessary to consider its setting: this includes the surrounding seas, and also Papua–New Guinea. This country, which can be divided into the units shown in Figure 13, presents a quite different geomorphic picture.

The south-west area of the Fly Platform is virtually a continuation of the Australian plate, but the rest of the country has a separate development.

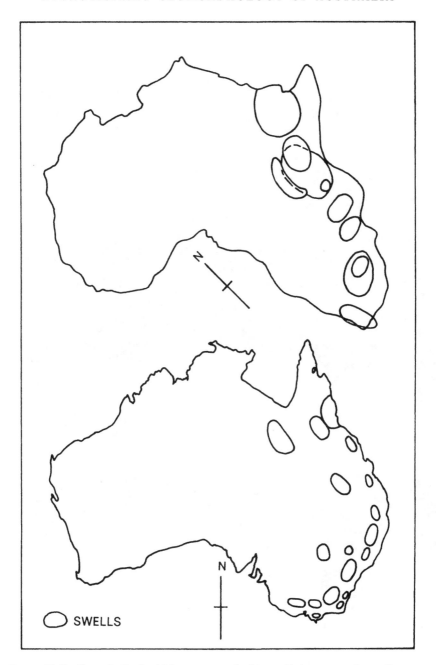

Figure 11 Swells and rifts in Africa compared with swells in eastern Australia.
Source: Le Bas, 1971.

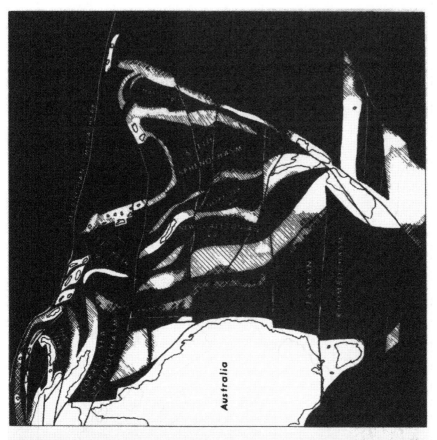

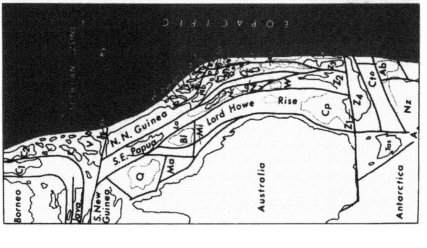

Figure 12 The fragmentation and drift of eastern Australia and its neighbours.
Source: Carey, 1976.

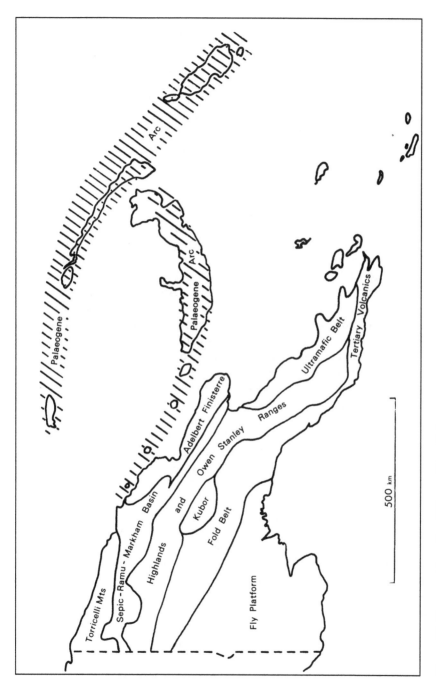

Figure 13 Geomorphic units of Papua–New Guinea.

The backbone of the country is a belt of mountain ranges which consist of two parts—the Central Highlands and the Owen Stanley Range with its continuation into the D'Entrecasteaux Islands—which may have formed separately. To the south is the Papuan Fold Belt of mainly Mesozoic rocks. The Sepik-Ramu-Markham depression is largely filled with Quaternary sediments. The Torricelli Range consists of folded upper Tertiary sediments (mainly Miocene and Pliocene) on a basement of metamorphic rocks. In the Adelbert and Finisterre Ranges folded upper Tertiary sedimentary rocks overlie unmetamorphosed lower Tertiary sediments. The Papuan Ultramafic Belt is believed to be a slice of old sea floor that has been thrust over the continental rocks of the Owen Stanley Range. The New Britain and New Ireland volcanic arcs have a Palaeogene origin.

The mosaic of geological patches of which New Guinea is built did not originate in place. Various fragments of New Guinea have drifted and collided, and their place of origin is a matter of speculation.

The New Guinea area is also subject to a great deal of shearing, and lateral displacement has brought together many slivers that may have been originally adjacent to eastern Australia. It is not possible to draw the palaeogeography of New Guinea like that of Australia, using the present outline as a base. Reconstruction, which in the present state of knowledge calls for flair and insight as well as familiarity with the geology and geomorphology, requires lateral movements of the various components.

Figure 12 shows the reconstruction suggested by Carey (1976). This model required only lateral movement and shearing of fragments, with expansion of new sea floors between the fragments. In this connection it is interesting to note that there is evidence for spreading of all the seas around Papua New Guinea. The Coral Sea sphenochasm opened between 60,000,000 and 50,000,000 years ago (Gardner, 1970): the Woodlark sphenochasm has been opening for the past 3,000,000 years and perhaps since 20,000,000 years (Luyendyk, MacDonald and Bryan, 1973; Milsom, 1970, 1974): the Bismarck Sea is currently spreading (Connelly, 1974) and the New Britain basin is probably an active spreading site. Subduction is not required in Carey's model, for he advocates an expanding earth.

Figure 14 shows two stages in an eight-stage sequence determined by Crook and Belbin (1978). They have a more conventional plate-tectonics model, with creation of new sea floor at some spreading sites, and destruction of sea floor at subduction sites. Nevertheless their reconstruction brings together fragments of Papua–New Guinea that were originally separated by about ten degrees of latitude.

One of the most interesting things about Papua–New Guinea is the vast amount of geomorphic development achieved in a short time. There are numerous Plio-Pleistocene granites now exposed at the ground surface, the youngest of which, at Ok Tedi, has a K-Ar date of 1,100,000 years. Much of the uplift of the highlands occurred in the Pleistocene. In Goodenough

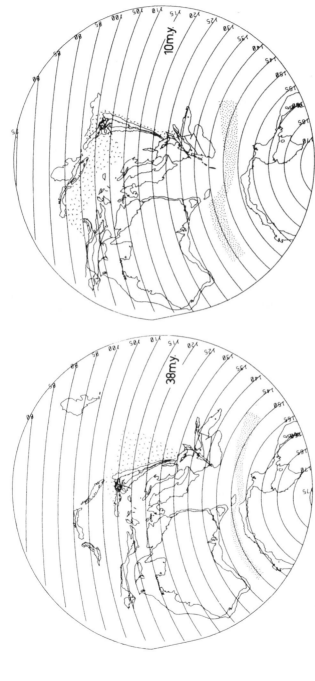

Figure 14 The southwest Pacific 38,000,000 years ago (left) and 10,000,000 years ago (right) (from Crook and Belbin, 1978). By 10,000,000 years ago the Finisterre New Britain block has met mainland New Guinea, but New Ireland–Manus, the Solomon Islands, and the Ontong Java plateau were still widely separated.
Key to stipple: dot pattern: sea floor formed by spreading during the preceding interval; dot-dash pattern: crust that will be consumed during the succeeding interval.

Figure 15 Planation surface cut across varied steeply dipping Palaeozoic rocks, Bungonia Gorge, New South Wales.

Island a granite-cored gneiss dome over 2,000 m high has emerged tectonically as a surface landform in Pleistocene times (Ollier and Pain, in press). In the Finisterre Peninsula raised coral reefs extend to over 2,500 m, and indicate an uplift rate of about 2 m/1,000 years (Chappell, 1974). This makes Australia look even more stable.

But despite these great differences in some of the rates of geomorphic processes, the total geomorphic histories of Australia and Papua–New Guinea must be related because they are geographically adjacent and they share some tectonic units.

A unified tectonic synthesis is something like this: the stable plate (which includes south-west Papua) has drifted north with a certain amount of warping and faulting that may herald a future splitting. A former extension of this plate to the east has broken into fragments that have moved independently. Some moved north faster than Australia and were then swept westwards at the Melanesian megashear by the western movement of the Pacific plate. With the complications of collision with younger island arcs and incorporation of younger sediments this complex chain of events created Papua–New Guinea. Thus the geomorphic histories of Australia and New Guinea are

Figure 16 Multiple planation surfaces, Coober Pedy, South Australia. The upper
surface has a silcrete cap and weathering profile with precious opal. The
very extensive lower surface, cut across the same Cretaceous rock at
the upper surface, usually has less than a metre of disturbed or transported
material over bedrock.

related through a common tectonic setting, and the longer time-scale is
required to understand even the relatively young geomorphology of Papua–
New Guinea.

Geomorphology is now on the same time-scale as global tectonics,
sea-floor spreading, continental drift and biological evolution, and geomorphologists must become involved with palaeogeography of land areas in the
way stratigraphers have long been concerned with the palaeogeography
of sea floors. Geomorphology, like stratigraphy and geophysics, provides
input in defining and solving major tectonic problems.

The significance of Australia's geomorphic history

Most of the prevailing theories or models in geomorphology were evolved
under the influence of the dominant idea that most landforms are Quaternary or modern, and that a relatively short time-scale is appropriate for

139

geomorphic concepts. The long time-scale demonstrated for Australian geomorphology, and the great variation in rates of landform development suggest that these ideas may not be appropriate in Australia, and may need revision or extension elsewhere.

Process studies

It is clear that whatever processes may be acting on the ground surface at present, the gross features of the Australian landscape were formed long ago, under conditions very different from those of today. Process studies may be interesting in their own right and tell us what is happening at present. They provide a basis for speculation, but they cannot be extrapolated to tell us what happened in the past. In Papua–New Guinea too, much of the landscape is inherited despite the relative youth of the country, and process studies are of limited value (Pain, 1978).

Uniformitarianism

In Australia the evidence suggests that the present is not invariably the key to the past, so process studies cannot be extrapolated on the scale of long-term geomorphology which applies to this continent. Things that were different in the past certainly include palaeolatitude, distance to the sea, climate, altitude (in many places), ocean currents and wind systems, vegetation cover and soil-forming processes. Other, and more fundamental, variables may have been in gravity, day length and the earth's radius.

Uniformitarianism remains our chief guide to the past, but it begins to fail when we deal with large spans of time. We have to relax present-day conditions to account for the past, and the general feeling is that the solution that requires least relaxation is the best. But in the end we can only discover what the past was like from consistent internal evidence, not from comparison with places that are thought to be somewhat similar today.

Dynamic equilibrium

Most of Australia is very old and is not in equilibrium with the present conditions of base level, tectonics or climate. There are parts of the landscape, especially those with angular ridges and valleys, where there is a rough equilibrium between various processes and landforms, especially in Papua–New Guinea, but the bulk of the landscape is a historic relic.

Climatic geomorphology

When so much of the landscape is of an age measured in geological periods the landscapes have inevitably experienced a variety of climates, but the

140

mark these climates leave is not overwhelming and is often negligible. Certainly the present-day climates have little to do with the major features of old landscapes. Throughout the Mesozoic, when the South Pole was just south of Tasmania, the climate was warm and temperate. There may have been slight cooling in the Palaeocene, but humid tropical or subtropical conditions prevailed through the Oligocene and most of the Miocene. A main period of laterite formation was the Oligocene, though laterites of earlier age are known. Aridity gradually spread from south-central Australia from Oligocene times on.

Climatic geomorphology, so far as it is used in Australia, involves the use of landforms to deduce past climates, rather than the application of climatic geomorphology theory to explain landforms.

Cyclical theories

Cyclical theories, whether Davisian or of any other kind such as the pedimentation cycle of King, do not fit the Australian scene very well. The existence of multiple planation surfaces indicates a succession of planation events, but as yet we are not sure that these correlate from place to place, or to what extent they can be regarded as evidence of 'cycles' rather than merely a succession. In the Pleistocene Australia was affected by a succession of climatic changes, with corresponding changes in groundwater levels, lake levels, run-off, dune mobility and even glaciation. There may be a vague cyclicity in these changes, but not on the scale of Davisian cycles. The Tertiary was the great time for cycles, if there ever were any. The Tertiary marine basins around Australia reveal a common succession which is roughly as follows:

1 Littoral sands and shales of Palaeocene to Eocene age (low sea level).
2 Marine conditions to the Upper Eocene (high sea level).
3 Low sea level in the Oligocene.
4 High sea level in the late Oligocene or early Miocene.
5 Low sea level in the late Miocene.
6 A brief and limited high sea level in the Pliocene, followed by retreat of the sea. Pliocene deposits are rare.

All these changes of sea level should be conducive to rejuvenation and formation of new cycles of erosion on the Davisian scheme, and to some extent some sort of cyclical landscape development did occur in the Tertiary. The situation is complicated by Tertiary tectonic movement, and further complicated by the fact that many landforms even pre-date these Tertiary changes.

There is little indication of cyclicity in pre-Tertiary geomorphology and geology. The Permian ice age was a unique event in the history of Australian landscape formation, as was the Cretaceous flood.

Nor do weathering phenomena show any cyclic patterns. One must assume that weathering goes on all the time, but there does seem to have been one major event (Mesozoic) when deep weathering profiles were formed and preserved for a very long time. In some places the succession of planation surfaces related to the partial stripping of this regolith; elsewhere there are typical features due to stripping of an irregular weathering front; and in still other places very deep weathering (hundreds of metres) is still preserved.

One of the major events that should initiate a new cycle of erosion is the formation of a new coastline by the splitting apart of continents. The splitting around Australia was progressive, and is recorded in the basins of deposition that border the continent: in Western Australia the Bonaparte Gulf and Canning Basins have lain at the continental margin facing an open ocean for 600,000,000 years, whereas the Perth Basin lay in the interior of Gondwanaland until it split apart in the Cretaceous. The Tasman sea opened off eastern Australia about 80,000,000 years ago, and the Eucla Basin originated with the separation of Australia and Antarctica in the Eocene. The disruption of the continent did indeed lead to the formation of new lowlands, like the Eastern Lowlands, separated from the upwarped plateaus of the Eastern Highlands by steep erosional escarpments. But the break-up was not all at the same time, we cannot distinguish different peneplains associated with particular breaks, and the time-scale involved seems to be much greater than that envisaged by Davis.

In Papua–New Guinea the accretion of land masses and the high rate of vertical movements makes cyclical notions inappropriate, although some planation surfaces do exist.

There is some basis for cyclic ideas in the Tertiary, but on the whole the geomorphic history of Australia has too many unique events to be regarded as cyclic.

Evolutionary geomorphology

Dynamic equilibrium, climatic geomorphology and process studies have all been shown to have little application to the geomorphology of a place like Australia that has a geomorphic history hundreds of millions of years long. If we also reject cyclic ideas and even uniformitarianism, what have we left? The answer is evolutionary geomorphology. By this phrase I do not mean that landforms evolve through a sequence of stages such as youth, maturity, old age, but that the earth's landscapes as a whole are evolving through time. This concept is perhaps easiest to see in relation to the concept of an evolving earth, a concept brought out in some recent geology books such as that by Windley (1977).

The whole tectonic and geomorphic system was very different in the distant past. Wynne-Edwards (1976) believes that the Proterozoic was a time of ductile flow rather than brittle fracture and failure characteristic of Phanerozoic time.

Other great changes to the system may result from a supposed increase in the crustal thickness of cratonic shields. Chapman and Pollack (1977) predict a lithospheric thickness of a few tens of kilometres in young oceans and continental orogenic provinces, in contrast to over 300 km in shield areas. This suggests greater viscosity beneath shields, offering an explanation for what they claim is an observed retarded motion of plates with shields. If in the future the lithosphere continues to thicken, the shields will become viscous anchors, and we may expect plate motion to diminish and eventually cease, bringing to an end the plate tectonic phase of earth evolution.

Carey (1976) has suggested a massive deviation from uniformitarianism in proposing that the earth has expanded, with creation most of the ocean floors, since the Jurassic. The period of expansion is roughly the same as that of landscape evolution, and if such expansion occurred it must surely have had a large and progressive effect on the evolution of geomorphic processes.

Even on a less controversial basis it is clear that several revolutions have occurred in the geomorphic system.

The early earth had a reducing atmosphere, and geomorphology would have been very different after the evolution of an oxygen-rich atmosphere. The early earth would have had a greater abundance of radioactive elements, especially those with relatively short half-lives, so generation of heat would have been greater, presumably affecting such things as tectonic and volcanic activity. The oceans evolved through time, and the amount of water falling on the land was probably very different in the past. The nature of landscape evolution would have been very different before the Devonian, when growth of a terrestrial cover of vegetation changed the geomorphic system. Another important biological change was the spread of grassland in the Cretaceous.

The prevailing tectonic view at present is that continents drift and occasionally collide to make larger land masses, which then split along new cracks from which new continents drift apart again. Island arcs and geosynclines may form at new continental edges, the rocks of which may ultimately be welded to the old continental fragment—a process called cratonization. Some theories suggest that continents are conserved by underthrusting at the edges of continents, others suggest that continents have become thicker as the earth evolved, which would have all sorts of geomorphic side-effects resulting from isostatic rise of the continents.

The origin of precambrian orogenic belts located within continents and bounded by older cratons can be examined by palaeomagnetic data. For the period 2,300–1,900,000,000 years data from the West African and Kaapvaal cratons form a coherent set with poles of similar age from each craton falling consistently on a combined apparent polar wander path constructed for this time interval. For the interval 1,100–700,000,000 years data from the Kalahari and Congo cratons likewise form a coherent set. Data from Australia for the time interval 2,500–1,100,000,000 years also form a coherent set irrespective of the craton from which they were derived. Data from the Superior, Churchill, Nain, Bear and Slave Provinces of North America form a coherent set for the interval 2,600–1,400,000,000 years.

The consistency of the data strongly suggests that the cratons were not previously widely separated and then converged to form orogenic belts. Indeed the palaeomagnetic data preclude such plate tectonic models with convergence of cratons to explain orogenesis. Precambrian orogeny must have resulted from a different mechanism, and it seems that the plate tectonics regime of today must have originated at some time between the Upper Precambrian and the Mesozoic. Even the longer estimate does not give time for many splits and collisions, so the geomorphology associated with plate tectonics is fairly novel in the long history of the earth.

Whatever may be true in the long term it seems certain that the present era is one of continental drift which started in about Jurassic times, though some spreading sites are younger and some are currently splitting, like the Red Sea rift. Certainly Gondwanaland was a supercontinent before the Jurassic, and much of Laurasia—the northern continental assemblage—was also attached to this vast landmass. Geomorphology on a supercontinent would be very different from that of today, resulting from the existence of vast inland areas at great distance from the sea, longer rivers, more inland deposition, and very different climatic patterns. The break-up of Gondwanaland would have important effects on each fragment, with rivers having shorter courses to the sea, rejuvenation and increased erosion on new continental edges, and associated tectonic features such as uplift at continental rims or the development of island arcs. The amalgam of fragments of land as in Papua–New Guinea brings together areas with very different geomorphic histories.

Each fragment of Gondwanaland would have its own individual and distinct history, with many unique events such as formation of new continental edges, biological isolation and local evolution, changes in latitude and the development of new climatic patterns which would depend on size and shape of the fragment, its latitude and orientation, and the location of inherited or newly formed highlands, and the effects of newly formed seas.

Australia is one of the Gondwanaland fragments, and its history must be seen on the extended time-scale that is appropriate for such major topics as continental drift, mountain building, and biological evolution. In this context some of the theories and fashions of geomorphology—process studies, dynamic equilibrium, and even cyclical theories—appear to have limited application and importance. The sort of geomorphology we see in Australia is evolutionary geomorphology, which is part of the story of an evolving earth.

References

ANDREWS, E.C. (1910) 'Geographical unity of Eastern Australia in Late and Post-Tertiary time', *J. Proc. R. Soc. N.S.W.* 67, 251–350

CAREY, S.W. (1976) *The expanding earth* (Amsterdam)

CHAPPELL, J. (1974) 'Geology of coral terraces, Huon Peninsula, New Guinea: a study of Quaternary tectonic movements and sea-level changes', *Geol. Soc. Am. Bull.* 85, 555–70

CHAPMAN, D.S. and POLLACH, H.N. (1977) 'Regional geotherms and lithospheric thickness', *Geology* 5, 265–8

CONNELLY, J.B. (1974) 'A structure interpretation of magnetometer and seismic pro-filer records in the Bismarck Sea, Melanesian Archipelago', *J. geol. Soc. Aust.* 21, 459–69

CROOK, K.A.W. and BELBIN, L. (1978) 'The southwest Pacific area during the last 90 million years', *J. geol. Soc. Aust.* 25, 23–40

GARDNER, J.V. (1970) 'Submarine geology of the western Coral Sea', *Geol. Soc. Am. Bull.* 81, 2599–614

JENNINGS, J.N. (1978) 'Reconstruction of Quaternary environments at Lake George', *Aust. Q. Newsletter* 12, 29–30

LE BAS, M.J. (1971) 'Per-alkaline volcanism, crustal swelling, and rifting', *Nature Phys. Sci.* 230, 85–87

LUENDYK, B.P., MACDONALD, K.C. and BRYAN, W.N. (1973) 'Rifting history of Woodlark Basin in the southwest Pacific', *Geol. Soc. Am. Bull.* 84, 1125–34

MCELHINNY, M.W. and MCWILLIAMS, M.D. (1977) 'Precambrian geodynamics—a palaeomagnetic view', *Tectonophysics* 40, 137–59

MILSOM, J.S. (1970) 'Woodlark Basin, a minor center of sea-floor spreading in Melanesia', *J. geophys. Res.* 75, 7335–9

MILSOM, J.S. (1974) 'East New Guinea', in SPENCER, A.M. (ed.) *Mesozoic-Cenozoic orogenic belts, Geol. Soc. Lond. Spec. Publ.* No. 4, pp. 463–74

OLLIER, C.D. (1977) 'Early landform evolution', in JEANS D.N. (ed.) *Australia: a geography* (London)

OLLIER, C.D. (1978) 'Tectonics and geomorphology of the Eastern Highlands', in DAVIES, J.L. and WILLIAMS, M.A.J. (eds) *Landform evolution in Australasia* (Canberra) pp. 5–47

OLLIER, C.D. and PAIN, C.F. (in press) 'Surficial gneiss domes of Papua New Guinea', *J. geol. Soc. Aust.*

PAIN, C.F. (1978) 'Landform inheritance in the Central Highlands of Papua New Guinea', in DAVIES, J.L. and WILLIAMS, M.A.J. (eds) *Landform evolution in Australasia* (Canberra) pp. 48–69

TAYLOR, T.G. (1911) 'Physiography of Eastern Australia', *Bull. Bur. Met. Aust.* 8

VAN DE GRAAFF, W.J.E., CROWE, R.W.A., BUNTIN, J.A. and MACKSON, M.J. (1977) 'Relict early Cainozoic drainages in arid Western Australia', *Z. geomorph.* 21, 379–400

WELLMAN, P. and MCDOUGALL, I. (1974) 'Cainozoic igneous activity in eastern Australia', *Tectonophysics* 23, 49–65

WILLIAMS, G.E. and GOODE, A.D.T. (1978) 'Possible western outlet for an ancient Murray River in South Australia', *Search* 8

WINDLEY, B.F. (1977) *The evolving continents* (London)

WRIGHT, R.L. (1963) 'Deep weathering and erosion surfaces in the Daly River basin, Northern Territory', *J. geol. Soc. Aust.* 17, 39–51

WYNNE-EDWARDS, H.R. (1976) 'Proterozoic ensialic orogenesis: the millipede model of ductile plate tectonics', *Am. J. Sci.* 276, 927–53

6

CONCEPT OF
THE GRADED RIVER

J.H. Mackin

Source: *Bulletin of the Geological Society of America* 59 (1948): 463–511.

Abstract

Grade is a condition of equilibrium in streams as agents of transportation. The validity of the concept has been questioned, but it is indispensable in any genetic study of fluvial erosional features and deposits. This paper modifies and extends the theory of grade originally set forth by Gilbert and Davis.

A graded stream is one in which, over a period of years, slope is delicately adjusted to provide, with available discharge and the prevailing channel characteristics, just the velocity required for transportation of all of the load supplied from above. Slope usually decreases in a downvalley direction, but because discharge, channel characteristics, and load do not vary systematically along the stream, the graded profile is not a simple mathematical curve. Corrasive power and bed rock resistance to corrasion determine the slope of the ungraded profile, but have no direct influence on the graded profile. Chiefly because of a difference in rate of downvalley decrease in caliber of load, the aggrading profile differs in form from the graded profile; the aggrading profile is, and the graded profile is not, asymptotic with respect to a horizontal line passing through base level. It is critical in any analysis of stream profiles to recognize the difference in slope-controlling factors in parts of the overall profile that are (1) graded, (2) ungraded, and (3) aggrading.

A graded stream responds to a change in conditions in accordance with Le Chatelier's general law:—"if a stress is brought to bear on a system in equilibrium, a reaction occurs, displacing the equilibrium in a direction that tends to absorb the effect of the stress." Readjustment is effected primarily by appropriate modification of slope by upbuilding or downcutting, and only to a minor extent or not at all by concomitant changes in channel characteristics. Paired examples illustrate (1) the almost telegraphic rapidity with which the first phases of the reaction of a graded stream to a number of artificial changes are propagated upvalley and downvalley and, (2) the more or less complete readjustment that is effected over a period of thousands of years to analogous natural changes.

The engineer is necessarily concerned chiefly with short-term and quantitative aspects of the reaction of a graded stream to changes in control, while the attention of the geologist is usually focused on the long-term and genetic aspects of the stream's response to changes. But the basic problems are the same, and a pooling of ideas and data may enable the engineer to improve his long range planning of river control measures and permit the geologist to interpret, in quantitative terms, the deposits of ancient streams.

Introduction

The concept of grade, as a condition of equilibrium in streams as agents of transportation, has been the fundamental basis for the understanding of fluvial landforms for the last half century. The geologic literature contains, however, a number of markedly different definitions of the concept, and many geologists have been troubled by its defects and inconsistencies. An analysis of some of these difficulties leads Kesseli (1941) to conclude that the views of Gilbert (1877) and Davis (1902) regarding the equilibrium relationship are untenable and that the concept of grade must be abandoned. This article is an outgrowth of studies of stream planation surfaces in Wyoming (Mackin, 1936, 1937), was started several years before Kesseli's critique was published, and is a revision of the concept rather than a defense of the writings of Gilbert and Davis.

The engineering literature provides a counterpart for the concept of grade in the idea of the "adjusted" or "regime" condition in streams. The engineer is concerned primarily with short-term reactions of adjusted streams to damming, shortening, and deepening operations and other river training measures. The geologist sees erosional and depositional features in valleys as records of the long-term response of the graded stream to various natural changes in conditions controlling its activity. These natural changes in control are in many instances closely comparable with those introduced by man. Because they are a good test of the concept of the graded or adjusted condition, a number of paired examples of long- and short-term reactions of streams to analogous changes are brought together here; citations are drawn about equally from geologic and engineering writings.

There is much of common interest in this type of synthesis, but the geologist and the engineer differ widely in background and habits of thought, and an attempt to bridge the gap requires certain compromises in use of terms and manner of treatment. General policies are as follows:

(1) Future advances in knowledge of stream processes will certainly be based increasingly on quantitative measurement and mathematical analysis. But the quantitative aspects of transportation by running water are controversial and are not essential for an evaluation of the concept of grade; the treatment here is qualitative. If, by clarifying some of the genetic aspects of the problem in qualitative terms, or focusing attention on them, the article clears the way for more rapid quantitative advances, it will have served part of its purpose.

(2) There are two possible approaches to the study of streams as agents of transportation, (A) in terms of relationships between slope, discharge, channel form, and the size of grains comprising the load, and (B) in terms of energy transformations. Preferably, the two should not be combined. But they *are* combined in most of the papers cited, and, while the thesis of this article depends wholly on the first approach, some discussion of energy transformations is necessary. The manner in which the term energy is used is well established in the literature; it may be regarded by the specialist as loose, but he will be merely irritated rather than misled.

(3) Transporting power is considered to be a function of *velocity*, rather than the *depth-slope* (tractive force) *relationship* that forms the basis for many mathematical treatments of transportation. This usage has the advantage of simplicity and, for present purposes, the differences are negligible (for analysis of these alternative theories see Rubey (1938) and discussion of Kramer (1935) by outstanding engineers, especially Matthes and Straub (p. 867–868).

(4) Partly in deference to the inveterate equation-skippers, but chiefly because critical differences between causes and effects do not appear in an equation, mathematical methods of expressing relationships are generally avoided.

Acknowledgments

After the 15-year period during which the views outlined here were developed it is difficult for me to distinguish between ideas that were arrived at independently and those gleaned from reading, discussions with numerous geologists and engineers, and lectures in the classrooms of Douglas Johnson and W.M. Davis. An effort has been made to credit other workers with specific points made by them, but the 80-odd citations certainly do not cover all of the cases in which the same thoughts have been expressed before, especially in the writings of Baulig (1926), Davis (1902), Gilbert (1877, 1914), and Rubey (1933, 1938) among the geologists, and Lane (1937), Salisbury (1937), Schoklitsch (1937) and Sonderegger (1935) among the engineers. Early papers, chiefly of historic interest, are not included in the bibliography (in this connection see Baulig, 1926).

I am indebted to W.W. Rubey and Lee Stokes (U. S. Geological Survey), Stafford C. Happ and Allen S. Cary (U. S. Army Engineers), and Robert C.

Hennes (Engineering, University of Washington) for critical comments on the manuscript.

Velocity and load

General statement

This section is a brief review of certain general principles of stream transportation, drawn chiefly from the works of Gilbert, Rubey and Hjulström. The principles are are based largely on laboratory studies and apply equally to graded streams and streams that are not graded. As outlined here they provide a basis for understanding observed behavior of graded streams; the concept of grade depends, not on any particular theory of transportation nor any special manner of apportioning energy losses, but on the form of the longitudinal profile developed by the debris-carrying stream under stable controlling conditions, and on profile changes that automatically readjust the stream to any change in controls.

Energy and velocity

The energy of a stream between any two points is proportional to the product of the mass and the total fall between the two points. This is, hereafter, the "total energy"; it increases with increase in discharge or slope but is increased also, negligibly for present purposes, by the presence of debris in motion in the water.

The energy is dissipated largely, or in some circumstances wholly, as heat developed by viscous shear within the stream. A rather artificial but useful distinction can be made between (1) energy dissipated in friction along the wetted perimeter of the channel (external frictional losses), (2) energy dissipated in friction between the diverse threads of the turbulent current (internal frictional losses), and (3) energy consumed in the transportation of load. The external and internal frictional losses occur whether or not the stream is engaged in transportation; these losses increase with increase in roughness of the channel, with irregularity in the trend or alignment of the channel, and with any departure from the ideal semicircular cross-sectional form that provides the shortest length of wetted perimeter per unit of cross-sectional area. Roughness, alignment, and cross-sectional form are referred to as "channel characteristics"; they determine the "hydraulic efficiency" of the channel. On the basis of an analysis of Gilbert's experimental results and other data, Rubey estimates that the frictional energy losses account for 96% to 97.5% of the total energy in some debris-carrying streams, and that the remaining energy is utilized in transportation (Rubey, 1933, p. 503). The point emphasized here is that the share of the total energy that is utilized in transportation is very small.

Transportation of boulders and pebbles that move only if they are rolled or dragged along the stream bed, and of smaller grains that must be lifted again and again by turbulent currents consumes energy. These pebbles and grains move slower than the water—the energy required to put them in motion and keep them in motion varies with grain size and quantity. Transportation of ultra-fine or colloidal particles with negligible settling velocities (in still water) does not tax the energy of the stream.

If the energy in a given segment were not utilized within that segment an acceleration in the rate of flow would result. Since this is usually not the case it appears that the energy in most segments is equal to the energy dissipated within those segments (Gilbert, 1877, p. 106). This conclusion taken together with the fact that the energy dissipated in internal and external friction is overwhelmingly greater than that consumed in transportation means that, total energy determined by slope and discharge remaining the same, relatively slight changes in the channel characteristics cause very marked changes in transporting power. The practical engineer concerned, for example, with design of non-silting and non-eroding canals is well aware of this relationship (Lane, 1937). It is not given due emphasis in geologic textbook discussion of stream transportation.

The several factors that enter into this energy balance in streams may be recast in terms of velocity:

Velocity increases with increase in slope of the water surface.

Increase in discharge is accompanied by increase in (a) the cross-sectional area and (b) the wetted perimeter of the channel. Since the natural channel is approximately rectangular in section the cross-sectional area increases approximately as the product of width and depth, while the wetted perimeter increases approximately as the sum of the width and twice the depth. Cross-sectional area therefore increases relatively to wetted perimeter with increase in discharge, and this change results in a relative decrease in frictional retardation of flow. Primarily for this reason, velocity varies with discharge.

As the channel departs from the ideal cross-sectional form, or as the floor and walls vary from smooth to rough, or as the trend varies from straight to tortuous, there is an increase in external frictional retardation of flow due to increased length of the wetted perimeter relative to cross-sectional area, and also an increase in internal frictional retardation of flow due to increased turbulence. For these reasons, velocity varies with variation in the channel characteristics.

Velocity is, then, a measure of the energy content of the stream. It varies with any change in the total energy resulting from change in slope or discharge, and, total energy remaining the same, it varies with any change in the energy dissipated in external or internal friction, as defined earlier. To complete the picture of energy-velocity interrelations for the case of the debris-laden stream we have Gilbert's experimental data indicating that velocity varies inversely with the amount of energy consumed in the transportation of load (1914, p. 225–230).

Competence

Competence is defined by Gilbert as a measure of the ability of the stream to transport debris in terms of particle size; the familiar statement is that the weight of the largest particles moved by a stream varies as the sixth power of the velocity. It is well known that there are notable differences in velocity in different parts of the cross section of a stream; the term, as used in the expression above, is usually interpreted as the average velocity. Rubey states that competence actually varies as the sixth power of the "bed velocity", and that the bed-velocity formula gives "reasonably close estimates of the maximum size of particles transported by some large natural streams for which adequate data are available" (Rubey, 1938, p. 137). For purposes of the present discussion, the significance of Rubey's analysis and of the experimental data presented by Gilbert and numerous other workers in this field is simply that velocity required for transportation of detritus increases very markedly with increase in particle size.

The "sixth-power law" was formulated to express the velocity requirements of that fraction of the load of a stream which moves by sliding, rolling, and bouncing along the stream's bed, *i.e.* the tractional load or bed load. But a large part of the load of most natural streams is transported in suspension. The size of the largest particles that can be carried in suspension depends, not on velocity directly, but upon the intensity of turbulence within the stream (Leighly, 1934). Turbulence itself is, however, a function of velocity among other factors; intensity of turbulence increases with increase in velocity. Thus, while Rubey's sixth-power law does not apply to the transportation of suspended load, the decrease in velocity requirements with decrease in grain size probably continues through the range of the larger grain sizes that are normally carried in suspension. A considerable fraction of the suspended load may consist of ultra-fine or colloidal particles with negligible setting velocities; maintenance of these ultra-fine materials in suspension depends only negligibly upon velocity.

Capacity

"Capacity", as defined by Gilbert (1914, p. 35), refers to "the maximum load a stream can carry". The experimental data on which the competence principle was based demonstrate also that, in a stream of given discharge, the velocity required for transportation varies with the quantity of any one grain size, the velocity requirements increasing with increase in the quantity or total weight of the material shed into the stream.

Gilbert used "capacity" in discussing data relating to transportation of weighed amounts of particles in the sand gravel size range under laboratory conditions. In spite of his explicit warning that his concept of capacity does not necessarily apply to natural streams (Gilbert, 1914, 223–230; *see also* Quirke, 1945) there has been a tendency so to apply it; the expressions "loaded to capacity", or "fully loaded", or "saturated with load" are frequently used in discussion of the graded condition in streams. These expressions are of course meaningless unless accompanied by some statement of the grain sizes or range in grade sizes which constitute the load. A stream "loaded to capacity" with coarse sand and pebbles could carry an enormously greater tonnage of material without change in velocity if the materials making up the load were crushed to silt size.

The capacity principle, like the competence principle, probably does not apply to the transportation of very small particles. Since the maintenance in suspension of ultra-fine clay particles and colloids (particles with negligible settling velocities) does not depend upon velocity, there is no theoretical upper limit to the amount of these materials that a stream of a given size and velocity can carry. A stream "loaded to capacity" with exceedingly fine particles would be a mud flow (Hjulström, 1935, p. 344–345).

There probably is in nature every gradation between normal streams and mud flows. Even the low concentrations of colloidal and ultra-fine particles that occur in normal streams undoubtedly tend slightly to increase carrying power by increasing the specific gravity of the water, and tend slightly to increase external and internal frictional energy losses by increasing the viscosity of the water. With higher and higher concentrations of these materials, particles of silt and sand and finally pebbles and boulders come to have negligible settling velocities in the medium until, in a mud flow, great blocks of rock can be carried buoyantly in a plastic mass that may move only a few feet an hour. These effects are negligible in normal streams.

The total load

Depending upon the lithologic characteristics, relief, and erosional processes in its drainage basin, and on processes in operation within the stream itself (as sorting), the range in grain sizes in the total load supplied to a given segment of a stream may vary widely. Moreover, the proportions of the several grain sizes in the total load may differ markedly in streams in which the range in grain size is the same. There is always, in normal streams, a decided "deficiency" in the supply of colloidal and ultra-fine materials.

Frequently the alluvial materials beneath and marginal to a stream channel include such an assortment of grain sizes that, as the velocity of the stream increases with seasonal increase in discharge, it is free to put in motion progressively coarser grain sizes up to the limits of its competence. There will be in this case a reasonably close relationship between the *quantity* of the debris in motion and the *largest grain sizes* that are in motion at a given time. Partly on this basis, and using an expression for the "average settling velocity of all of the debris particles being transported", Rubey has developed a means for evaluating both competence and capacity relationships in, terms of bed velocity: "in a stream free to pick up much sand and gravel as its velocity is increased, the unit width load will vary roughly as the third power of the 'bed' velocity" (Rubey, 1938, p. 139).

Rubey points out that this third-power principle applies only approximately in even those streams to which a fair proportion of the several movable grain sizes (excepting the ultra-fines) are available. In this case both competence and capacity might be said to be a function of velocity. But a stream completely adjusted to the transport of a large amount of sand and silt may carry no pebbles at all, due to a deficiency in supply, although gravel sizes are well within its competence. Gilbert's experimental proof that the quantity of load increases as the grain size decreases suggests that in this special case the total load per unit width of stream varies as a power of the bed velocity higher than the third power; it might be said that "capacity" is the critical factor in this case. A stream may, on the other hand, be supplied with a load consisting predominantly of pebbles and boulders, with a notably small proportion of sand. It appears that in this circumstance the total load per unit width of channel may vary as a power of the bed velocity lower than the third power; competence might be said to be the critical factor. (For an example of the significance of this point in a practical problem of design *see* Whipple's discussion of Missouri River slope, 1942, p. 1191–1200, 1212–1214.)

These numerical values, as such, are not important for purposes of the discussion to follow. But the possibility of notable variations in the proportions of the several grain sizes making up the total load, and the bearing of these variations (qualitatively) on the velocity requirements in the transporting stream, are important. Hereafter the expression "increase (or decrease) in load" means increase (or decrease) in the quantity and average grain size, in accordance with the case treated by Rubey. The expression "increase (or decrease) in calibre of load" means increase (or decrease) in particle size, the total load remaining the same.

The concept of grade

A graded stream is not then, strictly speaking, one in which there is "a balance between total energy and the work given the stream to do", or in which "energy supplied equals energy consumed"; a non-accelerating flow of water carrying no load in a flume or a bed-rock channel fulfills these requirements, but would hardly be considered graded in the geologic sense. It is not a stream in which "slope is adjusted to load"; the carrying power of a stream is a function of velocity, and slope is only one of the factors which bear on velocity. One of the attributes of a graded stream is a "balance between erosion and deposition", but definition of the condition of grade in terms of this balance, and emphasis on the "constant shifting" of the balance, is unfortunate because it focuses attention on incidental short-term changes in the activity of the stream and loses sight of the long-term balance which is the distinctive characteristic of the stream at grade. A graded stream is not a stream "loaded to capacity" because streams never carry a capacity load (by Gilbert's definition). These definitions are partly or basically sound, but all of them include half-truths that are sources of confusion.

A graded stream is not in any sense a stream which is unable to abrade its bed because "all of its energy is used in transportation", or because "transporting the load requires all the energy that was formerly (during youth) applied to downcutting". The particles comprising the load are the tools used in abrasion, and since abrasion does not involve a dissipation of energy independent of that consumed in the propulsion of the tools, abrasion may be regarded as an incidental result of the bouncing, sliding and rolling motion of the particles.

A graded stream is one in which, over a period of years, slope is delicately adjusted to provide, with available discharge and with prevailing channel characteristics, just the velocity required for the transportation of the load supplied from the drainage basin. The graded stream is a system in equilibrium; its diagnostic characteristic is that any change in any of the controlling factors will cause a displacement of the equilibrium in a direction ghat will tend to absorb the effect of the change.

By *stream* we mean, of course, that particular segment with which we are directly concerned; many rivers have both graded and ungraded parts. The expression *over a period of years* rules out seasonal and other short-term fluctuations on the one hand and, on the other, the exceedingly slow changes that accompany the progress of the erosion cycle. *Load* and *discharge* deserve the prominence given in the definition not because they are the only or even necessarily the most important factors controlling slope, but because they are the only factors which are, *in origin*, wholly independent of the stream. *Slope* stands alone because it appears to be the only factor in the equilibrium which is automatically adjustable by the stream itself in such a direction as to accomodate changes in external controls that call for changes in velocity.

The balance involved in the condition of grade can be stated in an equation, but this method of expression is inadequate for present purposes because the terms of an equation are transposable. As set up in an equation, for example, load is a function of velocity. In answer to a query as to which is the cause and which is the effect, the average engineer will assert that velocity controls or determines the load that is carried by a stream; and he may have misgivings as to the sanity of the party who raised the question. In a flume or rock-floored torrent velocity does, in a sense, determine the load that can be carried. But, over a period of years, the load supplied to a stream is actually dependent, not on the velocity of the stream, but on the lithology, relief, vegetative cover, and erosional processes in operation in its drainage basin, and, in the graded stream, that particular slope is maintained which will provide just the velocity required to transport all of the supplied load. In this very real sense velocity is determined by, or adjusted to, the load. In the graded stream, load is a cause, and velocity is an effect: this relationship is not transposable.

The sections that follow approach the question raised by Kesseli as to the validity of the concept of grade by considering (1) typical examples of streams at grade, (2) factors that control the slope of the profile under stable conditions, and (3) reactions of graded streams to natural and artificial changes.

Examples of streams at grade

A 50-mile segment of the Shoshone Valley east of Cody, Wyoming, contains a striking assemblage of river terraces, ranging from a few feet to several hundred feet above the stream. Inter-terrace scarps and the valleys of cross-cutting tributaries provide linear miles of exposures indicating that each terrace tread consists of a channeled and fluted rock floor, essentially flat in cross-valley profile, mantled by a uniformly thin (15 to 25 feet) sheet of alluvium. The rock floor of each terrace bevels inclined strata of varied types (Fig. 1). The mantle is made up largely of stream rounded pebbles wholly different lithologically from the local bedrock, and identical in composition with the detritus now being handled by the Shoshone River. The terrace surfaces and their planed rock floors exhibit smooth, concave-upward longitudinal profiles similar to that of the flood plain (Mackin, 1937, p. 825–837).

The manner of origin of the terraces is indicated by the present activity of the stream, which is meandering on its valley floor. Some curves are slicing laterally at the base of vertical to overhanging rock walls, and some are shifting downvalley from similar cut banks. As each meander shifts it leaves behind a gravelly surface exposed at low-water stage—it is safe to infer that the thickness of this channel gravel is at least equal to the depth of the channel. It is evident that the gravel deposit grows by lateral accretion as the stream shifts, and if, as seems likely, it rests on a rock floor, then this floor must have been cut by the shifting stream, *pari passu* with the deposition of the gravel. The gravel sheet represents the bed load; it is soon covered by fine silt and sand representing the finer fractions of the suspended load deposited by slow-moving or ponded overbank waters, and later by slope wash and side-stream alluvial fans. We do not *know* that the valley floor gravel sheet rests on bedrock because we canot see its base. But the gravel sheet on each terrace does rest on beveled bedrock, and edges laterally against the base of scarps with the same systematic curvature in plan as those being cut by the stream (Fig. 1).

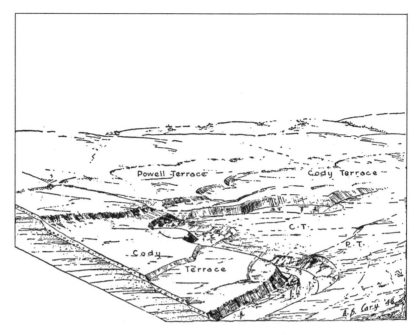

Figure 1 River terraces near Cody, Wyoming.
Drawn by Allen S. Cary from photographs taken from Cedar Mountain looking eastward down the Shoshone Valley. The alluvial veneer shown on the front of the block includes Shoshone River channel gravel and overbank silts, and side stream fan deposits; these materials are wholly different in origin and appearance but cannot be distinguished on the scale of the drawing. Note truncation of bed rock structure along the terrace scarps. The rear scarp of the Powell Terrace is about 90 feet high, but is largely covered by alluvial fans and slope wash.

These relations, taken together, indicate that the terraces are remnants of valley floors cut in bedrock by the lateral planation of the Shoshone River during earlier periods of very slow downcutting or pauses in downcutting (Mackin, 1937). They are altogether different in origin and structure from the equally valid type of terrace formed by partial filling of a valley and later drenching of the fill. And they most certainly were not formed by incidental deposition of gravel (as by floods?) on surfaces produced by other erosional process (successive downstepping peneplanes?); in that they were formed by the same agency at the same time, the gravel veneers are related directly, not incidentally, to the planed surfaces on which they rest.

Individual terrace remnants in the Shoshone Valley are more than half a mile wide and a higher Shoshone valley floor (Pole Cat Bench) is over 2 miles wide and essentially flat in cross profile. In the opening out of valley floors of such great breadth the river must have shifted repeatedly from side to side, trimming back first one valley side and then the other. As indicated above, the stream is now engaged in the same activity on a valley floor with an average declivity of more than 30 feet per mile; the slope of the earlier valley floors is (and was) of the same order of magnitude. Since streams of similar discharge and channel characteristics are now vigorously cutting downward in rock with much lower slopes, the question arises as to what held in check the downcutting of the high-gradient Shoshone during the very longperiod of planation. Bedrock resistance can be readily ruled out as a controlling factor, for the valley is underlain by sandstones and shales, and such contrasts in resistance as do occur are not reflected in the profiles of the present stream or the terraces. Even more compelling is the fact that the river has repeatedly opened out very broad valley floors by lateral corrasion in the same bedrock during periods when its downcutting was negligible.

The Shoshone failed to trench its valley floor during the terrace-cutting stages, and the present stream fails to trench its present valley floor because its high gradient is perfectly adjusted to provide, with available discharge and with the prevailing channel characteristics, just the velocity required for transportation of a large load of coarse rock waste continuously supplied to it from ramifying headwaters in the rugged Absaroka Range. Adjacent ephemeral streams that head on the arid floor of the Bighorn Basin and are supplied only with fine-textured detritus maintain lower slopes than the master streams that head in the mountains, although their discharge is only a small fraction of that of the master streams (Mackin, 1936; see also Rich, 1935, and Hunt, 1946, for parallel cases in Utah). Water diverted from the Shoshone River and freed of bed load must be conducted down the terrace surfaces in concrete canals or in canals interrupted by concrete dams; it would otherwise entrench itself below the headgates of the laterals. The "velocity required for transportation" is such that the river rolls and bounces 8 to 12 inch boulders along its bed. The river bed is so efficient a grinding mill that boulders of such rock types as dense andesite are reduced in diameter by one-half within a few tens of miles. But the velocity requirements for transportation are so definitely fixed that the river, flowing over sandstone and shale, could not lower its slope by downcutting during the planation stages, and is not able to cut down at the present time. If the slope were altered, as by warping, the river would be forced to restore it by cutting or filling as the case might be. The Shoshone east of Cody was during the planation stages, and probably is at the present time, a typical graded stream.

Valleys of tributaries of the Columbia River system, particularly those of the Clark Fork and Spokane rivers, illustrate the same additional relationships even more strikingly than the Shoshone Valley. These Columbia Basin valleys were partly filled with glacial, glacio-fluvial, and glacio-lacustrine deposits during the Pleistocene and have since been partly re-excavated. Here again, we look to terrace remnants of higher valley floors because exposures in the dissected terraces supply morphological data that could be obtained from the present valley floor only by hundreds of borings. The postglacial stream terraces (not to be confused with a wide variety of other types of terraces produced during the period of ice occupancy) usually consist of a sheet of channel gravel, 10 to 30 feet thick, overlain by typical overbank silt plus loess and slopewash from higher valley sides. The gravel sheet rests on a channeled and fluted surface which truncates disordered structures in till, lake clays and silts, older gravels, and bedrock; deposition of the gravels accompanied the cutting of the surface on which they lie. The longitudinal slopes of the terraces and of the present valley floors range from 5 feet per mile upward and the larger common pebble sizes in the terrace gravels and the present river bars approximate 6 inches in diameter. It is useful to consider how rapidly, in so far as scouring power is concerned, large streams carrying coarse gravel on these high slopes could trench downward in silt, and at the same time to note that the streams opened out broad valley floors by lateral planation in silt and bed rock in adjoining segments of their valleys, without trenching. The high longitudinal slopes were maintained during the planation stages simply because these slopes were required to provide the velocity needed for transportation of detritus continuously supplied to the streams; for emphasis through hyperbole, one might say that they would have been so maintained had the subjacent materials been cream cheese.

These high-gradient streams were selected as examples to indicate at the outset that the term grade carries no connotation of low declivity. The low-gradient Illinois River is, as Rubey states, an excellent example of a stream in equilibrium (1931) but it is no more excellent than the Columbia tributaries, the Shoshone, the Mesa-stage Rock Creek (Montana) with a slope of about 90 feet per mile (Mackin, 1937, p. 848–850), or many Southwestern pediment streams with much higher slopes (Bryan, 1922). The classic examples are, of course, the wet-weather streams that carved Gilbert's planation surfaces around the flanks of the Henry Mountains in Utah (1877). If, as suggested by some engineering articles, the adjusted stream is one that is stable in channel form and position (Pickels, 1941, p. 166) then the (engineering) adjusted stream is only one special type of the (geologic) graded stream, which is stable only in slope.

The Columbia tributaries are useful also as examples because these streams are at many points locally superposed from the fill onto rock knobs and spurs, some of which have caused falls or rapids now and during earlier planation stages. These streams consist, in other words, of graded segments separated by segments that are not graded, but this circumstance is certainly no defect in the theory of grade.

The graded stream as a system in equilibrium

The idea that a balanced or adjusted condition in streams is an expression of an equilibrium relationship, and that the graded profile is a slope of equilibrium is one of the oldest and most useful of geological concepts relating to streams. Many geologists seem to try to make their treatment of the balanced condition conform to the rigid definition of equilibrium used in the sciences of physics and chemistry. This point of view is reflected in the stock statement that the equilibrium is constantly shifting, approached, but rarely or never attained in the seasonally varying stream. It reaches its logical climax in Kesseli's argument that since discharge, velocity, and other factors are not literally constant in natural streams, no equilibrium can exist.

The requisite conditions for chemical equilibria (as between water and water vapor in a closed container) are; (1) absolute constancy of external controlling conditions (as temperature), and (2) a literally perfect balance between opposed tendencies (as the hail of molecules leaving and returning to the surface of the water). If the water-water vapor apparatus is housed in a laboratory where it is affected by constantly varying diurnal temperature changes, then, strictly speaking, the system rarely or never attains the perfect equivalence between opposed processes which is the essential mark of chemical equilibria. There is, moreover, in the precise chemical sense, a shifting between different states of equilibrium, but there is no such thing as a shifting equilibrium.

(1) *Constancy of controlling conditions* is certainly absent in any segment of a graded stream is attention is focused on its activity during any short period of time, as a year or part of a year. All natural streams vary in discharge, and in many the ratio of high-water to low-water discharge is several hundred to one. In some fully graded stream-transportation systems (in the geologic sense) there may be no discharge at all for most of the year. Velocity, load and all the other factors which enter into the economy of the balanced stream vary markedly with variations in discharge.

(2) *The perfect balance between opposed tendencies*, as an interchange between particles at rest on the bed and in motion in the stream, is not maintained in natural streams. In general, a stream flowing over alluvial materials within its competence tends to enlarge the channel during high-water stages, not only by increase in the height of the water surface, but also by scouring the bed. With decrease in discharge and slackening in velocity as the high-water stage recedes, the stream deposits that part of the load which it is no longer able to carry. Indeed, the same change in controlling conditions may give rise to opposite changes in different parts of the same channel at the same time; increase in discharge and velocity usually causes, for instance, scouring on bends and filling on "crossings" in meandering streams (Straub, 1942, p. 619).

Kesseli, implying (1941, p. 580) that Davis was not aware of seasonal variations in discharge and velocity in natural streams, misses the point. Davis considered the stream as an agent of transportation over a period of years—he was concerned with the forest rather than the trees. Over a period of years sufficiently long to include all the vagaries of the stream, the two independent controls (discharge and supplied load) may be essentially constant. Whatever the conclusions from *a priori* reasoning as to whether constancy of these conditions *should be* maintained in nature, *a posteriori* reasoning based upon the existence of widespread corrasion surfaces of the type represented by the Shoshone terraces indicates that they *are* so maintained long enough to produce distinctive land forms.

Scouring and filling with seasonal fluctuations in discharge and velocity occur in all streams; it is the peculiar and distinctive characteristic of the graded stream that after hundreds or thousands of such short-period fluctuations, entailing an enormous total footage of scouring and filling, the stream shows no change in altitude or declivity. Here again the extensive stream-planed rock surfaces of the Shoshone Valley, with their thin veneers of alluvium, are a case in point. In this long-term sense, there is an equivalence of opposed tendencies in the graded stream.

The concept of equilibrium is the basis for modern quantitative treatments of stream transportation; Rubey's mathematical analysis of "capacity" is appropriately entitled *Equilibrium conditions in debris-laden streams* (1933). The origin and significance of slope variations in the longitudinal profile of the graded stream under stable conditions can be understood only in terms of equilibrium relations. In its sensitivity to change, and its tendency to readjust itself to the changed conditions, the graded stream exhibits the chief and diagnostic mark of a system in equilibrium. It is, in other words, useful and necessary to consider a graded stream as a system in equilibrium, and it is altogether proper so to consider it, provided that it is stated explicitly that the type of equilibrium is different in mechanism and detail from the types treated by the chemist and the physicist, and from the equally valid types recognized by the zoologist and the botanist.[1]

Recognition of these differences eliminates the need for apologetic statements, seemingly made in deference to the chemical usage, to the effect that the stream shifts with every short-period fluctuation from one state of equilibrium to another, or toward another which it never quite attains. To the extent that this view has been the vogue, Kesseli's statement that the condition of grade is "elusive" is amply justified. Distribution of discharge over a short period of time, whether essentially uniform or largely concentrated in rare floods, has much significance with regard to the characteristics of the stream, and it is true that the greater part of the work of all streams both in erosion and transportation is accomplished during high-water stages. But whether the slope of the profile is determined during a brief annual period of high water (Baulig, 1926, p. 59) during the longer period of low water, or during some "bedforming stage" (Schaffernak, cited by Schoklitsch, 1937, p. 144) need not concern us here. It is not the particular stage of the stream in flood or low water, but the stream operating "over a period of years" that is the natural unit; the balance automatically maintained in this unit is, in its own way, quite as perfect as that of the most delicate equilibria dealt with in the "precise sciences".[2]

The shifting equilibrium

The expression "over a period of years" was used advisedly in the statement above regarding the essential constancy of controlling conditions; all of the conditions are subject to change over a period of geologic time. The changes may be sudden, or they may occur at a rate corresponding with the slow progress of the erosion cycle. Other things being equal, the manner in which the stream responds to changes is determined by the rate at which they occur.

A once-graded stream may, in response to a change tending to cause downcutting, (1) lower itself so slowly that each of its slightly lower profiles is maintained in *approximate* adjustment to that phase of the slowly changing conditions in existence at the time of its formation or, (2) be transformed by a relatively sudden change into a wholly unadjusted series of waterfalls and rapids, and re-establish a graded profile at a lower level only after a considerable lapse of time.

Similarly, a graded stream may (3) respond to slow uplift of a barrier across its path by upbuilding, each of its successively higher profiles being in *approximate* adjustment to the conditions at the time of its formation. Rate of uplift of the barrier may, on the other hand, (4) so far outstrip the rate of filling that a lake basin is formed. The stream will in this case develop a new graded profile only after a period of delta building following cessation of uplift.

In cases (2) and (4) the streams were clearly ungraded or out of equilibrium during the transitional periods. The sharply contrasted condition of the streams in cases (1) and (3) may be thought of as representing a *shifting equilibrium*. Use of this expression to indicate maintenance of approximate adjustment to a long-term change in control is justified by the fact that it describes what actually occurs.

In the discussion to follow it will be necessary to return again and again to the contrast between processes in operation in the stream in which the condition of equilibrium is maintained, the stream in which the equilibrium is shifting, and the stream in which there is no semblance of equilibrium. The landforms and deposits associated with these three conditions show differences that are of special significance to the geologist, but his terminology includes no simple and definitive terms for distinguishing them. It is therefore suggested, in accordance with Davis' (1902, p. 107) original proposal that "graded" be used specifically for the stream in which equilibrium is maintained, and that "degrading" and "aggrading" be restricted to cases of the shifting equilibrium. "Degrading" is downcutting approximately at grade, in contradistinction to such self-explanatory terms as trench or incise. "Aggrading" is upbuilding approximately at grade. "Regrading" refers to alteration in the form of the longitudinal profile by simultaneous aggrading and degrading in different parts (Johnson, 1932, p. 662). The term "degrade" is still available, of course, to describe the modeling of waste slopes in interstream areas.

There is no justification or need for using either "aggrade" or "degrade" to describe short-period variations in stream activity, that is, as synonyms for "filling" or "scouring" (of a channel), or for the more general terms "erosion" and "deposition". The following quotations from the last edition of an outstanding and most influential textbook illustrate, from the point of view of the present paper, a misuse of terms. The numbers in brackets are inserted for convenience in reference.

> (1) As downcutting reduces the gradient . . . a time comes when the increasing burden of transporting the load requires all of the energy that was formerly applied to downcutting. . . . The long profile has become a *profile of equilibrium* and the stream is said to be *graded.* (2) When a part of a main stream reaches grade, the local tributaries soon become graded with respect to it. (3) Any change in gradient, discharge, or load would upset the graded condition by altering the rate of erosion. A flood, for example, might convert the graded stream into one actively degrading, but with subsidence of the flood the graded condition would be restored. (4) Again, great increases in load are known to have converted graded streams into actively aggrading ones; for example, when glaciers appearing in their headwater regions poured great additional quantities of rock waste into them.
>
> (Longwell, Knopf and Flint, 1939, p. 64–65)

The view expressed in (1) has been discussed earlier; downcutting, or abrasion in general, does not involve an expenditure of energy independent of that consumed in friction and transportation of load. In (2) the term grade is used in what the present writer regards the proper long-term sense, but in (3) it is used in a wholly different short-term sense. In (3) the term degrading is synonymous with deepening of the bed by scouring (this usage involves a situation in which the *surface of the water is raised* when the stream is said to be degrading!). If a stream responds to a flood by degrading it presumably restores the graded condition by aggrading when the flood subsides. But in (4) aggrading is used to describe the response of a stream to a completely different long-term change in controlling conditions.

While no importance attaches to the terms, as such, it is the writer's opinion that the usage illustrated by these quotations is at least partly responsible for the confusion that forms the basis for Kesseli's attack on the theory of the graded river. The usage suggested here emphasizes the contrast between seasonal fluctuations in stream activity (or, indeed, the equally striking diurnal changes in certain proglacial streams) and such true shiftings of the equilibrium as those represented by epicycles of valley cutting and filling in the Southwest (Bryan, 1940, Bailey, 1935). The distinction is analagous to that made by the meteorologist between weather and climate, and it is just as fundamental. In addition, the proposed usage differentiates between the equilibrium that shifts in response to long-term change and the equilibrium that is maintained long enough to permit the stream to produce the distinctive landforms mentioned earlier. A question may arise as to the exact line of demarcation, in terms of feet of degradation or aggradation during so many thousands or millions of years, between the graded stream and the stream that is degrading or aggrading very slowly. Argument on this score leads nowhere—we classify natural phenomena not to assign each member of a series to a numerical pidgeonhole, but to clarify our understanding of their interrelationships. The distinction between the graded and the slowly degrading stream must, and should properly, depend on the nature of the problem and the point of view of the investigator.

Factors controlling the slope of the graded profile

General statement

Longitudinal profiles of graded streams are often considered to be smooth, "concave upward" curves, that is, curves that decrease systematically in slope in a downvalley direction. Systematic downvalley decrease in slope is, however, by no means an essential or necessary attribute of the graded profile. The stream receives contributions of water and debris from every part of its drainage basin, but the additions are concentrated largely at tributary junctions and the ratio of water to debris varies markedly from place to place, from the high-water ratio of a tributary issuing from a lake or other natural settling basin to the high-debris ratio represented by a talus slide. Superposed on, and in part the result of, the changes in load and discharge are changes in the channel characteristics; these affect the hydraulic efficiency of the channel and hence the slope of the stream.

Such changes along its length might seem to count against considering the graded stream a system in equilibrium, or to indicate that it should be regarded as a type of shifting equilibrium. Indeed, comparison of the manner in which a stream accomodates itself to changes in control from segment to segment under stable conditions with its reaction to a change in conditions (as warping or a climatic change) is a useful mental exercise. But these two types of changes are completely different in origin, and it would be fatal to confuse them in analysis of a given longitudinal profile. While the velocity of each unit segment of a graded stream under stable conditions differs from that of adjoining segments (being kept in balance with local variations in velocity requirements by appropriate adjustments in slope) the close interdependence between all of the segments is such that they are parts of one well-defined system. Factors bearing on the slope of the longitudinal profile that is *maintained* without change as long as conditions remain the same are considered in this section.

Downvalley increase in discharge

It is a matter of observation that large graded streams usually have lower slopes than smaller graded streams. Similarly, a graded stream formed by the confluence of two graded streams usually has, below the junction, a slope lower than that of either of the confluents. The essential reason for these relations, mentioned earlier, is that with increase in size of the channel there is usually an increase in cross-sectional area relative to wetted perimeter and a consequent relative decrease in frictional retardation of flow. A result is that large streams commonly have higher velocities than smaller streams with the same slope, or, stated in terms of the profile of a single stream, a downvalley segment with large discharge can maintain a given velocity on a lower slope than an upvalley segment with small discharge. In other words, a mere downvalley increase in discharge requires (or permits) a corresponding downvalley decrease in the slope of the graded profile.

Downvalley increase in ratio of load to discharge

Trunk streams often head in regions of high relief and flow in their lower portions through regions of relatively low relief. Under these circumstances the ratio of total load to discharge in the contributions of tributaries may be larger in the lower than in the upper parts of the stream. For the same reason, the caliber of the load supplied to the stream by tributaries, slope-washing, creep, and talus fall commonly decreases from head to mouth. Downvalley decrease in total load relative to discharge and/or downvalley decrease in the caliber of load shed into the stream, to the extent that they occur, require a corresponding downvalley decrease in the slope of the graded profile.

Downvalley decrease in ratio of load to discharge

The load of a graded stream may increase in a downvalley direction relative to its discharge because of evaporation or subsurface loss of water, or as a result of the entry of heavily loaded tributaries and various slower types of mass movement from its valley sides. Kesseli particularly emphasizes the latter process as being incompatible with the concept of grade, his statement being that if a stream be "fully loaded" it is manifestly impossible for it to acquire additional load as, for instance, the material caving from banks undercut in the process of valley floor widening (1941, p. 578). This statement is roughly equivalent to the contention that a given saturated solution, in

165

the presence of excess of the solute, cannot take more of that substance into solution. The solution can and must, of course, become more concentrated if any change in control, as an increase in temperature, displaces the equilibrium in the proper direction. Similarly, the graded stream can accommodate itself to the transportation of increased load at any point; it usually does so by a local increase in declivity.

Steepening of the Missouri profile at and below the junction of the Platte River is a case in point. The average declivity of the Missouri for 31 miles above the mouth of the Platte was .74 feet per mile as measured in 1931, and the average slope for 44 miles below the junction was 1.24 feet per mile (Whipple, 1942, p. 1185—in the original report by Straub, cited by Whipple, slopes for unspecified distances above and below the junction are given as .68 and 1.16 feet per mile, respectively; Straub, 1935, p. 1145). The slope of the lower part of the Platte is 3.2 feet per mile. Steepening of the Missouri profile is ascribed by Straub and Whipple to entry of the heavy gravel bed load of the Platte into the Missouri, which carries chiefly sand and silt above the junction. On the basis of an extended study of Missouri slopes and load, Straub generalizes as follows: "As is to be expected, the steepest part of the Missouri River below the point of confluence of the Yellowstone is in the vicinity of the mouths of the tributaries adding the largest bed load" (Straub, 1935, p. 1145).

It should be emphasized that entry of the Platte gravels into the Missouri is not due to any recent change in conditions. The local steepening of the Missouri profile is *not* a matter of "deposition" of gravels by the Platte at its mouth because the Missouri is unable to carry the load. If this were so the streams would be upbuilding rapidly, which is not the case. The Missouri profile below the junction is just steep enough to permit the stream to carry *all* of the added load. The profile break has been and will be maintained without change as long as conditions remain the same, which is the same as saying that the Missouri is graded above and below the junction.

Additions to the load of a graded stream resulting from various types of mass movements of the type mentioned by Kesseli are accomodated in the same way. The fact that nearly all streams receive detritus from these sources and, nevertheless, usually maintain smooth concave-upward profiles past the individual caving banks means simply that these additions are usually so small, relative to the great bulk of rock waste in process of transport along the channel in any given period of time, that their effects on the longitudinal profile are usually lost to view in the general downvalley lowering of declivity resulting from the other changes discussed in this section.

Increased slope of the Missouri at and below the Platte junction is required to provide increased velocity needed for transportation of the increased bed load, but this is not the whole story. The Missouri, a meandering river above the Platte junction, is characterized at and below the junction by a broad irregular channel with numerous bars. This change in habits, a result of the added bed load, almost certainly increases frictional retardation and, hence, calls for a steepening in slope to permit development of any given velocity by the Missouri. In other words, the effect of the Platte is two-fold—the total steepening of the main stream profile represents the adjustment required to accomodate both the direct and the indirect effects of the influx of Platte gravels.

Downvalley decrease in caliber of load

Graded profiles usually decrease in slope in a downvalley direction between tributary junctions, chiefly because of a downvalley decrease in caliber of load due to processes within the stream. The principle involved is that the velocity (and, other things being equal, the declivity) required for the transportation of the coarser fractions of a stream's load decreases with decrease in grain size, the total amount of the load remaining the same. The operation of this principle in natural streams is best illustrated by consideration of a graded segment without tributaries, in which additions and loss of water and detritus are negligible.

The lower portion of the Greybull River in the Bighorn Basin, Wyoming, approximates these ideal conditions. The river issues from the Absaroka Mountains and flows through the arid lowlands of the Basin to its junction with the Bighorn River, receiving its last perennial tributary (Wood River) about 50 miles above its mouth. A peculiar drainage pattern that delivers most of the intermittent drainage from immediately adjacent lowland areas to the Bighorn River by independent streams and an analysis of available discharge records provide reasonable assurance that there is no significant increase in the discharge of the Greybull in the 50-mile segment below Wood River. Throughout this segment the stream is meandering on a valley floor wider than the meander belt. Since the stream is neither aggrading nor degrading at a rate that would be appreciable over a period of years, the total load passing through all parts of the 50-mile segment in a given interval of time must be essentially the same, or may increase slightly downvalley as a result of bank erosion and other processes. In other words, variations in ratio of discharge to total load are of minor importance.

The profile of the Greybull valley floor decreases in slope from about 60 feet per mile at the head of the 50-mile segment to about 20 feet per mile at the lower end; that is, the profile is strongly "concave upward". Stream-cut rock terraces ranging from 50 to 1,200 feet above the present stream show a similar eastward (downvalley) decrease in slope. The river was certainly in essentially perfect adjustment during the long periods of lateral planation recorded by the terraces and probably still is. The downvalley decrease in slope must in this case be ascribed largely to a decrease in caliber of load in transit; pebbles of the valley floor and terrace gravel sheets decrease notably in size in a downvalley direction (Mackin, 1937, p. 858–862).

Discussions of the form of the graded profile often neglect downvalley decrease in caliber of load as a control, or, what is worse, imply that this decrease is the *result* of the decrease in slope. Discussion of this point belongs in a later section; it is sufficient to state here that downvalley decrease in caliber of load is an important cause of downvalley decrease in the slope of the graded profile, and that, in the graded stream, the decrease in caliber is due primarily to attritional comminution of particles comprising the load.

The bearing of decrease in caliber of load due to attrition on the slope of the graded profile is particularly emphasized in the European literature. Schoklitsch, for example, cites Sternberg to the effect that certain central European rivers show a systematic decrease in the weight of particles comprising the bed load as a function of distance traveled. He then points out (1) that "An examination of the profiles of natural watercourses reveals the striking fact that, with few exceptions, the slope (like the size of the bed-sediment particles) decreases from source to mouth"; (2) that it is therefore "quite logical to attempt to ascertain a relation between this law [the Sternberg law of decrease in particle weight] and the shape of the profile, at least in the stretches in which the reduction in size of the particles is due to abrasion"; and finally, (3) that "study of a number of river profiles showed that the slope . . . is proportional to the particle size". On this basis Schoklitsch develops an equation for the river profile (1937, p. 153). Many other writers have advanced the theory that graded or adjusted river profiles are mathematical curves, often without Schoklitsch's qualifications as to the effect of tributaries. The most recent contribution in this country (Shulits, 1941) presents a so-called "Rational equation of river-bed profile". The theory that longitudinal profiles closely approximate simple logarithmic curves has been utilized in geomorphic studies, chiefly in England. (*See*, for example, Jones, 1924; and Green, 1936; for critical discussion see Miller, 1939; and Lewis, 1945.)

Analysis of the mathematics of the graded profile and its important geomorphic implications lies beyond the scope of this article. The following will indicate how the theory applies in this qualitative treatment of the concept of grade:

Rubey, discussing the Shulits article, points out (1) that the Shulits equation is empirical rather than "rational"; (2) that many factors other than caliber of load bear on the slope of the profile, and (3) that the profile of a stream could vary directly as a power of bed-load diameters only if there were some "complex and as yet unformulated interrelationships among the many other variables" (p. 630).

The fact is, of course, that there can be no such interrelationships because there is no interdependence of all the factors bearing on slope in a trunk stream. The contribution of each tributary to a main stream is conditioned by the rocks, relief, and climate in the tributary drainage basin and is not systematically related to relationships in the main stream above the point where the tributary happens to enter. Changes in slope such as that caused by the Platte occur elsewhere on the Missouri, as indicated by Straub's generalization; in high-gradient western streams with which the writer is familiar these changes are even more striking than on the Missouri. The effect of tributaries is only one of several types of change along the length of the stream which are neither interdependent nor systematic. Mathematics can be an exceedingly useful tool in the study of river profiles, but it seems to the writer that the attitude that considers the job to be done because an *approximate* overall fit is obtained in the matching of curves is basically wrong; slurring over the irregularities gives a false impression of simplicity. Some types of numerical values, as the percentages of the several grain sizes in samples of sand and gravel, yield significant averages. But precise altitudes of points along a river profile are not subject to sampling errors; insofar as any understanding of its origin is concerned, the *breaks* in a semi-log plat of the profile are the significant elements, and a single straight line drawn through scattered points has little meaning. Downvalley decrease in caliber of load by reason of attrition is more nearly systematic than any of the other factors bearing on the slope of the graded profile. But caliber of load does not vary systematically in graded streams joined by tributaries, nor in graded streams which the rock types in the load differ notably in resistance to attrition; even if it did, caliber of load is only one of a number of partly or wholly independent factors controlling the graded slope.[3]

Relationship between channel characteristics and slope

As indicated earlier, the hydraulic efficiency of a channel varies with the channel characteristics; self-evident theoretical considerations and experimental data establish this point so clearly that it needs no discussion here. Close relationship between efficiency, as determined by channel characteristics, and slope of the graded stream is demonstrated compellingly by the reduction in slope from about .96 to about .69 feet per mile in a few years, as a result of artificial smoothing of tortuous curves and narrowing, and consequent automatic deepening, of parts of the Missouri channel (Whipple, 1942, p. 1199). The overall length of the channel was not significantly changed by the channel improvement measures. In terms of velocity the essential reason for the reduction in slope is that a reduction in frictional retardation of the current in the corrected channel permits the stream to develop the velocity required for transportation of its load on a lowered slope. In terms of energy the explanation is that a reduction in "internal" and "external" energy losses in the improved channel causes an automatic adjustment of slope of such nature as to reduce the total energy of the stream, the energy utilized in transportation remaining the same.

Because different segments of graded streams vary widely in channel characteristics, these variations certainly bear on the overall form of the longitudinal profile. But before discussing these effects it is necessary to consider a fundamental question regarding the theory of grade.

The concept of adjustment in section

The channel characteristics of a graded stream, like its slope, are developed by the stream itself. Both slope *and* channel characteristics vary from segment to segment, and any change in external controls usually results in changes in both of these variables. Because of the nature of his work the attention of the geologist is usually focused on slope; he knows, for example, that a graded stream responds to changes in load due to waxing and waning of glaciers in its drainage basin by appropriate adjustments in slope effected by upbuilding or downcutting. The attention of the engineer is, on the other hand, usually focused on the channel characteristics; he can, for example, enable an "adjusted" stream to transport an influx of mine waste that greatly increases its load by appropriate channel-improvement measures, without change in slope. A question arises, then, as to whether the foregoing definition which describes the graded stream as one "in which *slope* is delicately adjusted", etc., should not be revised to read, "in which *slope and channel characteristics* are delicately adjusted," etc.

That the question is a very real one is indicated by recent discussion of transportation by running water in the engineering literature which emphasizes especially the concept of the "adjusted", or "stable", or "regimen" (in the sense of equilibrium) cross-sectional form.[4] The discussion centers around problems of design of nonsilting and noneroding canals of various types. (Lane, 1937, with bibliography and discussion by 10 writers; important British papers include Griffith, 1927; and Lacey, 1930); but theoretical aspects of the relationship between cross-sectional form and transportation necessarily apply to natural streams. For example, Griffith observes (A) that natural streams with heavy bed load tend to flow in broad, shallow channels. He concludes (B) that the broad, shallow channel is the type of cross section best adapted for the transportation of heavy bed load. The general attitude of mind that makes (B) follow from (A) is expressed as follows: "A river fully charged with silt [meaning, in the geologic usage, debris without regard for particle size] *must obviously tend to adopt that form of section which will give it a maximum silt-carrying capacity*" (Griffith, 1927, p. 251) (italics mine).

There is perhaps deductive ground for believing that all the characteristics of natural channels, if they are to be "permanent", must somehow contribute to the ability of the stream to transport debris. It might be argued in the same deductive vein that the principle of "least work" would lead one to expect that changes in channel characteristics caused by a change in controlling conditions must be "adjustments" in that they must be of such nature as to adapt the stream to the new conditions. Analysis of the extent to which channel characteristics *are* adjustable in this sense may begin by considering channel-modeling processes in a straight channel in which movement of coarse-textured debris depends on velocity.

Adjustment in section in the straight channel

The semicircular section that would provide the least frictional retardation of flow for clear water or for water carrying only ultra-fine or colloidal particles is rarely or never developed or maintained by flowing water charged with coarse debris for many reasons, the more important of which are: (1) a large bed load requires high bed velocity and widening by bank erosion that must continue until velocity at the banks is reduced to the point where the resistance to erosion of the bank-forming materials equals the erosive force applied to them. (2) Shoaling by deposition will accompany widening of the narrow channel by erosion because the particles of the bed load tend to lodge, and move, and lodge again, the velocity required each time to set them in motion being greater than that required to keep them in motion, and because a higher velocity is required to set in motion a particle on the bed than one on the sloping banks. The operation of these processes results in a channel that is, under different circumstances, semielliptical (Lacey, 1930, p. 273; Lacey in discussion of Lane, 1937, p. 160) or parabolic (Pettis, in discussion of Lane, 1937, p. 149–151) in section, and this section, once developed by automatic modification of an originally too narrow or too wide section, will be stable, or in adjustment, or regimen as long as conditions remain the same. But it does not follow that the channel so formed will necessarily provide the "maximum silt-carrying capacity."

Widening causes (1) reduction in bed load moved per unit width of bed by reason of decrease in velocity that accompanies decrease in depth, and, at the same time, (2) increases the length of the cross section (that is, the number of width units) through which the bed load is moved. Tendencies (1) and (2) are opposed insofar as transportation of bed load is concerned; especially because of (2) the cross-sectional form that provides maximum efficiency for transportation of debris is wider and shallower than the semicircular section that gives maximum efficiency for movement of water. The form of the cross section most efficient for transportation varies with slope, with amount and caliber of the load, and especially with the proportions of the total load that are carried in suspension and moved along the bed. These factors are partly interdependent, and they certainly influence the form of cross section that is developed and maintained by the stream. But the form of the cross section depends also, for reasons indicated above, on a factor wholly independent of the stream, namely, resistance of the banks to erosion. For any one set of slope-debris charge factors there will be one critical degree of erodibility of the bank-forming materials such that an originally too narrow channel will quickly develop a cross section with depth-width relations that provide "the maximum silt-carrying capacity". If the bank-forming materials are less erodible than this critical degree the stream may tend to develop the maximum efficiency section over a period of time; that is, the final stable channel section, however long delayed, may approximate the ideal form for the given slope-debris-charge factors. But if the bank-forming materials are more erodible than the critical degree the stream will adopt and maintain a section that is wider and shallower than the ideal transportation section. In the case of the high-slope stream carrying coarse gravel between banks of incoherent sand, widening may continue until the channel disintegrates into a plexus of split channels and gravel bars that is the antithesis of efficiency for transportation. In this case, and generally, the operation of the "least work" principle is merely a matter of expediency and compromise with local conditions—the river braids because an arrangement of minor channels and bars is somewhat less inefficient than a single exceedingly wide and uniformly shallow channel.

It appears, therefore, that a statement to the effect that a flow of water charged with debris must necessarily develop for itself that form of section that will give it a "maximum debris-carrying capacity" is an invalid generalization because it ignores erodibility of the banks; the influence of this factor increases with increase in slope and caliber of debris. The cross-sectional form developed by a stream may be "stable" in that it is not subject to modification as long as conditions remain the same, but "stability" in section is no guarantee of maximum efficiency for transportation. There is no need to labor this point with examples; it suffices to say that in many instances on record, efficiency for transportation in streams and canals has been greatly increased by artificial modification of self-adopted sections.

Adjustment in section in the shifting channel

The tendency for erosional widening along both banks in the straight channel is, in the curving channel, localized and greatly accentuated at the outside of the curves. Whatever the type of meandering involved (Melton, 1936), the inner parts of the curving channel are bar-ridden shoals if the meanders are actively shifting. The meandering channel is usually deepest at the outside of the curves, but even here, if lateral shifting is rapid and the subjacent materials are resistant, the channel does not continue to operate in one place long enough to permit deepening to the potential depth of flood-stage scour. If outward shifting were stopped and if the shallows were filled so as to concentrate the flow in a narrowed channel, deepening would result. Efficiency for transportation would be further increased if the tortuous curves of the channel were smoothed.

The first point to be made, then, is that lateral shifting is one of the most important factors responsible for the inefficiency of the natural channel.

This holds for the meandering stream, and it is true also for the braided stream. In general, efficiency for transportation varies inversely with the rate of lateral shifting.

The second point is that, other things being equal, the rate of lateral shifting in the graded stream increases with velocity and load; high-velocity, gravel-carrying streams on piedmont slopes of semiarid mountains shift laterally far more rapidly than low-velocity, silt-carrying streams of humid lowlands.

Consider now how a meandering stream will react to a change in controlling conditions, as for example, an increase in load. The engineer might be able to accommodate the stream to the increased load by artificially increasing the efficiency of the channel without increase in slope. Under natural conditions there will usually be both change in the channel characteristics and increase in slope. The increase in slope will be effected automatically by aggradation, and aggradation will continue until the slope is steep enough to provide the velocity required to transport, with available discharge and the prevailing channel characteristics, *whatever they may be*, all the debris delivered to the stream. Modification of the channel characteristics, considered separately as a mechanism of readjustment, will, on the other hand, be self baffling because increase in velocity required by increase in load will itself entail an increase in lateral shifting which will, in turn, tend to decrease the efficiency of the channel, and hence the velocity of the stream.

A graded stream may react to an increase in load by a more drastic change in the channel characteristics, namely, by a change from a meandering to a braided habit. Braiding involves the choking of each functional channel by bar building; the resulting maze of shifting minor channels has a total overall proportionate depth much smaller than that of the corresponding meandering channel. Here again, the effect on velocity of a change in the channel characteristics is precisely the reverse of that called for by the original change in external controls. Eventual readjustment to the new conditions (including increased load *and* notably decreased channel efficiency) will be achieved by increase in slope effected through aggradation.

These examples are certainly not intended to establish a general rule to the effect that, with a change in control, channel characteristics necessarily shift in a manner opposite to that required to bring the stream into balance with the new conditions. In some instances the effects of changes in the channel characteristics may be negligible, and in other instances they may contribute notably to the readjustment. But the examples do serve to bring out a basic difference in the role of adjustments in section and adjustments in slope in the equilibrium of grade; while adjustments in section may or may not accommodate the effect of a change in control, slope is always modified, by the stream itself, in such a manner as to absorb the effect of the stress. This relationship, the tendency for one of a number of partly interdependent variables to act as the outstanding counterbalance in effecting a readjustment to new conditions, is familiar in many types of equilibria. In the case of grade it means simply that the stream normally reacts to a change in controls calling for an increase or decrease in energy required for transportation by increasing or decreasing the total energy through modification of slope rather than by effecting economies in the energy dissipated in friction. A graded stream is a system, prodigiously wasteful of energy at every bend and shoal, kept in a constant state of balance under stable conditions, and brought back into balance after any change in controls, primarily by appropriate adjustments in slope.

Effect of variation in channel characteristics on the graded profile

It has been indicated that changes in ratio of load to discharge, in caliber of load, and other factors, occur in the graded stream, and that, while none of these changes is necessarily systematic, they usually combine to cause a downvalley decrease in the slope of the profile. Variation in the channel characteristics from segment to segment under stable conditions is due, in part directly, to changes in these other factors—for example, decrease in the caliber of the load that is moved along the stream bed tends, other things being equal, to be accompanied by an increase in proportionate depth. But changes in the channel characteristics are due largely to change in the resultant of these other factors, namely, velocity, and, in turn, rate of lateral shifting. The paragraphs below are intended to show the nature of the changes that may be expected in an ideal case and how these will affect slope.

In the upper parts of the graded stream,[5] velocity, and therefore the power of the stream to cut laterally, is high, but the valley floor does not greatly exceed the width of the stream itself, and the rate of lateral shifting is inhibited by confining rock walls. In the absence of well-developed meanders, the channel is relatively straight. For these reasons the actual rate of lateral shifting in any representative segment is relatively slow, and the proportionate depth relatively large.

The middle parts of the graded stream are characterized by fully developed meanders and by decreased velocity and therefore decreased corrasive power. But the actual rate of lateral shifting may be increased because the tendency for lateral shifting increases with decrease in the radius of curves, and especially because the stream now operates largely in unconsolidated alluvium on a wide valley floor. Other things being equal, proportionate depth decreases with increase in the rate of lateral shifting.

Finally, in the lower parts of the stream, velocity and corrasive power may decrease until there is a notable decrease in the rate of lateral shifting even in alluvial materials, with a corresponding increase in proportionate depth. Degree of sinuousity of the channel may remain the same (as in the middle parts) or may decrease.

It follows from the earlier discussion that, in the measure that these changes in the channel characteristics affect velocity, they will result in departures from the theoretical profile adjusted to discharge and load, but with uniform channel characteristics throughout. The effect will be to decrease the slope required to provide the velocity needed for the transportation of load with available discharge in the upper parts, to increase the slope required etc., in the middle parts, and to decrease the slope required etc., in the lower parts.

Deductions as to the effects on slope of the contrasts between the upper and middle sets of conditions are verified by relations described by Gilbert on the Yuba River in California:

> Where the Yuba River passes from the Sierra Nevada to the broad Sacramento Valley its habit is rather abruptly changed. In the Narrows it is narrow and deep; a few miles downstream it has become wide and shallow. Its bed is of gravel, with slopes regulated by the river itself when in flood, and the same material composes the load it carries.
>
> In the Narrows the form ratio during high flood is 0.06 and the slope is 0.10 percent. Two miles downstream the form ratio is 0.008 and the slope is 0.34 percent. Thus the energy necessary to transport the load where the form ratio is 0.008 is more than three times that which suffices where the form ratio is 0.06; and it is evident that the larger ratio is much more efficient than the smaller.
>
> (Gilbert, 1914, p. 135)

Gilbert's "form ratio" is depth over width; it increases with increase in proportionate depth. While Gilbert does not say so, the abrupt change in depth-width relations on the Yuba is associated geographically with, and is certainly due in large part to, a change from a relatively low rate of lateral shifting in the rock-walled "Narrows" to a rapid rate of shifting on the piedmont alluvial plain. This downvalley *steepening* in slope at the point where a stream issues from a gorge is precisely the reverse of what we normally expect, and it is the reverse of what we usually find, because decrease in caliber of load, loss of water through infiltration and evaporation, and other factors, usually outweigh the effects of lateral shifting on channel characteristics and of channel characteristics on slope.

Factual relations at other California canyon mouths described by Sonderegger (1935, p. 296–300) indicate that the Yuba is not an isolated case; Sonderegger's reasoning as to the cause of the slope contrasts corresponds closely with that advanced here. These examples were not known to the writer when the deductions were set down in their present form—the examples therefore are "verification", by prediction of extraordinary or unique relationships, of the theory from which the deductions were drawn.

The effect of rate of lateral shifting on channel characteristics, and, in turn, of channel characteristics on slope in the middle and lower sets of conditions is strikingly illustrated by the Illinois River. This stream seems to exemplify the logical climax or end stage of the hypothetical sequence of downvalley changes in slope and channel characteristics outlined above.

According to Rubey the lower Illinois channel does not shift perceptibly, is much narrower but somewhat deeper than that of the adjacent part of the Mississippi, and is deepest at the inside rather than at the outside of its bends. The river has a slope of less than 2 inches per mile, actually lower than that of the Mississippi from Memphis to the Gulf.

An explanation of the remarkable habits of the Illinois, taken in part from Rubey's writings (Rubey, 1931a and the unpublished report mentioned earlier) and in part rationalized from factual relations described by him and shown on the topographic sheets, is as follows: The river flows in a valley formed by a much larger stream which served as the outlet of Lake Michigan during late-glacial times. Now, after several hundred feet of aggradation due in part to post-glacial aggradation of the Mississippi (which it enters), the lower Illinois is neither aggrading nor degrading at an appreciable rate. It is essentially graded (Rubey, 1931a, p. 366) and, as such, has a velocity adjusted to the transport of all of the debris shed into it. The slope of the stream is, in other words, adjustee to the present load under the prevaling conditions. The main stream is partly or wholly laked above the mouth of the Sangamon at Beardstown, and minor lateral tributaries below this point supply little coarse clastic debris. Under these circumstances the velocity required for transportation of the debris handled by the Illinois below Beardstown is very low—so low, in fact, that inertia fails to counteract the tendency of the principle current to follow the shortest and steepest route, which, in all curving streams, lies along the *inside* of the bends. This introduces a new factor tending to inhibit or halt lateral shifting (which would be very slow in any case because of low velocity) with the result that the stream maintains a relatively narrow and deep channel. The efficiency of this channel is so great, relative to that of actively shifting streams, that the Illinois develops the velocity required for the transportation of the load supplied to it on an exceptionally low slope.

The characteristics of the Illinois River are due largely to fortuitous circumstances (glacial drainage diversion, etc.), but the same type of regimen is approached in the lower flat reaches of many streams. Flattening of slope in the lower reaches is, for example, accompanied by a decrease in the rate of lateral shifting and an increase in proportionate depth on the Mississippi River (*see* Humphreys and Abbott, 1876, p. 107, 122) and the Brazos River in Texas (Barton, 1928, p. 622). It seems likely that this type of regimen, exceptional during the present epoch of crustal unrest and high-standing continents, may have been during earlier geologic periods a prevalent type in the lower parts of sluggish trunk streams draining areas of exceedingly low relief in the penultimate stages of the Davisian cycle of erosion.

Local variation from mean slope

Detailed surveys based on reading of closely spaced gauges at the same river stage usually reveal what may be called local variations in the slopes of graded streams. These local departures from mean slope in any short segment may be fixed in position over a period of years, or they may shift along the stream; they usually vary in position with variation in discharge. They are in most cases clearly related to local variations in proportionate depth or detailed roughness, or to sharp bends, split channels, and other irregularities in trend which increase frictional retardation of flow. The fact that some of the local variations are obviously not associated with changes in load has led one student of streams to publish this rather remarkable nonsequitur: "the declivity of the adjusted stream is not a function of load". The fact is, of course, that declivity is not controlled by load alone.

These local variations in slope merit only brief mention here for the same reason that seasonal variations in discharge, velocity, and load were left largely out of account in the discussion of equilibrium relationships. Local changes in slope of the water surface, as such, are usually symptoms of some local "defect" in the channel; from the point of view of the geologist they are usually negligible because they are not reflected in the slopes of the valley floors produced by streams.

Backwater and draw-down effects

The usual downvalley decrease in the slope of the graded stream has led to the suggestion that the profile tends to be asymptotic with respect to a horizontal plane passing its base level. Similarly, the profile of a tributary is sometimes supposed to approach the slope of the main stream near the junction. It will be shown later that in a special case (aggradation) there is such a tendency. But it is the essence of the concept of grade that declivity is controlled by velocity or energy requirements and, in the graded stream, base level, as long as it does not change, has no bearing on velocity or energy requirements. Base level controls the *level* or *elevation* at which the profile is developed, but it does not influence the *slope* of the profile. For this reason a generalization to the effect that the graded profile approaches base level asymptotically is not valid.

In detail, profile relations in the vicinity of downvalley control points (that is, either general or local base levels) differ markedly (1) where the stream enters still water, (2) where the downvalley control is the lip of a waterfall, and (3) where a tributary enters a trunk stream. In (1) the lowermost part of the profile may show the "back-water effect" to a greater or less extent depending upon a number of factors; additional factors are involved if the "still water" is tidal, and/or if outbuilding of a delta is in progress. In (2) the lowermost part of the profile will show a "draw-down curve," that is, a steepening in the slope of the water surface resulting from a decrease in cross-sectional area due to acceleration in velocity toward the point of free fall. In (3) the backwater curve may affect either tributary or trunk stream or both depending upon their relative velocity and discharge and on the angle (in plan) between the streams at the confluence. These three different types of "base level" and their contrasted effects on a transitional zone in the lowermost part of the profile qualify the statement made in the last paragraph with respect to the relation of the overall profile to base level. But these relationships are local details having no direct bearing on the concept of grade.

Effects of differences in rock resistance

Differential abrasion by streams tends to bring into relief differences in bedrock resistance; resistant rocks often form falls or rapids separating adjoining graded reaches. But if the stream be graded across the barrier, differences in bedrock resistance have no direct influence on its slope; theoretically and actually, as indicated earlier, graded streams cross belts of such contrasted rock types as quartzite and shale without change in slope at the contacts. Differences in the rock types traversed by the graded stream may, of course, cause changes in slope if associated contrasts in topography or lithology alter the amount and caliber of the load supplied to the stream, or if associated contrasts in valley-floor width affect rate of lateral shifting or details of trend or cross-sectional shape of the channel.

Summary

The longitudinal profile of a graded stream may be thought of as consisting of a number of segments, each differing from those that adjoin it but all closely related parts of one system. Definition of the unit segment (in terms of length, permissible slope variation, etc.) depends on the purpose of the investigation.

Each segment has the slope that will provide the velocity required for transportation of all of the load supplied to it from above, and this slope is maintained without change as long as controlling conditions remain the same. The graded profile is a slope of transportation; it is influenced directly neither by the corrasive power of the stream nor bed rock resistance to corrasion.

Some changes from segment to segment in factors controlling the slope of the graded profile are matters of geographic circumstance that are not systematic in any way; these include the downvalley increase in discharge, and the downvalley decrease in load relative to discharge, that characterize trunk streams flowing from highland areas through humid lowlands. Other changes, as downvalley decrease in caliber of load by reason of attrition, may be more or less systematic between tributary junctions. Still others, as change in channel characteristics, are partly dependent on changes in load and discharge; the channel characteristics are determined chiefly by caliber of load and rate of lateral shifting of the channel, and the rate of channel shifting is itself dependent on velocity and erodibility of the banks.

These changes are usually such as to decrease slope requirements in a downvalley direction but, because none of them is systematic, the graded profile cannot be a simple mathematical curve in anything more than a loose or superficial sense. We can proceed toward an understanding of the graded profile, not by "curve matching", but by rigorous analysis of adequate sets of data for unit segments of natural and laboratory streams numerous and varied enough to reveal the effect of variation of each of the factors separately. An essential prerequisite for efficiency in the gathering and analysis of the data is recognition of the difference between the graded profile that is maintained without change, and the ungraded profile that is being modified by upbuilding or downcutting.

Response of the graded stream to changes in control

General statement

The response of a graded stream to any change in control is systematic in that it is predictable in terms of Le Chatelier's general law: "If any stress is brought to bear on a system in equilibrium, a reaction occurs, displacing the equilibrium in a direction which tends to absorb the effect of the stress." This section outlines the manner in which a graded stream, as a system in equilibrium, reacts to "stresses" by considering the nature of its response to changes in some of the controlling conditions which were, in the preceding section, held constant.

The method of presentation adopted to some extent above but used more particularly here involves deduction, from the general concept, of specific reactions that should be expected as results of a number of changes in control, and the matching of these "expected consequences" with field examples. This method of testing the theory is neither superior nor inferior to the laboratory model method; it is simply different from the experimental method which it supplements but certainly cannot replace. Its validity as a test depends on (1) whether the deductions are logical, (2) whether the effects are specifically related to the stated causes, and (3) whether the examples are representative. Obviously, in some cases, the reasoning is inductive; the deductive method of presentation (Johnson, 1940) is not followed rigorously in most of these instances. But it is worth nothing that many of the reactions were in fact predicted purely on the basis of deduction, and later verified by reference to the record, and that a survey of factual relations set forth in the literature has failed to discover any effect of a given cause that does not fit the concept.

Examples are not cited, or are mentioned briefly, where relationships are clear cut or generally familiar. In some instances scores of examples bear out the deductive analysis, each differing from the others in nice detail; some selection was therefore necessary. In general, examples were selected in which a given cause can be related to a definite effect with the least explanatory argument. If alternative examples occur to the reader that illustrate a point more clearly than those cited, that is good; if, on the other hand, there are cases that fail to conform to the general thesis a description of them will be a contribution to our understanding of streams.

Engineering examples are usually intended to show the almost telegraphic rapidity with which preliminary reactions are propagated upvalley and/or downvalley from the point where a change has been introduced by man; some of the geologic examples show the nature of the response to analogous natural changes over a period of time in which more or less complete readjustment may be attained.

Discussion in the paragraphs below is confined to the general mechanism by which readjustment is effected. Contrasted methods of aggradation, the bearing of certain secondary cause-and-effect couples on the slope of the final readjusted profile, and the contrast between the form of the adjusted (or readjusted) profile of the graded stream and the disadjusted profile of the aggrading or degrading stream are treated later.

Increase in load

A once-graded stream responds to an increase in load primarily by steepening its declivity below the point of influx. The steepening is accomplished by deposition of part of the excess load in the channel at the point of influx with a consequent upbuilding of the channel at that point and the formation of a steepened part immediately below.[6] Steepening of any segment permits increased transport of load through that segment to the next segment which is in turn the site of deposition and steepening. Thus the effect of an increase in load is registered by the downvalley movement of a wave of deposition, large or small depending on the rate and manner of addition of the load, and deposition must continue throughout the stream below the point of influx until the slope is everywhere adjusted to the transport of all of the debris delivered to it.

The classic example of marked and immediate response of streams to increase in load associated with works of man is the aggradation and resulting widespread destruction of agricultural lands along the eastern side of the Great Valley of California caused by hydraulic mining on the western slopes of the Sierra Nevada between 1855 and 1884, when court decisions halted discharge of mining debris into the Sierra streams (Gilbert, 1917). Various surveys to 1894 are summarized as follows:

the deposit . . . was 20 miles long, had a maximum width of three miles, covering 16,000 acres and containing 600,000,000 cu. yd. It was 20 feet deep at the river's mouth, 35 feet deep at the edge of the foothills, and 80 feet deep 5 miles higher up on the Yuba River. The grade of the original bed was 5 feet per mile. After the fill was made, the grade per mile was 2[1/2] feet at the mouth, 10 feet at the middle zone, and 20 feet on the upper reaches.

(Waggoner, in discussion of Stevens, 1936, p. 271)

The effect of increase in load due to natural causes is most strikingly exemplified by the aggradation that commonly occurs when river valleys are invaded by glaciers. Automatic steepening of the declivity of the proglacial stream is not necessarily proof that great additional *quantities* of debris are being shed into them. The detritus carried from upvalley and delivered to the stream at the terminus of the glacier is usually much coarser than that formerly delivered to the same point from upvalley by running water, and increase in caliber may be more important than increase in quantity of load as a cause of the profile steepening.

Because glaciation is usually a relatively brief episode in the life history of the drainage basin, with the main cycle of advance and recession interrupted by numerous minor pulsations, and with continual change in topographic relations governing discharge of water and debris from the ice front, proglacial streams rarely or never attain adjustment. Their profiles therefore differ from the more or less completely adjusted pre-glacial and post-glacial profiles not only in overall slope, but also in form. (*See*, for example, MacClintock, 1922, p. 575, 681.)

Decrease in load

A decrease in load may be thought of as occurring at any point in the stream. It may be stated for the time being that the stream simply makes up for deficiency in the load supplied from above by picking up additional load from its channel floor. The net result is downcutting, with a consequent lowering in declivity downvalley from the point where the change occurred. Downcutting must continue until the profile is reduced to that slope which will provide just the velocity required to transport the reduced load.

Because of the operation of the reservoir as a settling basin, the dam is the most common manmade cause of decrease in load. The downcutting that may result is typically shown by the Rio Grande below the Elephant Butte Reservoir and the Colorado below Lake Mead (Stevens, 1938) the Saalach River below the Reichenhall Reservoir (Schoklitsch, 1937, p. 157) and in many other cases (Lane and others, 1934). This effect of decrease in load due to damming is almost always complicated by elimination of peak discharges and velocities that results from the use of the reservoir as a water storage basin.

Terraces cut in earlier fill characterize Pleistocene outwash plains and valley trains. The usual downvalley convergence of terrace profiles cut during the period of deglaciation is probably due primarily to decrease in caliber of load reaching any given segment of the degrading stream as the distance between that segment and the receding ice front increases.

Davis (1902, p. 261) treated the very gradual decrease in stream slope that results from decrease in load due to reduction in relief during the humid erosion cycle; Johnson (1932) and others described analogous effects around the borders of shrinking desert ranges.

Changes in discharge

If a segment of a graded stream receives all or most of its load at the upper end, changes in discharge call for readjustments in the slope in much the same way as changes in load. A decrease in discharge requires an increase in declivity because the load, remaining the same, must move faster through a smaller cross section, and because, as indicated earlier, decrease in the cross-sectional area of the channel involves a relative increase in frictional retardation of flow and, hence, a decrease in velocity. The stream affected by a decrease in discharge, being unable to transport all the load supplied to it on its former slope, deposits some of the load and thereby steepens the slope, the process continuing until the reduced stream is able, by reason of increased velocity on the steepened slope, to transport all the load shed into it. The opposite adjustment occurs in the case of increased discharge.

A special case of the operation of this principle is described by Salisbury (1937) in the lower Mississippi Valley. Subsequent to diversion of part of the Mississippi discharge into the Atchafalaya channel in 1882 there has been, downvalley from the point of diversion, silting of the bed of the reduced Mississippi, and lowering of the slope of the augmented Atchafalaya. Clearing of rafts on the Red River (the upper Atchafalaya), confinement of both the Mississippi and the Atchafalaya between levees, and other works of man have altered the regimen of both streams since the diversion. For these reasons, and particularly because of the exceedingly low slopes of the streams involved, the effects of the diversion are revealed only by careful evaluation of the evidence by Salisbury (chiefly in terms of variation in gauge heights) and by Lane (in discussion of Salisbury's paper, chiefly in terms of variation in discharge).

The distance from the point of diversion to the Gulf is about 125 miles along the Atchafalaya, and about 310 miles along the Mississippi. Deterioration of the discharge capacity of the Mississippi channel since the diversion began and concomitant increase in discharge through the steeper Atchafalaya suggest that we are viewing a type of deltaic drainage change, set in motion in this instance by man, which must have occurred repeatedly in the past under natural conditions. Salisbury's demonstration of silting in the trunk channel below the diversion provides an explanation of a mechanism by which deep-channel, slow-shifting streams like the Mississippi may transfer themselves to different radial positions as delta growth proceeds.

The principle and the general mechanics of readjustment are precisely the same in the more general case of a main stream which receives notable additions of debris from tributaries, but the effects on the form of the profile may be very different because loss of discharge in the main stream calls for local readjustments of slope at each tributary junction, usually enough steepening to permit the reduced main stream to transport the load supplied by the tributaries.

Aggradation of the trunk stream at and below tributary junctions is now in progress on the Rio Grande and the Colorado River below the Elephant and Boulder dams, due chiefly to elimination of peak discharges that formerly moved detritus delivered to the main channel by flash floods on the tributaries (Stevens, 1938). In these cases the effect of reduction of peak discharges locally outweighs the general tendency for downcutting due to retention in the reservoir of the bed load supplied from the upper parts of the trunk stream.

Recent widespread incision of valley floors by streams in the Southwest is ascribed by Bryan (1925; 1940) and Bailey (1935) to increase in peak discharge resulting from an increased rate of runoff due to reduction in vegetative cover. Assuming that this diagnosis is correct, whether the cause be climatic change (Bryan) or overgrazing (Bailey) the recent arroyo cutting is of special interest because, as in the examples last mentioned, a change in distribution of discharge through the year produces effects similar to those caused by a change in discharge, and because the downcutting, a preliminary result of deterioration in vegetative cover, may revert to upbuilding when and if increased load resulting from accelerated erosion of the denuded slopes begins to affect the streams.

The most common geologic cause of decrease in discharge is drainage diversion, and the most evident effect is the growth of side-stream fans in the valley of the reduced stream. The classic case is the partial blocking of the valley of the Petit Morin after capture of its headwaters by the Marne and the Aube; the marsh of St. Gond was caused by detrital accumulations which locally reversed the slope of the valley of the Petit Morin (Davis, 1896, p. 603, 604). Local slope reversals on a larger scale are represented by Lake Traverse and other lakes in the valley now occupied by the Minnesota River; this small postglacial stream has been unable to maintain the low-gradient valley cut during the Pleistocene by the very much larger "Warren River", which served as the outlet of Glacial Lake Agassiz. The final step in the blocking process is illustrated by an enormous accumulation of tributary fan detritus, possibly as much as 700 feet thick, in the gap cut by the former Shoshone River through the Pryor Mountains in Montana; in this case the discharge of the Shoshone was so greatly reduced by headwater diversion that the direction of flow of the beheaded trunk stream was completely reversed by tributary fans (Mackin, 1937, Fig. 6).

It appears, therefore, that reduction in discharge in a trunk stream joined by tributaries tends in general to cause profile changes that express the increased relative importance of the tributaries in the economy of the drainage system, and that, depending on how drastic is the reduction of the trunk stream, and how vigorous the tributaries, the main stream may respond by appropriate modification of its profile during a period of disadjustment or may disintegrate into a series of lakes and reversed segments.

Rise of base level

A rise of base level is equivalent to the rising of a barrier across the path of the graded stream. Each unit of increase in the height of the barrier tends to flatten the declivity immediately upstream. The stream, unable because of decreased declivity to carry all of the load through the flattened segment, deposits in the segment, thus increasing the declivity and transferring the flattening upvalley. Continuation of the process results in upstream propagation of a wave or, better, of an infinite number of small waves of deposition.

If the barrier is raised slowly the stream may maintain itself in approximate adjustment during the process; its rate of aggradation is determined by the rate at which the barrier is elevated. If the barrier is raised rapidly or instantaneously a lake is formed; the distal part of the delta is then the "flattened part" of the profile, and the rate of aggradation is determined by the rate of delta building into the lake. In either case the successive profiles developed during the period of readjustment will differ markedly in form from the original profile and from the eventual completely readjusted profile. The final readjusted profile will tend toward parallelsim with the original profile, differing in this respect from the cases treated above. But, because of secondary effects of aggradation to be discussed later, precise parallelism will usually not be achieved; the only generalization that can be made is that the new profile will be everywhere adjusted to the new prevailing conditions.

Rapid upvalley propagation of a wave of deposition due to rise in base level is well shown by profile modifications brought about within a few years by erection of Debris Barrier #1 on the Yuba River (Gilbert, 1917, p. 52–63); Gilbert's figure 7 illustrates the typical wedging out upstream of the preliminary detrital accumulations. (See also Sonderegger, 1935, Fig. 2, p. 298).

The same effect, but on a much larger scale, is seen in aggradation on the Rio Grande above the Elephant Butte Reservoir (Eakin and Brown, 1939, p. 90–99). At San Marcial, a town near the head of the reservoir, the channel of the Rio Grande has been raised at least 10 feet since 1916, when the reservoir was completed, and the town site is largely buried in silt. Blaney states that surveys made in 1934 show a rise of the channel of 7 feet since 1918 at La Joya, and 2 to 4 feet at Albuquerque (Blaney, in discussion of Stevens, 1936, p. 266). La Joya and Albuquerque are about 50 and 100 miles (airline) above San Marcial, respectively, and Albuquerque is 500 feet above the level of the reservoir.[7]

The Elephant Butte Reservoir is about 40 miles long and the spillway crest is 193 feet above the original river bed. A statement to the effect that the eventual readjusted profile of the Rio Grande, long after complete filling of the reservoir, will be parallel with the original river bed and about 200 feet higher is, of course, wholly indefensible. It is, however, probably closer to the truth than the comfortable assumption that the final debris fill will wedge out to zero within a few tens of miles above the original head of the reservoir. The profile of the aggrading Rio Grande during the period of disadjustment will be less steep than the original profile, but the final readjusted profile may be less steep or steeper. It is not necessary to look so far into the future for trouble—a few tens of feet of upbuilding to the latitude of Albuquerque would destroy highways and railways, towns, irrigation works, and farmlands with an aggregate value that may exceed the cost of the dam.

Aggradation of the Kickapoo River in Wisconsin is ascribed by Thwaites and Bates chiefly to Pleistocene upbuilding of the Wisconsin River, which it enters; it is in effect, therefore, a case of aggradation caused by a "geologic" rise of base level. As in all such cases, aggradation undoubtedly altered conditions controlling slope on the Kickapoo; it is interesting, nevertheless, to note that the modern profile for the first 100 miles above the mouth corresponds closely in slope with the original profile determined by well logs (Bates, 1939; Thwaites, 1928, p. 628).

In some respects analogous to the more or less adjusted Kickapoo and serving especially as an antidote for any idea that the readjusted stream *must* be less steep than the original stream is the striking case of aggradation above a debris barrier on the Bear River in California; the approximately adjusted profile of this river above the barrier is notably steeper than the original profile of the stream (Stevens, 1936, p. 219).

Johnson and Minaker (1945, p. 904) cite an unpublished report by Kaetz and Rich to the effect that study of profiles above 22 debris barriers shows that the slopes of the deposits average from 37 to 49 per cent of the original slope—these cases are recognized as not having reached "equilibrium conditions". They state further that in model studies slopes up to 90 per cent of the original streambed slopes were obtained. Their interpretation of the difference in the form of the original profile and the profiles developed during aggradation differs in detail from those advanced later in the present article, but their factual data, reasoning, and conclusions conform in all respects with the general theory of grade.

Loss of storage capacity by silting is one of the outstanding engineering problems of the century; the problem is many-sided, and the literature is extensive. (*See*, for example, Brown, 1944; Stevens, 1936, 1945; Witzig, 1944; and papers cited therein.) Two practical implications of the theory of grade bearing on special aspects of the problem follow so directly from the views and factual relations cited above as to merit brief mention in passing.

It seems to be common practice to determine the "useful life" of a reservoir (usually the time required for filling with debris to spillway level at the dam, but varying depending on the purpose of the reservoir) by dividing the yearly increment of debris into the capacity of the reservoir for water, with due allowance for compaction and related factors. This method is evidently based on the tacit assumption that, when the delta front has advanced any considerable distance toward the dam, the river will carry its detrital load from the original head of the reservoir to the delta front on a surface of no slope (i.e., the water level). Mere statement of this basic assumption in these terms demonstrates its absurdity. Remedial measures costing large sums are in part contingent on predicted rates of storage depletion; these predictions, particularly on high gradient streams carrying a coarse bed load, are subject to gross errors when they fail to take into account the progressive increase in the proportion of the stream's load that is deposited on the valley floor above the reservoir as aggradation proceeds. (*See* Eakin and Brown, 1939, p. 6, 7; Happ, in discussion of Stevens, 1945, p. 1298–1300.) Research on what may be called "the form of the aggrading profile" is much needed, and the results must of course be quantitative to be useful. It is the writer's conviction that some qualitative understanding of the complex interrelationships of the factors at play in aggrading streams is the essential prerequisite for such quantitative studies if they are to yield general "laws" rather than a set of empirical constants of the type that lead some engineers to conclude that each individual stream has its own rules of conduct.

Study of the form of the aggrading profile may be expected to pay dividends not only by increasing accuracy of the storage depletion rate but also by developing methods by which the rate may be decreased. As indicated especially by the soil conservationists (Brown, 1944), a basic part of any comprehensive attack on the silting problem is reduction insofar as practicable of the debris shed into streams from uplands in the drainage area. The next step is to decrease the load delivered by the streams to the reservoir; the trend of thought in this connection seems to favor use of debris dams. But another method of holding debris out of the reservoir merits more attention than it has received. Normal upvalley aggradational processes may be accelerated by use of groins and other temporary and inexpensive structures so placed as to cause the stream to decrease its transporting efficiency (perhaps it would be better to say—so placed as to accentuate the natural tendency of the meandering or braided channel to be exceedingly inefficient). This method conforms with the well-recognized principle that, in dealing with rivers, better results may be achieved with less human effort by working with the water, rather than against it; river training to induce deposition is to impounding debris by damming as river training to maintain a navigable channel is to dredging. Training measures intelligently planned to increase the rate of upbuilding of the channel by deposition of bed load, and to increase the rate of vertical accretion on the floodplain by overbank deposition of suspended load, combined with suitable types of agricultural utilization of the valley floor as upbuilding proceeds, may provide a partial answer to the problem of reservoir silting. Quantitative data are needed to evaluate the practicability of controlled upvalley aggradation as alternative or supplementary to the debris dam system.

These suggestions apply to the simple case (as that of Lake Mead on the Colorado) where protection of the reservoir is the primary concern. The situation is quite different on the Rio Grande, where aggradation above the Elephant Butte reservoir will destroy valuable property on the valley floor. The engineer charged with corrective measures on the Rio Grande will face an interesting dilemma. He can, by increasing the efficiency of the channel, permit the stream to transport its load on a lower slope, thus delaying destruction of upvalley property and shortening the useful life of the reservoir, or he can, by decreasing the efficiency of the channel, force the river to develop a higher slope by aggradation, thus hastening destruction of upvalley property and lengthening the life of the reservoir.

Lowering of base level

Lowering in base level, is, insofar as the response of the stream is concerned, essentially the same as the lowering of a barrier in its path. Each small lowering of the control point steepens the gradient immediately upstream. Accelerated velocity in the steepened portion results in downcutting, and the steepening is propagated upvalley. Downcutting must continue until the slope is again completely adjusted to supply just the velocity required to transport all of the debris shed into the stream; as in the last case, and with the same qualifications, the final readjusted profile will tend to parallel the original profile.

A man-made change which corresponds with a lowering in base level is the local shortening of a stream by the elimination of meander loops; the general trend of the upvalley effects is indicated or, better, merely suggested by changes in the profile of the Mississippi River brought about by a series of artificial cutoffs and other channel improvements between 1929 and 1939. A generalized and more or less diagrammatic profile (Ferguson, 1939, p. 829) shows the new slope between the cutoffs essentially the same as the original slope, the changes in level resulting from the individual cutoffs being cumulative in an upvalley direction. At Arkansas City, at the head of the cutoffs, the river level was lowered about 15 feet. The effect was noted in 1939 at a gauge 107 miles above the head of the cutoffs, where there was a lowering of 2 or 3 feet in flood stage. The river has certainly not yet adjusted itself to the new conditions; the chief significance of the recorded profile changes to date is the sensitiveness of the stream to "lowering of base level", and the extremely rapid headward progression of the first effects of that lowering.

Macar, on the basis of studies of natural cutoffs in European and American rivers, concludes that the "steepened part" caused by a given cutoff moves headward in large streams several hundred times faster than the rate of downcutting of the stream; his profiles show subparallelism between the original and readjusted slopes (1934).

Classification of changes in control

Changes of the type discussed above may be usefully divided into two catagories: (1) "upvalley", and (2) "downvalley" changes in controls; the terms are used, of course, with reference to a particular segment under consideration. As indicated in an earlier section the slope of the graded profile is determined primarily by load and discharge, which are "upvalley" factors; change in either or both of these factors calls for changes in *slope*. The level at which the profile is developed is determined by base level, which is a "downvalley" factor; change in base level calls for a change in the level of the profile but does not, directly, call for change in its slope. A single change usually affects the stream differently above and below the point where the change occurs; for example, if the load delivered by a tributary is greatly increased the effect on the trunk stream below the junction is increase in load, and above the junction rise in base level. The distinction between upvalley and downvalley changes control is of special importance in interpreting the significance of parallel and converging terrace and valley-floor profiles.

Crustal movements, as tilting or warping of any segment of a graded stream, constitute a third category of changes. Such changes are neither upvalley nor downvalley, and they do not necessarily involve changes in load, discharge, or base level; they simply cause a forcible distortion of all earlier developed terrace and valleyfloor profiles within the segment affected. The stream responds by building or downcutting or both. Except insofar as base level or factors which control slope are changed, either in consequence of the crustal movements directly or by reason of the upbuilding or downcutting of the stream, the completely readjusted profile will be developed at the same level and will have the same slope as the original profile.

Regrading with progress of the erosion cycle

A stream may lengthen during its life history by headward erosion, delta growth, and the development of meanders. Lengthening by the slow extension of ungraded headwaters and by capture involves additions of discharge and load to the stream. The contrasted effects of these additions have been treated earlier; their net effect on the graded portion of the stream is determined by their relative importance. The change, with the progress of the erosion cycle, from the relatively straight course of youth to the meandering course of maturity involves a systematic lengthening of the stream. This lengthening, together with that resulting from delta building, tends to decrease declivity and hence calls for aggradation to maintain the slope required for the transport of load. Whether aggradation will actually occur and how far upvalley its effects may be felt at any stage depends upon the relative rate of an accompanying opposed upvalley change, namely, the tendency for slow lowering of graded declivity resulting from decrease in load with advancement of the cycle. Green has developed an interesting mathematical treatment of the change in the form of the profile by which, with certain initial assumptions made, the past and future shifting of the zones of aggradation and degradation can be determined (Green, 1936; *see also* Miller, 1939; Lewis, 1945).

These changes may be difficult or impossible to detect in large, low-gradient streams of humid regions, especially because they tend to be masked by or confused with the effects of subsidence in the deltaic area and eustatic changes in sea level. Their effects are more clearly seen in high-declivity graded streams flowing from semiarid ranges to closed desert basins under such conditions that base level is raised *Pari passu* with decrease in the declivity of the stream that accompanies reduction of the range. As shown by Johnson and others, the stream maintains its graded condition under these conditions by *regrading*, being engaged in aggradation in the lower levels at the same time that it is degrading on the rock-floored pediment, the line of demarcation between the contrasted zones shifting rangeward or basinward depending on the relative importance of the downvalley and upvalley factors (Johnson, 1932).

There is perhaps nothing seriously wrong with the statement that the stream maintains a graded condition while continuously lowering its slope during the process of the cycle of erosion; the statement is merely an oversimplification that is confusing because it is inconsistent. In a textbook discussion of the *cycle of erosion* it would seem preferable to state that, having developed an equilibrium slope under existing conditions at any given stage of the cycle, that is, having attained a condition of grade, the stream must continually alter its slope and its graded condition as controlling conditions change. The rate of change in controlling conditions is exceedingly slow, and the stream's adjustment to the controlling conditions is exceedingly delicate. The lag between change in conditions and adjustment is therefore so completely negligible that, viewed at any one time, the stream may be properly regarded as being graded. But if attention is focused on the orderly change in landforms during the erosion cycle, that is, *if the topic under discussion is the cycle*, then the emphasis should be not on the "static" equilibrium at any one time, but on the gradually shifting equilibrium over a long period of geologic time. In the course of the erosion cycle the stream maintains itself, not in any one graded condition, but in an infinite number of different graded states, each differing slightly from the last and each appropriate to the existing conditions.

If there is anything paradoxical or confusing in this relationship it escapes the writer. But even the best student may be hopelessly confused if the concept of grade and the wholly different concept of the erosion cycle are churned up together and administered in one dose. Grade must be explained in terms of channel processes and thoroughly understood as a condition of equilibrium in the stream as a transporting agent before the fourth dimension, geologic time, is introduced in a consideration of the role of the stream in the cycle.

An understanding of grade is essential for any theoretical analysis of the erosion cycle, but the principle geologic application of the concept is in interpretation of erosional and depositional landforms produced directly by the work of streams. The field worker in most continental areas finds that he must deal with streams that have at different times in the recent past engaged in episodes of downcutting and valley filling, both proceeding at varying rates and punctuated by pauses. The complex history of most modern stream valleys is undoubtedly due largely to the climatic fluctuations, crustal movements, and eustatic changes in sea level that characterize the Pleistocene. Valley floors and terrace remnants of several types and with various slopes in the same valley record the automatic tendency of the streams to adjust themselves to these changing conditions; even if equilibrium were never attained the theory of grade would be indispensable for any understanding of these strivings toward it. But, as shown by examples cited earlier, which could be multiplied a hundred fold, essential equilibrium is in fact often attained and maintained long enough for the production of distinctive landforms by long-continued lateral planation at the same level.

Short-term changes

After the meandering habit is fully developed in any segment the length of the stream is not significantly altered by continued shifting of the meanders; the shortening effect of occasional cutoffs is cumulative only in the pages of Mark Twain's "Life on the Mississippi". That these local shortenings are compensated, over a period of years, by continued slow growth of all of the loops is indicated by the fact that, although 20 cutoffs between 1722 and 1884 shortened the Mississippi by about 249 miles, the river was by 1929 about the same as the original length (Pickels, 1941, p. 339).

Disturbance of equilibrium relations in the stream by sudden local shortenings and slow general lengthening are not inconsistent with the concept of grade as defined here. If a cutoff results in a "significant" break in the longitudinal profile the stream may be properly considered to be ungraded at that point; the height of the break that is "significant" depends simply on the point of view of the investigator. Schoklitsch's analysis of profile readjustments in the vicinity of a single cutoff, too long and not sufficiently pertinent to be summarized here, is an excellent illustration of readjustment on a small scale (1937, p. 153, 154; see also Shulits, 1936). The effect of a number of cutoffs was cited above, also from the engineering literature, to illustrate the very rapid headward propagation of a local steepening in the longitudinal profile. But in most cutoffs in streams flowing in alluvium the resulting steepening in the bed is less than its normal relief from bend to crossing, the break in the slope of the water surface is less than the difference between high and low water levels, and the increased velocity in the steepened part is less than the seasonal variation in velocity. Because breaks of this type usually leave no recognizable record in the valley-floor features, they may in many types of geologic studies be properly neglected in considering whether the stream is graded. With other short-term fluctuations in velocity, discharge, and load, they are considered to be covered by the expression "over a period of years" which is an essential part of the definition of grade.

Deposits of graded and aggrading streams

The body of detritus that floors the valley of a graded stream usually consists of three types of material, unlike in origin: (1) sand and gravel originally carried as bed load and deposited in the shifting channel, (2) an overlying sheet of sand and silt deposited from suspension in slow moving or ponded overbank waters, and (3) detritus not directly related to the work of the main stream, as talus and slope wash from the valley sides, and loess and wind-blown sand.[8] The maximum potential thickness of the channel gravel sheet equals the maximum potential depth of scour in the shifting channel during high-water stages; a thickness of a few tens of feet does not necessarily indicate that the stream has aggraded or is aggrading. The maximum thickness attainable by the overbank silt is determined by the height to which it can be built by successive overspreadings of the valley floor by flood waters without long-term change in the river level; the natural levee, as such, is not an evidence of aggradation. There is no well-defined limiting thickness for deposits of type (3); for example, over 150 feet of fan wash accumulated over the main stream channel gravel sheet in the Shoshone Valley in Wyoming during a period when the Shoshone River was not aggrading (Mackin, 1937, Pl. 1, p. 827–833).

True aggradation, which involves a systematic long-term rise in the river level, may be effected by a thickening of deposits of types (1), (2), or (3), or any combination of them.

An aggrading stream may produce a thick fill that consists largely or wholly of channel deposits. Numerous examples occur in valleys marginal to glaciated ranges, where the fill was deposited by proglacial streams and subsequently trenched so that its internal structure can be seen; exposures in the frontal scarp of a great fill terrace in the Cle Elum Valley in Washington, for instance, show over 200 feet of uniformly bedded gravels. This preponderance of channel deposits certainly does not mean that the Pleistocene Cle Elum River carried no suspended load, nor does it mean that the successively higher valley floors formed during the period of aggradation carried no veneers of overbank silt.

Aggradation may, on the other hand, produce a fill made up largely of silt and lacustrine clay—the Oligocene White River deposits of the Great Plains are a case in point. The prevailing fine texture of the White River beds certainly does not represent the loads carried by the Oligocene streams; lenses and stringers of channel gravel, contrasting sharply with the overbank silts and clays, occur throughout the White River sediments (Osborn, 1929, p. 103–109).

The thick aggradational fill in the Kickapoo Valley in Wisconsin consists chiefly of slope wash derived from the local valley sides (Bates, 1939, p. 870–876). This is perhaps an exceptional case, but greater or less amounts of local wash, partly reworked and interfingering with main stream deposits, are to be expected in all valley fills.

Contrasts between aggradational fills of different types depend largely on contrasts in the rate of lateral shifting of the depositing stream relative to the rate of upbuilding of the deposit and, especially, on the mechanism of lateral shifting, whether by meander swing and sweep processes (Melton, 1936) or by avulsion as in the braided stream. These habits depend in turn on slope, discharge, load, channel characteristics, and the hydrographic regimen of the stream, resistance of the valley-floor materials to lateral cutting, vegetative cover on the valley-floor, distance from source of the detritus and from the mouth of the stream, relief and erosional processes on adjacent uplands, and other factors that cannot be evaluated here. The points to be made are simply: (1) that *deposits* formed by or associated with aggrading streams differ markedly from the *loads* carried by them, (2) that distinguishing between channel and overbank deposits is the first essential step in interpreting modern valley fills or ancient fluviatile sediments, and (3) that, even after this distinction is made it is virtually impossible to work directly from the grade sizes represented in the channel deposits to the characteristics of the depositing streams because there is no simple relationship between the deposits of an aggrading stream and such partly interdependent factors as slope, discharge, channel characteristics, velocity, and load. We cannot proceed directly from laboratory-determined laws relating to stream *transportation processes* to interpretation of ancient stream *deposits*. An alternative and promising route of attack on the problem is via study of deposits now being formed by natural streams of many types to determine whether the sum total of all of the characteristics of given deposits is uniquely related to the particular modern streams by which they are being formed, and to proceed thence to an understanding of the characteristics of ancient streams by comparison of their deposits with deposits of modern streams of known characteristics.

Here again, as in every phase of the study of streams, there is a broad field in which the interests of the geologist and the engineer overlap. For example, correct evaluation of the factors that control what proportion of overbank silts and channel gravels are incorporated in an aggrading valley fill under natural conditions may permit intelligent modification of these factors to alter, to the advantage of a reservoir, the ratio of bed load to suspended load in the debris charge delivered to the reservoir by the aggrading streams that enter it. The marked decrease in the rate of storage depletion in Lake MacMillan brought about by accidental introduction and development of a dense growth of saltcedar (tamarisk) on the aggradational flood plain above the reservoir (Eakin and Brown, 1939, p. 17–18; Walter, in discussion of Taylor, 1929, p. 1722–1725) illustrates the effectiveness of modification of only one of the factors controlling silt deposition on the valley floor.

Secondary effects of aggradation

Decrease in supplied load

Aggradation (for whatever reason) involves burial of the lower waste-shedding slopes of the valley sides, and a consequent reduction in the load supplied to the stream system as a whole. Usually far more important, back-filling in valleys tributary to the aggrading trunk stream tends to reduce the load delivered by the tributaries. In the extreme case, aggradation by the main stream may so far outstrip upbuilding by its tributaries that the lower portions of their valleys are ponded; settling basins formed in this manner serve to entrap all or most of the clastic waste carried by the tributaries. (For examples, *see* Lobeck, 1939.) In general, decrease in load calls for decrease in the slope of the main stream.

Decrease in discharge

A stream flowing in a valley cut in rock normally carries most of the runoff from its drainage basin as surface discharge. But in an aggraded valley a considerable part of the runoff may pass through the detrital filling as underflow, with a consequent reduction in surface discharge available for the transportation of load. In general, decrease in discharge calls for an increase in declivity.

Change in channel characteristics

Aggradation often involves a shoaling and widening of the channel, and it may cause very marked changes in the stream's characteristics, as from a meandering to a braided habit. These changes usually tend to reduce the efficiency of the channel and therefore to require an increase in declivity.

General statement

Changes of the types listed above are more or less incidental results of the reaction (aggradation) of a stream to a given "primary" change in control. Since they, in turn, call for modifications in declivity, they may be regarded as "secondary" controlling factors. They operate especially during the period of aggradation, but they may continue to affect the stream long after it has readjusted itself to the new conditions. Thus, while certain "downvalley" changes in control (as rise of base level) do not directly call for change in slope, the eventual readjusted profile may, under different circumstances, be steeper or less steep than the original profile because of secondary chain-reaction effects. Any attempt to predict the final form or slope of the readjusted profile by evaluating the effects of a given man-made primary change in controls which fails to take into account these secondary changes in slope-controlling factors is liable to serious error. In general, the net effect on declivity of any primary change in control will be the algebraic sum of the parallel or opposed effects of the primary change and the associated secondary changes.

Profiles of adjusted and disadjusted streams

Sorting in the graded stream

As the velocity of the graded stream increases with seasonal increase in discharge, progressively coarser grains are set in motion up to the limits of the competence of the current and the supply of detritus available in the bed and banks. As velocity decreases with passing of the high-water stage, the materials in motion are thrown down in order of decreasing grain size. Every part of the channel deposit that veneers the valley floor is worked over again and again as the stream shifts laterally, each time with selective deposition of the coarser materials and a winnowing out of the fines. For this reason the valley-floor channel deposits in any part of the valley contain notably higher proportions of coarse materials than the bed load in transit through that segment of the valley in any representative period of time. There is, in this sense, a sorting process in operation in the graded stream.

Because the finer grains in the bed load are set in motion earlier, move faster while in motion, and are retained in motion longer during each seasonal fluctuation in discharge, there is a "running ahead of the fines" in the channel of the graded stream. But this "running ahead of the fines" does not cause a downvalley decrease in the caliber of the load handled by the stream. The fines move faster toward and into any given segment, but they also move faster through and away from that segment; the average grain size in the bed load moving through any segment is the same as though all the grain sizes moved at the same rate. The primary cause of the downvalley decrease in caliber of load in the graded stream is attrition, not sorting.

The same conclusion holds for the channel gravel sheet which veneers the valley floor. Every boulder or pebble that is lodged in the valley-floor alluvium in any segment of the valley at any period of time is likely to be moved and lodged again farther downvalley in successive intervals of time. Downvalley decrease in the grain sizes making up the channel gravel sheet is due, not to sorting, but to attrition during stages of movement in the channel, possibly accentuated by weathering during periods of lodgement.

Sorting in the aggrading stream

It has been noted earlier that aggradation may be effected either by a downvalley movement of a wave or waves of steepened declivity, or by an upvalley propagation of a wave or waves of flattened declivity. The essential relationship in both cases is that of a given segment shedding debris into an adjacent (downvalley) segment that is not quite adjusted to the transportation of that debris. Under these circumstances the operation of the seasonally expanding and contracting channel combines with a deficiency in velocity to assure that the coarsest fractions of the load entering any disadjusted segment will be deposited in that segment. Since the channel of the stream is raised as aggradation proceeds, the materials so deposited will not be subject to reworking and continued downvalley movement. There is in the aggrading stream, in other words, a "permanent withdrawal from circulation" of the coarser fractions of the bed load, and the "running ahead of the fines" in this case makes for a real and substantial downvalley decrease in the caliber of the load in transit, and in the materials deposited by the stream as it aggrades.

Exchange in graded and aggrading streams

The channel of the graded stream, shifting back and forth in a valley-floor gravel sheet made up largely of channel deposits which may be somewhat coarser than the average bed load in transit, has little chance to decrease the caliber of its load by the process of exchange. But the channel of the aggrading stream may shift laterally in alluvium consisting essentially of contemporaneous overbank deposits. In this case the materials added to the load of the stream by scour and caving at the outside of a shifting bend may be largely fine silts and sands, but the coarser fractions of the bed load supplied from upvalley tend to be deposited and left behind as the channel shifts. The term exchange as used here does not imply any equivalence in weight or bulk of the material set in motion and deposited; the stream simply cuts at the outside of its bends because inertia normally holds the strongest current against the outside bank; it deposits the coarsest fractions of the load in transit on its channel floor with every seasonal fluctuation and leaves some of these deposits behind as the channel shifts. In the aggrading stream, in contrast to the graded stream, this type of exchange may result in a notable downvalley decrease in the caliber of load in transit.

Effect of downvalley change in caliber of load
on the form of the profile

It has been repeatedly emphasized that the declivity of the graded stream is controlled by load (and other factors); the declivity is adjusted to furnish just the velocity required for the transportation of all the load supplied to the stream. In the aggrading stream the supplied load does not control declivity in the same degree because, by definition, *all* of the supplied load is not transported. But it is important to recognize that the amount of

195

material moved through any segment of the channel of the aggrading stream in any interval of time is enormously greater than the amount deposited, and that even in the aggrading stream the declivity of each segment is *approximately* adjusted to the load in transit through that segment. In general, with decrease in the discrepancy between the supplied load and the load in transit, the aggrading stream approaches the graded condition.

No generalization can be made with regard to the average steepness of the profiles of graded and aggrading streams as such; both may vary from a small fraction of a foot to hundreds of feet per mile. But an exceedingly useful generalization can be made with regard to a contrast in the *form* of the profiles of graded and aggrading streams. Since declivity is in general adjusted to caliber of load in transit, and since the downvalley decrease in caliber of load in aggrading streams (by attrition, sorting, and exchange) is much more rapid than the downvalley decrease in caliber of load in graded streams (by attrition), it follows that declivity should decrease in a downvalley direction much more rapidly in the aggrading stream than in the graded stream under otherwise similar conditions. The profile of aggradation should be, in other words, more strongly "concave upward" than the graded profile. Thus, while the profile of the graded stream usually shows no tendency to be asymptotic with respect to a horizontal plane passing through a downvalley control point, the profile of the aggrading stream should and usually does show a definite tendency in this direction.

The writer has found two "rules" that follow from the discussion above to be useful tools in field study and interpretation of terraces of many types in stream valleys: (1) If there is any considerable length of stream upvalley from a given segment, aggradational channel deposits in that segment are so consistently finer in grain size than earlier or later deposits formed when the stream was at grade that variation in grain size and sorting serves as a criterion, for example, in distinguishing between channel deposits laid down in a valley-filling stage and the channel gravel sheet that mantles terraces cut in the fill during a subsequent degradational stage. (2) Aggradational profiles (recorded by terrace remnants) are usually steeper than earlier or later graded profiles in the upper parts of proglacial valleys, but the contrast in slope decreases in a downvalley direction and may be reversed, so that the aggradational profile is less steep than the graded profile in the vicinity of a downvalley control point.

Conclusions

Let us suppose that a stream endowed with a constant volume of water, is at some point continuously supplied with as great a load as it is capable of carrying. For so great a distance as its velocity remains the same, it will neither corrade (downward) nor deposit, but will leave the slope of its bed unchanged. But if in its progress it reaches a place where a less declivity of bed gives a diminished velocity, its capacity for transportation will become less than the load and part of the load will be deposited. Or if in its progress it reaches a place where a greater declivity of bed gives an increased velocity, the capacity for transportation will become greater than the load and there will be corrasion of the bed. In this way a stream which has a supply of debris equal to its capacity, tends to build up the gentler slopes of its bed and cut away the steeper. It tends to establish a single uniform slope. . . .

Every segment is a member of a series, receiving the water and waste of the segment above it, and discharging its own water and waste upon the segment below. If one member of the series is eroded with exceptional rapidity, two things immediately result; first, the member above has its level of discharge lowered, and its rate of erosion is thereby increased; and second, the member below, being clogged with an exceptional load of detritus, has its rate of erosion diminished. The acceleration above and the retardation below, diminish the declivity of the member in which the disturbance originated; and as the declivity is reduced the rate of erosion is likewise reduced.

But the effect does not stop here. The disturbance which has been transferred from one member of the series to the two which adjoin it, is by them transmitted to others, and does not cease until it has reached the confines of the drainage basin. For in each basin all lines of drainage unite in a main line, and a disturbance on any line is communicated through it to the main line and thence to every tributary. And as any member of the system may influence all the others, so each member is influenced by every other. There is interdependence throughout the system.

These paragraphs, taken with minor changes in wording from the Henry Mountain report, dated 1877, set forth the essence of Gilbert's idea of grade in streams; additions and modifications discussed in the present paper are chiefly matters of qualifying detail. The principal conclusion, that the concept of the graded stream as a system in equilibrium is valid, is based on:

(1) Citation of broad valley floors cut by long-continued planation at the same lavel by high-gradient streams crossing rock types of varying resistance to corrasion;

(2) Analysis of the form of the longitudinal profile developed and maintained by the graded stream under stable conditions, demonstrating by citation of cases that, in each segment, slope is adjusted to provide, with available discharge and under prevailing channel conditions, just the

velocity required for transportation of all of the load supplied to that segment without regard for variation in resistance to corrasion in the subjacent materials; and,

(3) an outline of the manner in which graded streams readjust themselves to natural and man-made changes in controlling conditions of several types, demonstrating that the stream responds to such changes always so as to "absorb the effect of the stress", and thus exhibits the chief and diagnostic characteristic of the equilibrium system.

A critical point in connection with (2) is that, because in a trunk stream conditions controlling slope do not vary systematically from segment to segment, the longitudinal profile cannot be a simple mathematical curve. This conclusion is qualitative; if, in conformity with it, we cease to smooth out real departures from uniformity and center the attack on them, with an adequate understanding of the genetic relationships of the independent and interdependent factors involved, then mathematical analysis of longitudinal profiles will advance our knowledge of streams.

A second generalization, important because it has been so generally neglected in geologic writings, is that the slope of the graded profile is adjusted to, or controlled by, not only the classic "load and discharge" but also the cross-sectional form and alignment of the channel—the more efficient the channel, the lower the slope.

In connection with (3) the present study tends to confirm the standard geologic view that streams readjust themselves to new conditions primarily by adjustments in slope, and only in minor degree by modification of the channel section. This statement is so phrased as to avoid any semblance of a "law"—certainly no fetish attaches to slope, and each individual case must be judged on the basis of the evidence. But it does appear that, confronted by changed conditions that call for increased or decreased energy for transportation, the stream usually responds by increasing or decreasing its total energy by appropriate adjustments in slope rather than by effecting economies in the energy dissipated in friction.

Additional generalizations include the distinction between "upvalley" and "downvalley" changes in control and between "upvalley" and "downvalley" reactions of the stream to a given change, the contrast between the form of the disadjusted profile during the period of readjustment and the final readjusted profile, and the effect of secondary changes in control on the slope of the readjusted profile.

With a few minor lapses, this paper does not treat the practical implications of the concept of grade. In geology these ramify widely, ranging from the power of rivers to corrade laterally to interpretation of ancient fluvial sediments and the origin of unconformities beneath and within them. In connection with control of rivers by men, a safe general implication is that the engineer who alters natural equilibrium relations by diversion or damming or channel-improvement measures will often find that he has a bull by the tail and is unable to let go—as he continues to correct or suppress undesirable phases of the chain reaction of the stream to the initial "stress" he will necessarily place increasing emphasis on study of the genetic aspects of the equilibrium in order that he may work *with* rivers, rather than merely *on* them. It is certain that the long-term response of streams to the operations of the present generation of engineers will provide much employment for future generations of engineers and lawyers.

In this connection the most important point brought out by the study may well be the striking analogy between the streams' response to the works of man and to accidents and interruptions due to geologic causes. Nature has brought to bear on streams nearly all of the changes in controlling conditions that are involved in modern engineering works; the record of the long-term reaction of rivers to past geologic changes that is revealed by terraces and in dissected valley fills should contribute much to an understanding of the future of streams that man seeks to control, and will call for changes in design. Conversely, every advance in knowledge of erosional, transportation, and sedimentation processes deriving from engineering investigations will increase the geologist's ability to interpret the record of the past. As the engineer becomes more and more concerned with the genetic aspects of his especial problems (as he must), and as the geologist learns more about the quantitative aspects of his especial problems (as *he* must), it will become evident that the problems are in large measure the same. As Rubey (1931b) puts it, there is "a need for close cooperation among *students of stream-work.*"

Notes

1 For discussion of equilibrium as a "universal law" of wide application *see* W.D. Bancroft's Presidential Address to the American Chemical Society (1911). *See also* "The principle of dynamic equilibrium" as applied in Oceanography (Sverdrup *et al.*, 1942, p. 160).

2 It is useful in this connection to look again at the water-water vapor system, this time focusing attention on the smallest conceivable surface area of water. It will be noted that *two* molecules may leave the water surface and only one return to it in any exceedingly short period of time. The level of the water surface is lowered. In the next unit of time two molecules return and only one leaves; the surface of the water returns to its former position. The point is that time and space relationships must enter into any consideration of equilibria. The time and space relations of the balance in graded streams are of a wholly different order of magnitude from those that obtain in the chemist's laboratory, but the perfection of equivalence of opposed tendencies is none the less perfect.

3 Slopes of ungraded stream segments are determined by the depth to which the stream has cut downward. In headwater basins in bed rock fairly uniform in resistance to corrasion the ungraded profile tends to decrease in declivity in a downvalley direction, the reason being downyalley increase in discharge and therefore corrasive power. Ungraded profiles of this type are superficially similar to

graded profiles but they are completely different in origin; the ungraded profile is conditioned by the corrasive power of the stream, bed rock resistance to corrasion, and the length of time that the stream has been downcutting, while the graded profile is an adjusted slope of transportation that is influenced negligibly, if at all, by these factors. Mathematical analysis of a stream profile that fails to distinguish between graded and ungraded segments involves the fundamental error of mixing different types of data, and can lead only to frustration or to conclusions that are unsound.

4 I am indebted to W.W. Rubey for calling my attention to the need for somewhat more extended discussion of the concept of adjusted cross sections than was accorded that concept in an early draft of this article examined by him. He kindly loaned me an unpublished manuscript on the Hardin-Brussels quadrangles in Illinois, in which the channel characteristics of the Illinois River are analyzed mathematically (for published abstract, *see* Rubey, 1931a), and subsequent correspondence with him clarified and extended my own views. These views do not parallel his in all respects, and he is not responsible for them. But I would like to emphasize that everything that may have lasting value in this section is an outgrowth of his council.

5 The terms "upper", "middle", and "lower" are used here for convenience to designate contrasted sets of conditions that bear on actual (not potential) rate of lateral shifting of a graded stream. These contrasted sets of conditions usually occur in the geographic order suggested by the terms, but depending on the geology of the drainage basin, they may occur in any order along the part of the stream that is graded. The terms should not be confused with the upper (downcutting), middle (cutting and filling), and lower (upbuilding) parts of the overall profile recognized by many workers. These subdivisions of the profile as a whole are discussed later; they have no place in the present treatment of factors bearing on the graded slope.

6 A stream affected by a change in controls of any type "deposits part of its load" or "picks up more load" by appropriate modifications in the amount of material deposited and picked up in the course of its normal seasonal fluctuations, and the net differences are usually very small compared with the great bulk of material moved and relaid during these fluctuations.

7 Stafford C. Happ points out (personal communication) that increased sedimentary loads of tributaries in the last 50 years have probably caused aggradation in the Rio Grande Valley independently of the influence of the Elephant Butte Reservoir. If quantitative evaluation of adequate data indicate that all of the upbuilding noted by Blaney at La Joya and Albuquerque is due to these "upvalley" influences it will mean simply that the effects of the reservoir have not yet reached these points. For definite evidence that aggradation had extended to a point at least 15 miles above the reservoir by 1941 see Happ, in discussion of Stevens (1945, p. 1298).

8 For a more elaborate classification of valley floor materials *see* Happ, Rittenhouse and Dobsen, 1940, p. 22–31.

References cited

Bailey, Reed W. (1935) *Epicycles of erosion in the valleys of the Colorado Plateau Province*, Jour. Geol., vol. 43, p. 337–355.

Bancroft, Wilder D. (1911) *A universal law*, Am. Chem. Soc., Jour., vol. 33, p. 91–120.

Barton, Donald C. (1928) *Meandering in tidal streams*, Jour. Geol., vol. 36, p. 615–629.

Bates, Robert E. (1939) *Geomorphic history of the Kickapoo region, Wisconsin*, Geol. Soc. Am., Bull., vol. 50, p. 819–879.

Baulig, M.H. (1926) *La notion de profil d'equilibre, histoire et critique*, Cong. Inter. Geog. (Le Claire, 1925), C. R., vol. 3, p. 51–63.

Brown, Carl B. (1944) *The control of reservoir silting*, U. S. Dept. Agric., Misc. Pub. 521.

Bryan, Kirk (1922) *Erosion and sedimentation in the Papago Country, Arizona*, U. S. Geol. Survey, Bull. 730.

—— (1925) *Date of channel trenching (arroyo cutting) in the arid Southwest*, Science, vol. 62, p. 338–344.

—— (1940) *Pre Columbian agriculture in the Southwest as conditioned by periods of alluviation*, 8th Am. Sci. Cong., Pr., vol. 2, p. 57–74.

Davis, W.M. (1896) *The Seine, the Meuse, and the Moselle*, Nat. Geog. Mag., vol. 7, p. 189–202, 228–238.

—— (1902) *Base level, grade and peneplain*, Jour. Geol., vol. 10, p. 77–111.

Eakin, Henry, and Brown, Carl B. (1939) *Silting in reservoirs*, U. S. Dept. Agric., Tech. Bull. 524.

Ferguson, Harley B. (1939) *Construction of Mississippi cut-offs: Effects of Mississippi cut-offs*, Civil Eng., vol. 8, p. 725–729, 826–829.

Gilbert, G.K. (1877) *Report on geology of the Henry Mountains*, U. S. Geog. Geol. Survey Rocky Mountain Region, 160 p.

—— (1914) *The transportation of debris by running water*, U. S. Geol. Survey, Prof. Paper 86.

—— (1917) *Hydraulic-mining debris in the Sierra Nevada*, U. S. Geol. Survey, Prof. Paper 105.

Green, J.F.N. (1936) *The terraces of southernmost England*, Geol. Soc. London, Quart. Jour., vol. 92, p. LVIII–LXXXVIII (Presidential address).

Griffith, W.M. (1927) *A theory of silt and scour*, Inst. Civil. Eng., Pr., vol. 223, p. 243–314.

Happ, Stafford C., Rittenhouse, Gordon, and Dobson, G.C. (1940) *Some principles of accelerated stream and valley sedimentation*, U. S. Dept. Agric., Tech. Bull. 695.

Hjulström, Filip (1935) *Studies of the morphological activity of rivers as illustrated by the River Fyris* Univ. Upsala, Geol. Inst., vol. 25, p. 221–527.

Humphreys, A.A., and Abbot, Henry L. (1876) *Report on the physics and hydraulics of the Mississippi River*, U. S. Army, Corps of Eng., Prof. Paper 13.

Hunt, Chas B. (1946) *Guidebook to the geology and geography of the Henry Mountains region*, Utah Geol. Soc., Guidebook to the Geology of Utah, no. 1.

Johnson, Douglas (1932) *Rock planes of arid regions*, Geog. Rev., vol. 22, p. 656–665.

—— (1940) *Studies in scientific method; IV, The deductive method of presentation*, Jour. Geomorph. vol. 3, p. 59–64.

Johnson, J.W., and Minaker, W.L. (1945) *Movement and deposition of sediment in the vicinity of debris-barriers*, Am. Geophys. Union, Tr. 1944, p. 901–906.

Jones, O.T. (1924) *The upper Towy drainage-system*, Geol. Soc. London, Quart. Jour., vol. 80, p. 568–609.

Kesseli, John E. (1941) *The concept of the graded river*, Jour. Geol., vol. 49, p. 561–588.

Kramer, Hans (1935) *Sand mixtures and sand movement in fluvial models*, Am. Soc. Civ. Eng., Tr., vol. 100, p. 798–878.

Lacey, Gerald (1930) *Stable channels in alluvium*, Inst. Civil Eng., Pr., vol. 229, p. 259–384.

Lane, E.W. (1937) *Stable channels in erodible materials*, Am. Soc. Civil Eng., Tr., vol. 102, p. 123–194.

—— and others (1934) *Retrogression of levels in river beds below dams*, Eng. News-Record, vol. 112, p. 836–840.

Leighly, John (1934) *Turbulence and the transportation of rock debris by streams*, Geog. Rev., vol. 24, p. 453–464.

Lewis, W.V. (1945) *Nickpoints and the curve of water erosion*, Geol. Mag., vol. 82, p. 256–266.

Longwell, C.R., Knopf, A., and Flint, R.F. (1939) *Textbook of Geology, Part I, Physical Geology*, John Wiley and Sons, N.Y.

MacClintock, Paul (1922) *The Pleistocene history of the lower Wisconsin River*, Jour. Geol., vol. 30, p. 673–689.

Macar, Paul (1934) *Effects of cut-off meanders on the longitudinal profiles of rivers*, Jour. Geol., vol. 42, p. 523–536.

Mackin, J. Hoover (1936) *The capture of the Greybull River*, Am. Jour. Sci., 5th Ser., vol. 31, p. 373–385.

—— (1937) *Erosional history of the Big Horn Basin, Wyoming*, Geol. Soc. Am., Bull., vol. 48, p. 813–894.

Melton, F.A. (1936) *An empirical classification of flood-plain streams*, Geog. Rev., vol. 26, p. 593–609.

Miller, A. Austin (1939) *Attainable standards of accuracy in the determination of preglacial sea levels*, Jour. Geomorph., vol. 2, p. 95–115.

Osborn, Henry Fairfield (1929) *The titanotheres of ancient Wyoming, Dakota, and Nebraska*, U. S. Geol. Survey, Mon. 55.

Pickels, George W. (1941) *Drainage and flood-control engineering*, McGraw-Hill Book Co., N. Y.

Quirke, Terence T. (1945) *Velocity and load of a stream*, Jour. Geol., vol. 53, p. 125–132.

Rich, John L. (1935) *Origin and evolution of rock fans and pediments*, Geol. Soc. Am., Bull., vol. 46, p. 999–1024.

Rubey, W.W. (1931a) *The Illinois River, a problem in channel equilibrium* (Abstract) Wash. Acad. Sci., Jour., vol. 21, p. 366–367.

—— (1931b) *A need for closer cooperation among students of stream-work*, Am. Geophys. Union, Tr. 1931, p. 216–219.

—— (1933) *Equilibrium conditions in debris-laden streams*, Am. Geophys. Union, Tr. 1933, p. 497–505.

—— (1938) *The force required to move particles on a stream bed*, U. S. Geol. Survey, Prof. Paper 189E.

Salisbury, E.F. (1937) *Influence of diversion on the Mississippi and Atchafalaya Rivers*, Am. Soc. Civil Eng., Tr., vol. 102, p. 75–122.

Schoklitsch, Armin (1937) *Hydraulic structures* (*Der Wasserbau*—translated by Samuel Shulits), Am. Soc. Mech. Eng., N. Y.

Shulits, Samuel (1936) *Fluvial morphology in terms of slope, abrasion and bed-load*, Am. Geophys. Union, Tr. 1936, p. 440–444.

—— (1941) *Rational equation of river-bed profile*, Am. Geophys. Union, Tr. 1941, p. 622–629.

Sonderegger, A.L. (1935) *Modifying the physiographic balance by conservation measures*, Am. Soc. Civil Eng., Tr., vol. 100, p. 284–346.

Stevens, J.C. (1936) *The silt problem*, Am. Soc. Civil Eng., Tr., vol. 101, p. 207–288.

—— (1938) *The effect of silt removal and flow regulation on the regimen of the Rio Grande and Colorado Rivers*, Am. Geophys. Union, Tr., 1938, p. 653.

—— (1945) *Future of Lake Mead and Elephant Butte reservoir*, Am. Soc. Civil Eng., Tr. vol. 111 p. 1231–1342.

Straub, L.G. (1935) *Missouri River*, H. R. Doc. 238, 73rd Cong., 2nd Session, Appendix IV, p. 1032–1245.

—— (1942) *Mechanics of rivers*, in Meinzer, O.E. (editor) *Physics of the Earth, IX, Hydrology*, McGraw-Hill Book Co., N. Y., 712 p.

Sverdrup, H.U., Johnson, Martin W., and Fleming, Richard H. (1942) *The oceans*, Prentice-Hall, Inc., New York, 1087 p.

Taylor, T.U. (1929) *Silting of the Lake at Austin, Texas*, Am. Soc. Civil Eng., Tr., vol. 93, p. 1681–1735.

Thwaites, F.T. (1928) *Pre-Wisconsin terraces of the driftless area of Wisconsin*, Geol. Soc. Am., Bull., vol. 39, p. 621–642.

Whipple, William, Jr. (1942) *Missouri River slope and sediment*, Am Soc. Civil Eng., Tr., vol. 107, p. 1178–1214.

Witzig, Berard J. (1944) *Sedimentation in reservoirs*, Am. Soc. Civil Eng., Tr., vol. 109, p. 1047–1106.

7

ON THE SURVIVAL
OF PALEOFORMS

C.R. Twidale

Source: *American Journal of Science* 276 (1976): 77–95.

Abstract

The presence in the contemporary landscape of very old paleoforms, and especially paleosurfaces of low relief, is inconsistent with some of the commonly espoused models of landscape evolution. And although it is not completely at odds with others, the survival of paleoforms nevertheless poses difficulties.

Stabilization and persistence of the land surface are clearly possible in some circumstances. Various factors—structural conditions, positive diastrophism, localization of river erosion and various feedback mechanisms—operating singly or in concert are suggested in explanation of these survivals.

A model of landscape evolution involving increasing relief amplitude is proposed, not to the exclusion of all others, but as one of several possibilities each of which should be considered, and probably modified, in the light of field evidence.

Introduction

W.D. Thornbury (1954, p. 26) was probably expressing a widely held view when he wrote that "Little of the earth's topography is older than Tertiary, and most of it is no older than Pleistocene". In part this attitude reflects experience and consideration of landscapes profoundly modified either by Quaternary glaciation, or by tectonism, sealevel fluctuations, and climatic changes during that period (see for instance Firman, 1969, p. 206). It mirrors acceptance of W.M. Davis' (1909, p. 266 and following) belief that streamwork in one form or another extends over the entire land surface, and that the latter is subject to slow but continuous and inevitable destruction. It results from a rejection of the whole concept of multicyclic landscape development which is based on the realization that many of the alleged old peneplains described in the literature are little more than "lines in the sky" derived from theory and imagination and not from field evidence (Sauer, 1925; Robinson, 1963; Chorley, 1965).

But having admitted that many of the "dissected peneplains" recognized in many parts of the world are of this intangible quality, it must be asserted also that remnants of some very old paleosurfaces of low relief (paleoplains) and other paleoforms constitute an integral part of many contemporary landscapes. The remnants exist, and they can in many instances be dated within close limits: some studies in denudation chronology are firmly based in stratigraphy (see, for instance, Wooldridge and Linton, 1939; Wopfner and Twidale, 1967). Paleosurfaces constitute only minor elements of the landscape in some areas, but in others such as southern Africa (King, 1940, 1950a, 1960) and central and western Australia various paleoplains form a major part of the modern landscape.

Evidence

Linton (1957, p. 67) has claimed that "All the hill and mountain features we see today belong to Tertiary times unless the present cycle of erosion is exhuming some earlier land surface". And it is true that some elements of the contemporary land surface are paleoforms that have been resurrected from beneath sediments or volcanic rocks. Thus granite inselbergs of Precambrian age and exhumed type have been reported from the west and southwest of the Canadian Shield (Vogt, 1953a and b). Granite domes occur also in the western Murray Basin in South Australia[1], reexposed by the erosion of Miocene marine strata (Twidale, 1968, p. 121–123), and on west-ern Eyre Peninsula where they have been partially exhumed from beneath late Pleistocene aeolianite (Twidale, Bourne, and Smith, 1976). Resurrected paleoplains of Precambrian age have been identified in northern Greenland (Cowie, 1961), in the Canadian Arctic (Ambrose, 1964), and, in limited expos-ures, from north-eastern Eyre Peninsula and the eastern Gawler Ranges (Twidale, Bourne, and Smith, 1976). An exhumed surface of low relief varying in age from locality to locality between upper Jurassic and Lower Cretaceous is extensively exposed at the western margin of the Great Artesian Basin (Twidale, 1956a; Carter and Öpik, 1961; Wopfner, 1964; Wopfner and Twidale, 1967). It has been identified also on the eastern margin of the Basin (Twidale, 1956a) and elsewhere (Woodard, 1955; Hays, 1967). An unusual landscape exhumed from beneath the late Cretaceous Deccan Traps has been described from the Bari district of central India where step-faulted planation surfaces are being reexposed as a result of the erosional stripping of the volcanics (Choubey, 1972).

These and many other exhumed land surfaces are of considerable interest and in places comprise a significant part of the modern land surface, but

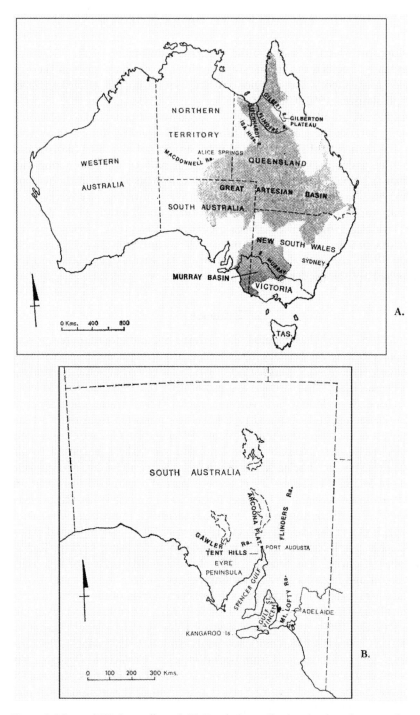

Figure 1 Maps of (A) Australia and (B) South Australia showing locations mentioned.

they have been preserved as a result of burial and have only recently been reexposed to subaerial attack. Their survival thus poses few problems.

But many paleoplains, which are evidently of considerable antiquity, have survived long periods of exposure to weathering and erosion. The extensive planation surfaces of southern Africa for example are of epigene origin, have never been buried, and yet are of great antiquity (King, 1950a, 1960). The African Surface which in places carries a laterite duricrust is of early-mid Tertiary age, and the higher Gondwana Surface or surfaces preserved above the Drakensberg escarpment developed during the Cretaceous. Moreover King believes there are small remnants of a slightly higher landscape of Jurassic age preserved on the Drakensberg (L.C. King, personal commun.).

These old land surfaces of southern Africa apparently have their equivalents in other continents, including Australia (King, 1950a, 1950b, 1960). Certainly a lateritized paleoplain of early-mid Tertiary age is widespread in northern Australia (Twidale, 1956a; Hays, 1967; Öpik, Carter, and Randal, 1973), and a siliceous duricrust of similar age-range is extensively preserved in central and southern areas of the continent (Wopfner, 1960; Sprigg, 1963; Stephens, 1964, 1971; Langford-Smith and Dury, 1965; Wopfner and Twidale, 1967; Wopfner, Callen, and Harris, 1974). Surfaces that predate the Cainozoic have been recognized in central and southern Australia where some of the ranges of the Alice Springs region are considered to be relics of a Cretaceous landscape (Mabbutt, 1965, 1967). The same worker has been quoted as stating that there are remnants of a surface predating the Cretaceous in the Macdonnell Ranges (J.A. Mabbutt, cited *in* Brown, Campbell, and Crook, 1968, p. 304). However, this is not the "only known contemporary land surface that has persisted unburied since pre-Tertiary times..." (Brown, Campbell, and Crook, 1968, p. 304). For example the lateritized summit surface of the Mount Lofty Ranges, Kangaroo Island, and southern Eyre Peninsula is older than the Middle Jurassic and probably of Triassic age (Daily, Twidale, and Milnes, 1974). Similarly the plateaus and domed plateaus that form the prominent summit surface of the Arcoona Plateau and its southerly extension in the Tent Hill region are of Cretaceous age (Twidale, Shepherd, and Thomson, 1970; Twidale, Bourne, and Smith, 1976).

There is no evidence to suggest that these paleoplains and other paleoforms have been buried. But many are of etch type (Wayland, 1934): the present surface represents either an exposed indurated horizon of the weathering profile, or the weathering front (Mabbutt, 1961a), that is, the base of significant weathering, or the junction between the regolith and the intrinsically fresh bedrock. Thus to take a simple example, the so-called "new Plateau" which occupies extensive areas of interior Western Australia is undoubtedly an etch plain formed by the stripping of the lateritic mantle, which developed beneath the higher "old Plateau," and the consequent exposure of the weathering front (Jutson, 1914, p. 96 and following; Mabbutt, 1961b). The laterite-capped surfaces, which give rise to plateau, mesa, and butte assemblages in many parts of southern and northern Australia (see, for example, Whitehouse, 1940; Twidale, 1956a; Wright, 1963; Hays, 1967; Stephens, 1964; Prider, 1966), are themselves etch surfaces, for the exposure of the ferruginous zone implies the stripping of the sandy A-horizon characteristic of modern laterite profiles. This A-horizon is commonly several meters thick, though it varies from place to place (McNeil, 1964; Maignien, 1966), and where present this porous and permeable, though friable, sand cover forms an absorbent cushion protecting the ferruginous horizon. So long as the sands remain vegetated they are washed away only near valleys and escarpments and act as a protective carapace to the land surface for long periods. In north Queensland, the Gilberton Plateau (Twidale, 1956a, 1966a) which forms a prominent divide of Cretaceous sandstone is not greatly dissected. The ferruginized zone of the mid-Tertiary laterite developed on the sandstone is exposed in bounding scarps and in the river gorges, but only a few score meters away the sands, under a eucalypt woodland, still cover the indurated horizon. Elsewhere in the same general region little or no sand remains above the pisolitic ironstone on the Normanton and Donors plateaus which are near the coast and close to the Flinders and Leichhardt rivers.

The silcrete-capped tablelands of central and southern Australia are probably of similar type, though as no modern analogue is known the character of the complete silcrete profile cannot be determined with certainty. A calcrete duricrust has formed on the late Pleistocene Wudinna Surface which occupies much of central and northern Eyre Peninsula (Twidale, Bourne, and Smith, 1976). It has a protective function, and though the A-horizon of calcareous silts remains in situ over wide areas there are also many localities where the calcrete horizon is exposed through the stripping of this cover.

Apart from these paleoplain remnants other landforms seemingly of considerable antiquity and also of etch type survive as integral parts of the contemporary landform assemblage. Thus flared slopes, tafoni, and platforms etched in granite and together indicative of former hill-plain junctions occur at various levels above the present piedmont zone on some of the higher bornhardts of northern Eyre Peninsula (Twidale and Bourne, 1975a). They have been correlated with paleoplain remnants of the adjacent plains and uplands, so that not only are various generations of granite inselberg tentatively identified, but the higher residuals are subdivided into elevational zones the age of which increases with height above the present hill-plain junction. In brief, many granite surfaces and forms date from middle Tertiary times, but some few may be of Mesozoic age. The minor granite landforms mentioned formed at the weathering front beneath the land surface. Hence it can be argued that the former regolithic cover has afforded a measure of protection and that the erosion of the mantle and exposure of the front may have taken a considerable time. But observations of modern erosion suggest that unconsolidated materials such as soil mantles are readily removed by stream action.

Even if it is accepted that estimates of the contemporary rate of degradation of land surfaces are several orders too high (Dole and Stabler, 1909; Judson and Ritter, 1964; see also Gilluly, 1955; Menard, 1961) to provide an accurate yardstick of erosion in the geological past, there has surely been ample time for the very ancient features preserved in the present landscape to have been eradicated several times over. Yet the silcreted land surface of central Australia has survived perhaps 20 m.y. of weathering and erosion under varied climatic conditions, as has the laterite surface of the northern areas of the continent. The laterite surface of the Gulfs region of South Australia is even more remarkable, for it has persisted through some 200 m.y. of epigene attack. The forms preserved on the granite residuals of Eyre Peninsula have likewise withstood long periods of exposure and yet remain recognizably the landforms that developed under weathering attack many millions of years ago.

Models of landscape evolution

The survival of these paleoforms is in some degree an embarrassment to all of the commonly accepted models of landscape development. Perhaps the best-known and most widely favored and applied model of landscape evolution is that of W.M. Davis (1899, 1909) whose deductive scheme involves stream incision accompanied by the gradual lowering of divides by rillwork, wash, and wasting. The end product of these combined activities is the peneplain. One of the several difficulties inherent in the Davisian cycle as propounded by its author is the inevitable elimination of all vestiges of former peneplains following relative lowering of baselevel, stream incision, and lowering of divides. The identification of remnants of uplifted and dissected paleoplains such as have been described from various parts of the world is a denial of the effectiveness of the agents cited as being at work on divides. Theoretically peneplanation and the lowering of interfluves it implies are inimical to the development of multicyclic landscapes and to the survival of paleoforms.

A similar difficulty attends the schemes involving steady state development or dynamic equilibrium such as that derived by Davis (1922) with his concept of the "old-from-birth" peneplain and the more elaborate developments of Hack (1960) and Chorley (1962, 1965). Steady state implies continuous and equal erosion of the entire land surface, once the system is in adjustment. This is contrary to the field evidence of the widespread survival of old land surfaces. Even if it were argued that such survivals denote temporary maladjustments of the system, it is reasonable to ask how long, in view of the great age of some of the paleoforms, can maladjustment prevail before it is considered the norm?

Models based on the retreat of slopes (Penck, 1924, 1953; King, 1953, 1957, 1960) better explain the survival of paleoplains at least in the short term, for the essence of the concept is that high plain or plateau surfaces are destroyed only as rapidly as the scarps heralding the approach of the new cycle and lower surface recede inland: new plains extend at the expense of the old. But the development of flared slopes at successively lower levels on granite inselbergs (Twidale and Bourne, 1975a) and other evidence of piedmont zones, which have changed location very little in time, argues against significant retreat in some areas. Moreover the great age of some of the paleosurfaces surely implies that scarps retreat only very slowly, and the persistence of very old yet steep slopes surely requires demonstration and, if possible, explanation. Some of the models deduced by Kennedy (1962) explain the survival of paleosurfaces through a dominance of stream incision over wasting of divides, but similar problems arise concerning the ineffectiveness of the processes at work on interfluves.

Thus the Davisian and steady state concepts offer no theoretical possibility for the survival of paleoforms, but the scarp retreat hypothesis and some of Kennedy's models do, provided the relative immunity of interfluves to weathering and erosion can be explained.

Survival factors and mechanisms

General statement

The survival of paleoforms must be due either to temporal or spatial variations in the distribution of erosional energy, or to the varied susceptibility of the land surface to erosion, or to both these factors. Estimates of the relative rate of erosion under different climatic conditions have been made (see, for example, Corbel, 1959), but the present is probably not typical of the geological past, and even if it were, difficulties are introduced everywhere by the realities of climatic change. In some regions, it is possible to say that climatic conditions since the formation of a particular form have been conducive to its survival. For instance the arid conditions, which have obtained for much of later Cainozoic time, have no doubt aided the survival of the silcrete surface in central Australia. But the relative significance of arid and humid phases cannot yet be evaluated, partly because the duration of the two is not known, partly because the relative effect of brief downpours and periods of river flow compared to more regular rainfall and flow is difficult to assess in absolute terms. Rather can it be argued that significant erosion takes place largely during periods of climatic change, when the land surface is in geomorphological disequilibrium.

But by and large no generalizations can as yet be made about climatic control of erosional rates in the past.

Various possible factors involving spatial variation are discussed principally in respect to paleosurface remnants that occur in South Australia, though reference is made to other regions. In broad terms these are diastrophism and structure, the nature of river work, and autocatalytic or reinforcement mechanisms. The significance of some of these was perceived as long ago as 1924 by Eleanora Bliss Knopf who published a paper concerned with erosion surfaces in the eastern Appalachians. In it she wrote (p. 667):

> The effect of (such) repeated uplift is to postpone indefinitely the total degradation of a region by constant addition to the area undergoing reduction and by setting back the progress of the work . . . after each interruption.

This statement is both perceptive and provocative. Perceptive because it is consistent with field evidence in many diastrophically disturbed areas such as the Flinders and Mount Lofty ranges, and provocative because by implication it calls to question the basic tenets of hypotheses of landform development which were as sacrosanct half a century ago as they are now.

Diastrophism and structure

Linton (1957, p. 67) was aware of the significance of diastrophic rejuvenation. He wrote that "it is unlikely that any mountains can endure as such for more than about twenty million years unless they are raised up anew." Both the Mount Lofty and the Flinders ranges have suffered recurrent uplift, mainly along old-established fracture zones, through the Phanerozoic (Campana, 1958a and b). Both are still seismically active (Sutton and White, 1968). Both uplands are characterized by intense marginal dissection, with complex valley-in-valley forms indicative of repeated phases of rejuvenation. On the western escarpment of the Mount Lofty Ranges perched benches indicate the former piedmont zone now carried high above the present hillplain junction by faulting. In both uplands there are high plains or remnants of paleoplains located in core regions: that in the Mount Lofty Ranges comprises residuals capped by the early Mesozoic laterite together with its etch-plain equivalent (Twidale, 1968, p. 315–316; Twidale and Bourne, 1975b); in the northern Flinders Ranges the high plain is demonstrably an exhumed surface which predates the Lower Cretaceous (Woodard, 1955; Twidale, 1969), but in the central and southern regions (Twidale, 1966, 1969) there is no evidence of Cretaceous sedimentation, and the high plain remnants there, which cut across folded argillaceous sediments, are more likely to be of epigene type eroded by rivers which graded to the Lower Cretaceous shoreline.

Thus in these regions the field evidence conforms to the theoretical consequences of repeated uplift. Yet in both the preservation of the paleoplain remnants involves more than repeated faulting. For example the central Flinders Ranges is essentially a simple denuded anticline from which at least

6,000 m of sediment has been eroded. Sandstone and limestone are prominent high in the stratigraphic sequence and are thus exposed near the margins of the structure where they form ridges and ranges. It is these outcrops and uplands, not fault zones, that delimit and delineate the uplands, and being marginally located they to some extent buttress the weaker sediments exposed between them by acting as local baselevels of erosion for the streams that drain radially from the Flinders. Profound erosion of the anticline has exposed the deep core of the structure which was in compression, where the joints are tight and where the strata are thus resistant to weathering and erosion. It is not by chance that the high plain surface is best preserved in the central Flinders Ranges where several structural factors combine with positive diastrophism to render the land surface less vulnerable to erosion.

The Mount Lofty Ranges are in many respects similar. The upland is developed on an anticline the flanks of which are faulted to form a complex horst. The bulk of the laterite plateau and high plain is centrally located close to the compressional zone but there are many remnants in marginal locations, near the fault scarps. In the west these survivals can partly be explained in terms of their being buttressed by sandstone and limestone outcrops, but in the east they are preserved on gneiss and schist which are not notably resistant to weathering and erosion.

Of course, the ferruginous zone of the laterite, being commonly pisolitic and vesicular, is permeable and porous and resistant to erosion. Moreover it retains the protective A-horizon in many places. But the kaolinized zone beneath the iron rich horizon is readily eroded, causing undermining and collapse of the whole slope. This indeed is the reason for the extensive development of the etch plain in the upland, and particularly in the east.

Localization of river erosion

In both the Flinders and the Mount Lofty ranges repeated diastrophism and structural factors are likely to be responsible for the core survivals in the uplands. In neither area can the preservation of paleoplains be attributed to interior location, for both are long narrow uplands no part of which is far removed from baselevel.

But in other places remoteness from the sea or from effective stream action is an important factor contributing to the preservation of paleoplains. Thus in southern Africa the land mass is so large that rivers only slowly extend their maximum influence in the interior, and it is significant that the old planation surfaces there are distributed in roughly concentric zones with the older surfaces located in the deepest interior (King, 1960). Again remoteness from river action can be cited in explanation of the survival of the prominent summit surface of etch character and Cretaceous age preserved in the Gawler Ranges (Twidale, Bourne, and Smith, 1976): the area is located inland and is surrounded by plains that are either climatically or lithologically arid (that is, underlain by limestone or other permeable rocks). The high plains of central and western Australia may be preserved for similar reasons.

But distance fom the ocean or from major drainage lines is not the complete explanation for such survivals. Ritter (1968) has pointed out that not only are there considerable variations in the present rate of erosion in the several major catchments of the continental United States but has also indicated that there are major contrasts between different parts of the same catchment. This suggestion is borne out by observations in the Flinders Ranges, for example, where lithological variations largely determine drainage density and hence degree of landscape dissection and paleoplain survival.

Why do some rivers develop long and deep valleys but extend their influence only a short distance from their channels? Knopf (1924, p. 667) was clearly aware of the contradiction between the evidence of some landscapes and the views expressed by the doyen of American geomorphologists and perhaps the most authoritative voice of the day, W.M. Davis. Davis considered that the movement of waste on divides and in rivers are "extreme members of a continuous series" (Davis, 1909, p. 267). He regarded both as moving mixtures of waste and water in variable proportions, and in consequence he saw the "river" as extending "all over its basin and up to its very divides" (Davis, 1909, p. 267). "Ordinarily treated" he wrote "the river is like the veins of a leaf; broadly viewed, it is like the entire leaf." Yet Knopf implied that the various processes at work on the valley side slopes and interfluves are not as effective in lowering the land surface as are trunk streams. And she was not alone in this view, for Crickmay (1932, 1959, 1968, 1969, 1971) has maintained much the same argument.

The reasons for such a contrast between the rate of lowering of stream beds and the intervening divides are surely related to the nature of water flow at the two sites. On the interfluves some precipitation infiltrates into the subsurface and makes its way underground to the stream lines. Part of the surface runoff takes the form of diffuse flow, and any linear flows are of low volume. All these are, however, brought together in the trunk streams which are thus of much greater capacity than any of the rills and streamlets at work on the divides. Moreover the catchments of the divide streams are very limited, whereas the trunk streams gather runoff from large and probably extending areas upstream.

The contrast in the rate of lowering of divides and of stream beds is most pronounced in areas of permeable and pervious rocks. Basalt is resistant to wash by virtue of its well developed jointing which permits water to infiltrate into the subsurface so that interfluves and indeed the major part of the basalt surface are eroded only very slowly. But Waters (1955, p. 676) reports that in the Yakima area of Washington State where there are major linear flows joint blocks are plucked away by the river in spate and it is not uncommon for deep gorges to be eroded in and through basalt formations.

Similar deep gorges incised below virtually undissected basalt plateaus occur also in north Queensland (Twidale, 1956b). Some sandstone and limestone areas are resistant by virtue of their ability to swallow water: they are lithological deserts. Thus in the area west of Sydney the high plains and plateaus underlain by flat-lying Mesozoic strata, including massive sandstone formations, are incised by such rivers as the Hawkesbury, Wollondilly, Colo, and Shoalhaven. The upland summit plain is of considerable antiquity, but whatever its precise age the valleys are deep and long but only narrow. Over much of its middle section, for example, the Shoalhaven is only 2 km wide but 500 m deep: a feature that can be termed a gorge, bearing in mind that few valleys are deeper than they are wide (Johnson, 1932).

The River Murray in South Australia is an exotic or allogenic stream which flows in a trench some 25 m deep and 1 km wide and is excavated essentially in flat-lying massively bedded Miocene calcarenites. The widening is due almost wholly to lateral corrasion by the river during post Pliocene times. During the last 12 to 15 m.y. wash, weathering, tributaries, and mass movements on the slopes have achieved little significant retreat of the valley sides (Twidale, 1968, p. 171-175).

Similar situations obtain in different structural settings. Many rivers have carved gorges in resistant members of folded sequences, such as quartzites, as a result of diversion, antecedence, superimposition, or valley impression (Twidale, 1972a). The adjacent ridges remain upstanding and bear testimony to the localized nature of river erosion, for some of the gorges can be correlated with planation surfaces of considerable antiquity (see, for instance, Twidale, 1966b). But even the very recently developed adjustments of drainage illustrate the same point: the famous antecedent drainage channel through the Shaur Anticline in western Iran has cut a minor gorge some 4 m deep in 1,200 yrs, but during that same period there has been no significant valley widening (Lees, 1955, p. 224).

Channels eroded in unconsolidated rocks display similar forms. Gullies cut in unconsolidated sediments such as clays and fanglomerates are invariably trough-shaped with precipitous and in places overhanging sides. This is so whether the gullies are 10 cm or 10 m deep and is independent of the presence of a capping of turf, calcrete, or gibber. In some instances the gullies are eroded in crumbly clays which crack and swell easily and which can offer only minimal resistance to erosion; yet the sidewalls stand in cliffs.

Thus both in cohesive and in unconsolidated rocks, there is strong suggestion that though stream beds are rapidly worn down bank erosion and valley side erosion are less effective.

There is also some evidence that river systems accomplish only slight erosion at the margins of their catchments and that significant back-wearing is restricted to the stream channels that penetrate to the periphery. In many areas, and in various structural situations, there are well-developed zones of scarp-foot weathering which surely argue an essentially stable location of the escarpment and of the scarp-foot for a considerable period of time. Thus in the Lincoln Gap area west of Port Augusta the flat-lying Precambrian strata have been dissected to form plateaus, mesas, and buttes. Three major planation surfaces have been identified in the region (Twidale, Shepherd, and Thomson, 1970). The youngest, of only limited extent, is related to present baselevel. The second, characterized by pediments carrying a discontinuous veneer of silcrete, stands 5 to 10 m above the first, is probably of mid Tertiary age, and is commonly separated from the backing scarp by a scarp-foot valley. The scarps lead up to the third and oldest surface, the prominent summit high plain of the Arcoona Plateau which is of Cretaceous age.

The dissection of the silcrete surface has taken place since the Miocene. The valleys are up to 3 km, but more commonly 1 km, wide so that lateral erosion due to all forces has proceeded at a maximum rate of about 0.15 mm per annum. During this time the backing scarps below the Cretaceous surface have receded by only a few score meters at most, for the scarp-foot valleys and associated weathering zones remain clearly in juxtaposition with the escarpments. There is no doubt that scarps have receded due to slumping, rock falls, and gullying, but it is localized and slow overall. The presence of silcrete in the marginal zones of drainage basins and particularly in scarp-foot situations (Bassett, 1954; Twidale, Shepherd, and Thomson, 1970; Hutton, Twidale, Milnes, and Rosser, 1972) is also indicative of topographic stability, for the evidence suggests that silcrete forms only very slowly and under essentially static conditions (Hutton, Twidale, and Milnes, 1976).

Thus although the case should not be overstated (for catchments have obviously been enlarged and stream piracy has occurred) there is much to suggest that headward stream erosion proceeds at a much faster rate than general scarp retreat.

Feedback mechanisms

Slope retreat is the dominant mode of slope modification in many areas. Particularly is this so where slope behavior is controlled by the presence of a caprock (Tricart, 1957; Twidale, 1960). In these conditions the rate of scarp retreat is largely determined by the activity of self-regulatory mechanisms. Though scarp-foot weathering induces slope instability, collapse, and hence recession (Twidale, 1960, 1967a and b), it is gullying that is in large measure responsible for the regrading of scarps. Coarse debris from the bluff accumulates in gullies, the beds of which are thus protected against further erosion. The low divides between gullies are however unprotected and are eroded; they in turn are lowered and become receptacles for coarse debris from the bluff above. This process of gully gravure (Bryan, 1940) involving the alternation of the locus of intense erosion through the protective effects of coarse debris derived from the bluff effectively slows the rate of overall retreat of the scarp.

Thus the stripping of the laterite capping of the southern Mount Lofty Ranges can be shown to have taken place at a maximum of 1 cm in 200 yrs or 0.05 mm per annum, if it is assumed that the laterite originally extended to the margins of the fault blocks, as the present distribution of remnants suggests, and if in order to maximize the rate of recession it is assumed that the drainage was established de novo on the laterite plateau. This is more comparable with rates of slope erosion derived from observations in hot arid regions than with those derived from the humid tropics. Scarps bordering the Grand Canyon have retreated at a rate of only 0.06 cm per annum (Schumm and Chorley, 1966), and sandstone cliffs in Colorado appear to recede as a result of rainwash at a rate of 0.005 mm a year (Schumm, 1964). These contrast strongly with contemporary rates of erosion in the humid tropics: almost 8 cm per annum on 40° to 50° slopes in Oahu (Wentworth, 1943); 1 cm in 7 to 10 yrs in one area and in 44 yrs in another in New Guinea (Simonett, 1967); and 0.1 to 10 cm (variable) per century over an entire catchment in central Japan (Yoshikawa, 1974).

This contrast may in part be explained by the self regulatory mechanism described above in the context of gully gravure. For in the humid tropics weathering is so efficient that little or no coarse debris survives to afford the ephemeral but significant protection.

A different type of feedback mechanism involving reinforcement or autocatalytic effects has also been invoked to explain the preservation of paleoplains on sandstone and other resistant massifs within fold mountain belts, in areas of horizontal strata (Twidale, Bourne, and Smith, 1974), and of paleoforms on granite and other inselbergs such as those described from northwestern Eyre Peninsula (Twidale and Bourne, 1975a).

The ridges and ranges of fold mountain belts are built mainly of intrinsically resistant rocks, but once they are in relief they shed water to the adjacent valleys or plains. The latter are eroded in weaker rocks anyway, but they receive a disproportionately high run-off and are thus more and more vulnerable to weathering and erosion. Thus there is a tendency not only for relief amplitude to increase, as has been suggested for the Flinders Ranges through the Cainozoic (Twidale, 1966b), but for remnants of paleoplains developed on sandstone for instance to be long preserved. Such residuals of the late Mesozoic exhumed surface also survive on quartzites of the fold sequence involving Precambrian rocks in the Isa Highlands (Twidale, 1956a). Plateaus capped by sandstone also shed water to the plains and themselves are little affected by it, though there is some sapping of the caprock and consequent collapse and recession of the bluffs. Granite residuals of various shapes and sizes owe their development fundamentally to their effectively massive structure and to their propensity for shedding water to the adjacent plains, but the nature of granite and its reaction to moisture must also be borne in mind. It has long been known that granite is susceptible to attack by moisture, but that it is virtually inert when dry (Barton,

1916; Bain, 1923; Griggs, 1936; Wahrhaftig, 1965). This compounds the tendency to the preservation of inselbergs and similar forms and the weathering and lowering of plains on compartments of more densely jointed granite.

Summary

The gravamen of this discussion has been that several factors, including lithology, structure, diastrophism, the unequal activity of rivers within their catchments, and self regulatory and autocatalytic mechanisms all contribute to the survival of paleoforms. Where several occur in conjunction, as they do on the western side of the Mount Lofty Ranges where there are massive sandstones, a lateritic carapace, recurrent positive diastrophism, protective alluvial aprons, and unequal stream activity, landforms, and land surfaces can survive long periods of weathering and erosion. Many of the paleosurfaces, which have not been exhumed, are admittedly of etch type, but their survival through long periods of geological time is nevertheless remarkable.

An alternative model

Models of landscape evolution either take insufficient or no account of the mechanisms and survivals outlined in this paper. The occurrence in the contemporary land surface of considerable areas of paleoplain assemblages is quite alien to some of these concepts, and though more conducive to survival than other models scarp retreat alone does not explain the continued presence of very old paleosurfaces in the landscape.

In an earlier discussion of the development of the Flinders Ranges (Twidale, 1966b, p. 26) the term "persistence of relief" was used to describe the longevity of the basic pattern of relief through the Cainozoic in the southern uplands despite profound erosion. It is not suggested that the ridges have suffered no lowering, but that they have been degraded so slowly that old land surfaces survive on them, and that relief amplitude has increased. Nor is it argued that the planimetric location of ridges and valleys has gone unchanged, for deep erosion of folds has brought about the migration of ridges and valleys (Twidale, 1972a) and even the advance of escarpments (Twidale, 1927b). But the basic pattern of relief has remained for some 70 m.y.

In similar fashion reinforcement mechanisms have been invoked in explanation of the survival of the various inselbergs of northwestern Eyre Peninsula through at least the Cainozoic and in some instances possibly through the last 200 m.y. of earth history (Twidale and Bourne, 1975a).

Such survivals of landforms are embraced in a model characterized by persistent and increasing relief, a model in which initial structural contrasts are accentuated by weathering and erosion of weaker zones (fig. 2). Continued exploitation of weaker zones is made possible by recurrent uplift, by

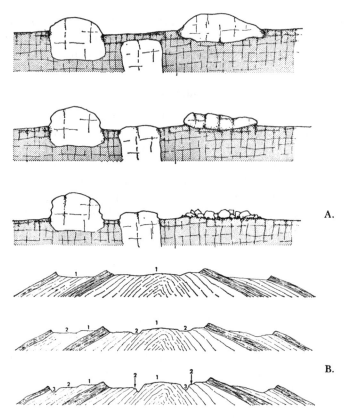

Figure 2 Suggested models of landscape development: (A) granite—stippled areas represent weathered granite; (B) fold mountain range—1. oldest paleoplain, 2. second paleoplain, 3. present surface of low relief.

restricted effectiveness of river systems, and by reinforcement effects. This model is offered not as one of universal application but as one of the several, outlined earlier, which should be considered in the light of the field evidence.

It may for instance be necessary to differentiate between landscape evolution in humid lands, where erosion of slopes is relatively active and where divides are seemingly being consumed at a measurably rapid rate (Wentworth, 1943; Simonett, 1967), and arid or semiarid environments (either climatic or lithological) where under comparable conditions of anthropogenically induced disequilibrium, slopes are being eroded only very slowly (see Schumm, 1964; Schumm and Chorley, 1966). In humid regions, paleoforms do not long survive; in arid zones, they may. Thus it is that the tropical and subtropical deserts are characterized by extensive survivals of paleoforms, while they persist in the humid lands only where structural conditions are favorable, as for instance on the chalk uplands of southeastern Britain (Wooldridge and Linton, 1939) and in regions, such as the Labrador Peninsula where coastal indentations are few, where recent dissection is marginal, and where the high plain which occupies much of the Peninsula is an etch surface representing a preglacial weathering front stripped of its regolithic cover by glacial erosion during the Pleistocene.

It has been said that "The essence of geomorphology is the discrimination of the ancient from the modern" (Bryan, 1950, p. 198). It may be that Bryan did not appreciate how ancient are some facets of the modern land surface, but certain it is that studies of denudation chronology at their best furnish a perspective on the evolution of the contemporary landscape without which our comprehension and appreciation would be sadly diminished. Whether such investigations are viewed as historical geology, as Bryan averred, and whether or not they are regarded as geomorphological, is, *pace* Chorley (1965), irrelevant. Even if the conclusions reached by many workers over the years are only partly correct, it is clear that remnants of paleoforms are an integral part of the modern land surface, particularly in tropical arid and semiarid regions. Their persistence provides valuable insights into the character of erosional processes. The hills are not everlasting as Jacob implied (*Genesis*, 49, 26), but they persist for much longer periods than has been generally conceded.

Acknowledgments

The author extends his thanks to Dr. Colin H. Crickmay of Calgary, Professor W.C. Bradley of Colorado, and to his Research Assistant, Jennie Bourne, for their useful and constructive criticisms of this paper. The opinions expressed are however the responsibility of the writer.

Note

1 For location map showing Australian locations mentioned, see figure 1.

References

Ambrose, J.W., 1964, Exhumed paleoplains of the Precambrian Shield of North America: Am. Jour. Sci., v. 262, p. 817–857.

Bain, A.D.N., 1923, The formation of inselberge: Geol. Mag., v. 60, p. 97–101.

Barton, D.C., 1916, Notes on the disintegration of granite in Egypt: Jour. Geology, v. 24, p. 382–393.

Bassett, H., 1954, Silification of rocks by surface waters: Am. Jour. Sci., v. 252, p. 733–735.

Brown, D.A., Campbell, K.S.W., and Cook, K.A.W., 1968, The geological evolution of Australia and New Zealand: Sydney, Pergamon Press, 409 p.

Bryan, K., 1940, Gully gravure—a method of slope retreat: Jour. Geomorphology, v. 3, p. 89–107.

—— 1950, The place of geomorphology in the geographic sciences: Assoc. Am. Geographics Annals, v. 40, p. 196–208.

Campana, B., 1958a, The Mt. Lofty-Olary region and Kangaroo Island, in Glaessner, M.F., and Parkin, L.W., eds., The geology of South Australia: Melbourne, Melbourne Univ. Press, p. 3–27.

—— 1958b, The Flinders Ranges, in Glaessner, M.F., and Parkin, L.W., eds., The geology of South Australia: Melbourne, Melbourne Univ. Press, p. 28–45.

Carter, E.K., and Öpik, A.A., 1961, Lawn Hill—4 mile geological series: Australia Bur. Mineral Resources Geology and Geophysics Expl. Notes, v. 21, 17 p.

Chorley, R.J., 1962, Geomorphology and general systems theory: U.S. Geol. Survey Prof. Paper 500-B, 10 p.

—— 1965, A re-evaluation of the geomorphic system of W.M. Davis, in Chorley, R.J., and Haggett, P., eds., Frontiers in geographical teaching: London, Methuen, 379 p.

Choubey, V.D., 1972, Pre-Deccan trap topography in central India and coastal warping in relation to Narmada Rift structure and volcanic activity: Bull. volcanol., v. 25, p. 660–685.

Corbel, Jean, 1959, Vitesse d'érosion: Zeitschr. Geomorphologie, v. 3, p. 1–28.

Cowie, J.W., 1961, Contributions to the geology of North Greenland: Medd. Grønland, v. 164, 47 p.

Crickmay, C.H., 1932, The significance of the physiography of the Cypress Hills: Canadian Field Naturalist, v. 46, p. 185–186.

—— 1959, A preliminary inquiry into the formulation and applicability of the geological principle of uniformity: Calgary, Crickmay, 50 p.

—— 1968, Some central aspects of the scientific study of scenery: Calgary, Crickmay, 36 p.

—— 1969, The art of looking at broad valleys: Calgary, Crickmay, 21 p.

—— 1971, The role of the river: Calgary, Crickmay, 30 p.

Daily, B., Twidale, C.R., and Milnes, A.R., 1974, The age of the lateritised summit surface on Kangaroo Island and adjacent areas of South Australia: Geol. Soc. Australia Jour., v. 21, p. 387–392.

Davis, W.M., 1899, The geographical cycle: Geog. Jour., v, 14, p. 481–504.

—— 1909, Geographical essays: Boston, Dover, 777 p.

—— 1922, Peneplains and the geographical cycle, Geol. Soc. America Bull., v. 33, p. 587–598.

Dole, R.B., and Stabler, H., 1909, Denudation: U.S. Geol. Survey Water Supply Paper 234, p. 78–93.

Firman, J.B., 1969, Quaternary period, in Parkin, L.W., ed., Handbook of South Australian geology: Adelaide, South Australia Geol. Survey, p. 204–233.

Gilluly, James, 1955, Geologic contrasts between continents and ocean basins: Geol. Soc. America Spec. Paper 62, p. 7–18.

Griggs, D.T., 1936, The factor of fatigue in rock exfoliation: Jour. Geology, v. 44, p. 783–796.

Hack, J.T., 1960, Interpretation of erosional topography in humid temperate regions: Am. Jour. Sci., v. 258A, p. 80–97.

Hays, J., 1967, Land surfaces and laterites in the north of the Northern Territory, *in* Jennings, J.N., and Mabbutt, J.A., eds., Landform studies from Australia and New Guinea: Canberra, Australian Natl. Univ. Press, p. 182–210.

Hutton, J.T., Twidale, C.R., and Milnes, A.R., 1976, Characteristics and origin of some Australian silcretes, *in* Langford-Smith, T., ed., Silcrete in Australia: Armidale, Univ. New England Press, in press.

Hutton, J.T., Twidale, C.R., Milnes, A.R., and Rosser, H., 1972, Composition and genesis of silcretes and silcrete skins from the Beda Valley, southern Arcoona Plateau, South Australia: Geol. Soc. Australia Jour., v. 19, p. 31–39.

Johnson, D.W., 1932, Streams and their significance: Jour. Geology, v. 40, p. 481–497.

Judson, S., and Ritter, D., 1964, Rates of regional denudation in the United States: Jour. Geophys. Research, v. 69, p. 3395–3401.

Jutson, J.T., 1914, The physiography (geomorphology) of Western Australia: Western Australia Geol. Survey Bull. 95, 366 p.

Kennedy, W.Q., 1962, Some theoretical factors in geomorphological analysis: Geol. Mag., v. 99, p. 304–312.

King, L.C., 1940, South African scenery: Edinburgh, Oliver & Boyd, 379 p.

—— 1950a, The study of the world's plainlands: Geol. Soc. London Quart. Jour., v. 106, p. 101–131.

—— 1950b, The cyclic land surfaces of Australia: Royal Soc. Victoria Proc., v. 62, p. 79–95.

—— 1953, Canons of landscape evolution: Geol. Soc. America Bull., v. 64, p. 721–752.

—— 1957, The uniformitarian nature of hillslopes: Geol. Soc. Edinburgh Trans., v. 17, p. 81–102.

—— 1960, Morphology of the earth: Edinburgh, Oliver & Boyd, 699 p.

Knopf, E.B., 1924, Correlation of residual erosion surfaces in the eastern Appalachian highlands: Geol. Soc. America Bull., v. 35, p. 633–668.

Langford-Smith, T., and Dury, G.H., 1965, Distribution and character and attitude of the duricrust in the northwest of New South Wales and the adjacent areas of Queensland: Am. Jour. Sci., v. 263, p. 179–190.

Lees, G.M., 1955, Recent earth movements in the Middle East: Geol. Rundschau, v. 42, p. 221–226.

Linton, D.L., 1957, The everlasting hills: Adv. Sci., v. 14, p. 58–67.

Mabbutt, J.A., 1961a, 'Basal surface' or 'weathering front': Geol. Assoc. London Proc., v. 72, p. 357–358.

—— 1961b, A stripped land surface in Western Australia: Inst. British Geographers Pub. (Trans. Papers) 29, p. 101–114.

—— 1965, The weathered land surface in central Australia: Zeitschr. Geomorphologie, v. 9, p. 82–114.

—— 1967, Denudation chronology in central Australia; structure, climate, and landform inheritance in the Alice Springs area, *in* Jennings, J.N., and Mabbutt, J.A., eds., Landform studies from Australia and New Guinea: Canberra, Australian Natl. Univ Press, p. 144–181.

Maignien, R., 1966, Review of research on laterites: Nat. Resources Research (UNESCO), v. 4, 148 p.

McNeil, Mary, 1964, Lateritic soils: Sci. Am., v. 211 (NN), p. 97–102.

Menard, H.W., 1961, Some rates of regional erosion: Jour. Geology, v. 69, p. 154–161.

Öpik, A.A., Carter, E.K., and Randal, M.A., 1973, Notes on the first edition Camooweal geological sheet, Queensland, 1961: Australian Bur. Mineral Resources Geology and Geophysics Rec., v. 83, 27 p.

Penck, W., 1924, Die morphologische Analyse: Stuttgart, Engelhorns, 283 p.

—— 1953, Morphological analysis of land forms, translated by Czeck, H., and Boswell, K.C.: London, Macmillan, 429 p.

Prider, R., 1966, The lateritized surface of Western Australia: Australian Jour. Sci., v. 28, p. 443–451.

Ritter, D.F., 1968, Continental erosion, in Fairbridge, R.W., Encyclopaedia of geomorphology: New York, Reinhold, p. 169–174.

Robinson, G., 1963, A consideration of the relations of geomorphology and geography: Prof. Geography, 15, p. 13–17.

Sauer, C.O., 1925, The morphology of landscape: Univ. Calif. Pub. in Geography, v. 2, p. 19–53.

Schumm, S.A., 1964, Seasonal variations of erosion rates and processes on hillslopes in western Colorado: Zeitschr. Geomorphologie, Suppl. 5, p. 215–238.

Schumm, S.A., and Chorley, R.J., 1966, Talus weathering and scarp recession in the Colorado plateaus: Zeitschr. Geomorphologie, v. 10, p. 11–36.

Simonett, D.S., 1967, Landslide distribution and earthquakes in the Bewani and Torricelli mountains, New Guinea, in Jennings, J.N., and Mabbutt, J.A., eds., Landform studies from Australia and New Guinea: Canberra, Australian Natl. Univ. Press, p. 64–84.

Sprigg, R.C., 1963, Geology and petroleum prospects of the Simpson Desert: Royal Soc. South Australia Trans., v. 86, p. 35–65.

Stephens, C.G., 1964, Silcretes of central Australia: Nature (London), v. 203, no. 4952, p. 1407.

—— 1971, Laterite and silcrete in Australia: Geoderma, v. 5, p. 5–52.

Sutton, D.J., and White, R.E.. 1968, The seismicity of South Australia: Geol. Soc. Australia Jour., v. 15, p. 25–32.

Thornbury, W.D., 1954, Principles of geomorphology: New York, John Wiley & Sons, Inc., 618 p.

Tricart, J., 1957, Misè au point: l'évolution des versants: L'inform. Geog., v. 21, p. 108–115.

Twidale, C.R., 1956a, Chronology of denudation in northwest Queensland: Geol. Soc. America Bull., v. 67, p. 867–982.

—— 1956b, Physiographic reconnaissance of some volcanic provinces in North Queensland, Australia: Bull. volcanol., v. 7, p. 3–23.

—— 1960, Some problems of slope development: Geol. Soc. Australia Jour., v. 6, p. 131–147.

—— 1966a, Geomorphology of the Leichhardt-Gilbert area, northwest Queensland: Melbourne, CSIRO, Land Research Ser., v. 16, 56 p.

—— 1966b, Chronology of denudation in the southern Flinders Ranges, South Australia: Royal Soc. South Australia Trans., v. 90, p. 3–28.

—— 1967a, Hillslopes and pediments in the Flinders Ranges, South Australia, in Jennings, J.N., and Mabbutt, J.A., eds., Landform studies from Australia and New Guinea: Canberra, Australian Natl. Univ. Press, p. 95–117.

—— 1967b, Origin of the piedmont angle as evidenced in South Australia: Jour. Geology, v. 75, p. 393–411.

—— 1968, Geomorphology, with special reference to Australia: Melbourne, Nelson, 406 p.

—— 1969, Geomorphology of the Flinders Ranges, in Corbett, D.W.P., ed., Natural History of the Flinders Ranges: Adelaide, Southern Australia Library Pub., p. 57–137.

—— 1972a, The neglected third dimension: Zeitschr. Geomorphologie, v. 16, p. 283–300.

—— 1972b, Landform development in the Lake Eyre region, Australia: Geog. Rev., v. 62, p. 40–70.

Twidale, C.R., and Bourne, J.A., 1975a, Episodic exposure of inselbergs: Geol. Soc. America Bull., v. 86, p. 1473–1481.

—— 1975b, Geomorphological evolution of part of the eastern Mount Lofty Ranges, South Australia: Royal Soc. South Australia Trans., v. 99, in press.

—— 1976, Age and origin of palaeosurfaces on Eyre Peninsula and the southern Gawler Ranges, South Australia: Zeitschr. Geomorphologie, in press.

Twidale, C.R., Bourne, J.A., and Smith, D.M., 1974, Reinforcement and stabilisation mechanisms in landform development: Rev. Géomorphologie Dynamique, v. 23, p. 115–125.

—— 1976, Age and origin of palaeosurfaces on Eyre Peninsula and the southern Gawler Ranges, South Australia: Zeitschr. Geomorphologie, in press.

Twidale, C.R., Shepherd, J.A., and Thomson, R.M., 1970, Geomorphology of the southern part of the Arcoona Plateau and the Tent Hill region west and north of Port Augusta, South Australia: Royal Soc. South Australia Trans., v. 94, p. 55–67.

Vogt, Jean, 1953a, Un problème morphologique du bouclier canadien: le relief granitique: Rev. Géomorph. Dynamique, 4th année, p. 85–95.

—— 1953b, Un nouvel exemple d'inversion de relief aux dépens d'un batholite granitique: Rev. Géomorph. Dynamique, 4th année, p. 231–235.

Wahrhaftig, C., 1965, Stepped topography of the southern Sierra Nevada, California: Geol. Soc. America Bull., v. 76, p. 1165–1190.

Waters, A.C., 1955, Geomorphology of south-central Washington, illustrated by the Yakima East Quadrangle: Geol. Soc. America Bull., v. 66, p. 663–684.

Wayland, E.J., 1934, Peneplains and some erosional platforms: Uganda Geol. Survey Annual Rept. Bull., p. 77–79.

Wentworth, C.K., 1943, Soil avalanches on Oahu, Hawaii: Geol. Soc. America Bull., v. 54, p. 53–64.

Whitehouse, F.W., 1940, Studies in the late geological history of Queensland: the lateritic soils of western Queensland: Queensland Univ. Papers, Dept. Geology, v. 2 (new ser.), no. 1, p. 2–22.

Woodard, G.D., 1955, The stratigraphic succession in the vicinity of Mt. Babbage Station, South Australia: Royal Soc. South Australia Trans., v. 78, p. 8–17.

Wooldridge, S.W., and Linton, D.L., 1939, Structure, surface and drainage in South-East England: Inst. British Geographers Pub. (Trans. Papers) 10, p. 1–124.

Wopfner, H., 1960, On some structural development in the central part of the Great Australian Artesian Basin: Royal Soc. South Australia Trans., v. 83, p. 179–193.

—— 1964, Permian-Jurassic history of the western Great Artesian Basin: Royal Soc. South Australia Trans., v. 88, p. 117–128.

Wopfner, H., Callen, R., and Harris, W.K., 1974, The lower Tertiary Eyre formation of the southwestern Great Artesian Basin: Geol. Soc. Australia Jour., v. 21, p. 17–51.

Wopfner, H., and Twidale, C.R., 1967, Geomorphological history of the Lake Eyre Basin, in Jennings, J.N., and Mabbutt, J.A., eds., Landform studies from Australia and New Guinea: Canberra, Australian Natl. Univ. Press, p. 118–143.

Wright, R.L., 1963, Deep weathering and erosion surfaces in the Daly River Basin, Northern Territory: Geol. Soc. Australia Jour., v. 10, p. 151–163.

Yoshikawa, T., 1974, Denudation and tectonic movement in contemporary Japan: Tokyo Univ. Geography Dept. Bull. 6, p. 1–14.

8

LANDSCAPE DISSECTION, ISOSTATIC UPLIFT, AND THE MORPHOLOGIC DEVELOPMENT OF OROGENS

*A.R. Gilchrist, M.A. Summerfield
and H.A.P. Cockburn*

Source: *Geology* 22 (1994): 963–6.

Abstract

We examine the morphologic factors that determine the isostatic response to landscape dissection in orogenic terrains in order to quantify the relative contributions of active tectonic uplift and passive isostatic uplift in generating mountain topography. We demonstrate that although peaks many times the elevation of an assumed initial plateau can theoretically be generated, only moderate amounts of peak uplift can be explained by isostatic rebound in real terrains. Analysis of topographic data from the European Alps and other orogens indicates that isostatic uplift in response to dissection can account for about one-fourth of the elevation of the highest peaks if local isostatic compensation is assumed and less if the lithosphere has significant flexural rigidity. We conclude that although isostatic uplift can be significant in orogens, high peaks are predominantly a consequence of tectonic processes in convergent settings.

Introduction

The isostatic response to denudation in orogenic terrains is important for several reasons. First, and most obviously, the pattern of isostatic response influences the overall morphology of an orogen. Second, denudational unloading leads to passive crustal uplift that must be distinguished from active uplift driven directly by tectonics in studies of the dynamics of mountain

225

building (Burbank, 1992a). Indeed, it has been suggested that episodes of enhanced rates of denudation associated with late Cenozoic climatic change could induce increased rates of crustal uplift that may have been previously misinterpreted as the result of active tectonic processes (Molnar and England, 1990). Third, isostatic uplift can lead to localized absolute increases in surface elevation of mountain peaks and ridges in dissected terrain while the regional mean altitude of an orogen is reduced by denudation. This raising of mountain crests, it has been argued, could, in turn, induce climatic cooling of the kind observed during the late Cenozoic (Molnar and England, 1990).

It has long been acknowledged that the dissection of a mountain belt can lead to an absolute increase in the elevation of its summits, but quantitative analyses have been lacking. Wager (1933) argued that the elevation difference between peaks in the eastern Himalayas and the Tibetan Plateau, with a mean elevation of ~5 km, could be accounted for by the isostatic response to incision of the plateau edge, although Holmes (1965) thought this effect alone to be insufficient to account for the highest peaks (~8,500 m). More recently, Molnar and England (1990) have suggested that the isostatic response to dissection may have raised the highest peaks in some orogens up to two or more times the mean regional elevation of preexisting low local-relief terrain. Our aims here are to assess the morphologic factors determining the isostatic response to dissection and to evaluate the importance of isostasy in generating high peaks in orogenic terrains. We focus on the central European Alps as an example of a small, deeply dissected orogen and briefly discuss the implications of this study for larger orogens such as the Himalayas and the Andes that include high, intraorogen plateaus.

Describing landscape dissection

Relief, or more accurately local relief, is dependent on the scale of observation and generally increases with area. The morphometry of landscapes can be described simply as a function of elevation by the minimum (e_{min}), mean (e_{mean}), and maximum (e_{max}) values within a spatial domain at a specified resolution. It is useful to consider topography in these terms because many digital data sets now available characterize terrain in such a format. Local relief (r) is the maximum depth of valley incision ($e_{max} - e_{min}$), and the mean depth of valley incision is here termed the depth of dissection (d), i.e., the amount of material required to infill a valley to the top of the surrounding peaks ($e_{max} - e_{mean}$). Clearly, d controls the isostatic response to dissection. In what follows, we consider that the landscape is initially a flat surface with no relief, which maximizes our estimates of denudational unloading by dissection. In addition, we assume that sea level is fixed and acts as the reference level for elevation and also the base level of erosion.

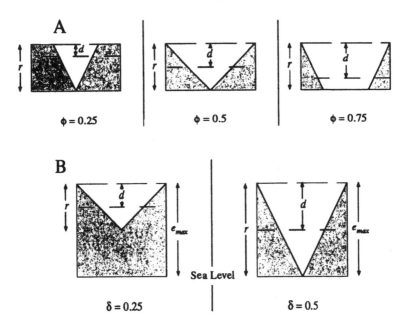

Figure 1 A: Effect of valley shape on ϕ. B: For given ϕ, in this case 0.5, effect of valley depth on δ. Each illustration represents valley cross section in different types of terrain. Upper dashed line is elevation of peaks (e_{max}); lower dashed line is mean level of terrain. Local relief (r) and depth of dissection (d) are shown.

Nondimensional parameters can also be used to characterize dissected landscapes and to compare features at different scales. Here, we define

$$\phi = d/r, \tag{1}$$

which equals the fraction of the local-relief volume occupied by valleys. The effect of different valley shapes on ϕ is illustrated in Figure 1a. Also,

$$\delta = d/e_{max}, \tag{2}$$

which is the proportion of the total landscape volume above base level that is occupied by valleys. Figure 1b shows that for different landscapes with the same ϕ, valley depth affects d, which is closely related to δ. Equations 1 and 2 can be combined to show how ϕ and δ are related (Fig. 2). The result indicates that for δ to approach unity—which represents the upper limit of d and hence the maximum isostatic response to dissection—there must be both deep valleys incised close to base level ($r/e_{max} \approx 1$) and also valleys with a cross-sectional area much greater than that of intervening peaks ($\phi \geqslant 0.5$).

227

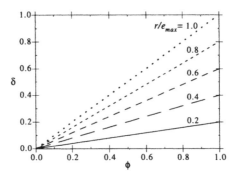

Figure 2 Relation between ϕ and δ for different values of ratio r/e_{max}.

Isostatic response to landscape dissection

The isostatic response to landscape dissection acts to increase the elevation of peaks that are not eroded, even though the mean elevation of the terrain decreases (Wager, 1933; Holmes, 1965; Molnar and England, 1990). Isostatic uplift is primarily determined by the isostatic response function (i), equal to ρ_c/ρ_m, where ρ_c is the density of material eroded from the top of the crust and ρ_m is the density of mantle at the depth of compensation; i is the amount of isostatic uplift per unit depth of dissection and is generally slightly less than the depth of dissection (~0.82).

The isostatic uplift of peaks due to landscape dissection can be defined as the ratio (χ) of maximum peak height in dissected terrain to an initial plateau of lower elevation that included the summits of now uplifted peaks. This definition assumes that the initial plateau had no local relief and no erosion has occurred from the top of uplifted peaks. The elevation of the initial plateau is unknown but can be estimated by calculating the isostatic depression of the landscape due to infilling the valleys between peaks. The relation between δ and χ is given by

$$\chi = \frac{e_{\mathrm{max}}}{e_{\mathrm{max}}(1 - \delta i C)}, \tag{3}$$

where C is the degree of isostatic compensation ($0 \leq C \leq 1$), dependent on the wavelength of unloading, that is important when the lithosphere has flexural rigidity (Turcotte and Schubert, 1982). Here we only consider the case where $C = 1$ for all wavelengths and the lithosphere has no rigidity (Airy isostasy). In reality, χ will likely be reduced due to flexural compensation ($C < 1$ for short wavelengths), and the following estimates are maximum values. The relation between δ and χ is shown in Figure 3 for several values of i. This indicates that peaks several times the elevation of an initial

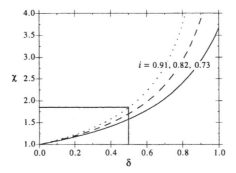

Figure 3 Relation between δ and χ for various isostatic response functions (i) assuming local compensation ($C = 1$). Range of χ for $\delta < 0.5$ is contained in lower left rectangle.

plateau can theoretically be generated if a landscape has a high value of δ. In addition, χ is greater when i approaches unity for a given δ. However, as we subsequently show, δ rarely exceeds ~0.5 in orogenic terrains and then not in the region of the highest peaks, and this reduces χ overall to <2 and significantly diminishes the actual magnitude of isostatically induced peak uplift.

Application to the European Alps

The European Alps are the morphological expression of the convergence between the African and Eurasian plates; the main orogenic phase began in the early Tertiary (Trümpy, 1980). Convergence has been accommodated by subduction of the Eurasian plate below the African plate and by crustal thickening in the Alpine region (Valasek *et al.*, 1991). The late Tertiary heralded a change in convergence direction that marked the end of extensive deformation. The altitudes of peaks in the European Alps reach >4,000 m; those in the selected study region of eastern Switzerland attain an elevation of ~3,500 m near the orogen axis and generally decrease in elevation across its flanks (Fig. 4). Landscape dissection is characterized by ridges and peaks separating deep valleys oriented both transverse and parallel to the orogen axis. The gross morphology of the European Alps differs from larger orogens such as the Himalayas and the Andes, which contain intraorogen plateaus, the Tibetan Plateau and Altiplano, respectively. In these cases, the highest peaks are located along the dissected edge of the plateaus, adjacent to the foreland.

The digital elevation model of the central European Alps (Fig. 4) has been used to determine variations in e_{min}, e_{mean}, and e_{max} across the range along a northwest-southeast profile, perpendicular to the orogen axis, by using the method of Burbank (1992b) and Fielding *et al.* (1994) for

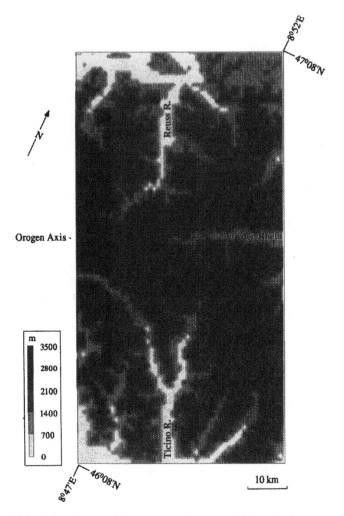

Figure 4 Digital elevation model for part of eastern Switzerland covering area of 53 × 100 km with grid resolution of 1 km. Long axis of grid is in approximately northwest-southeast direction, oriented perpendicular to orogen axis. Elevation data digitized from 1:100,000 scale topographic maps of Switzerland.

constructing swath profiles. The results (Fig. 5) show that e_{min}, e_{mean}, and e_{max} generally increase from either end of the profile, although the latter two variables decrease in value around the center because of the orientation of the Vorder Rhein Valley (Fig. 4). Both d and r reach their maximum values on either side of the profile center, approximately coincident with the location of the highest peaks.

230

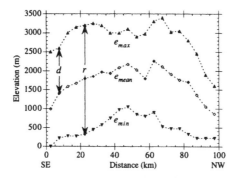

Figure 5 Profile across Swiss Alps derived from data in Figure 4 showing variations in peak (e_{max}), mean (e_{mean}), and valley (e_{min}) elevations. Local relief (r) and depth of dissection (d) are also indicated. The 100-km-long profile was divided into 5 km segments, and data contained in a 53-km-wide swath were projected onto profile. Local relief is dependent on horizontal length (L) over which relief is measured, and as L increases to ~10–30 km, local relief converges to constant value for variety of mountain belts, including European Alps (Ahnert, 1984). Therefore, swath width of 53 km is sufficient to determine characteristic local relief of region.

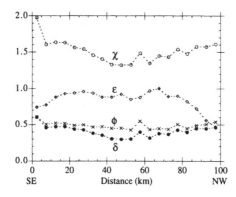

Figure 6 Variation in ϕ and δ calculated from Figure 5 and also χ, assuming local isostatic compensation, when i is 0.82. Maximum peak elevation normalized to vary between 0 and 1 (ε) is shown for comparison.

The evaluation of ϕ and δ along the profile across the orogen (from data in Fig. 5) enables an assessment of the importance of isostatically induced surface uplift (Fig. 6). The analysis shows that ϕ and δ are higher at the profile edges and lowest in the center. Values of δ attain ~0.5 at the edges of the profile and fall to ~0.3 in the center, the exception being in the region of the Vorder Rhein Valley (Fig. 4). This trend is the opposite of that of maximum peak altitude, which is higher toward the center of the profile and decreases at the edges. This indicates that χ is higher at the profile edges (~1.6) than around the center (~1.35), given the local isostatic compensation suggested by gravity data (Lyon-Caen and Molnar, 1989). The isostatic uplift of peaks is therefore relatively more important along the flanks of the orogen as valleys here are incised closer to base level than around the axis. Around the highest peaks adjacent to the axis, χ is ~1.5, so that the isostatic response to dissection can account for up to ~1,000 m of the elevation of the highest peaks, or about a quarter of their total elevation. Various studies have suggested that the lithosphere beneath the Alps has significant flexural rigidity (Karner and Watts, 1983; Watts, 1992). If so, $C < 1$, and the amount of isostatic uplift is reduced accordingly.

Implications for large orogens with plateaus

The analysis presented here can be readily applied to other regions, and we have examined morphologic data for the Himalayas (Fielding *et al.*, 1994) and the Andes (Burbank, 1992b) to determine the importance of isostatic uplift in these larger orogens. They are wider and have higher peaks than the European Alps and contain high plateaus, with the most elevated peaks being located along the dissected edge next to the low-lying foreland. In both cases, local relief is lowest across the plateau, increases toward the plateau edge, around the highest peaks, and decreases again across the orogen flanks to negligible values in the foreland. We estimate that values of δ for the highest peaks are ~0.3 for the southern margin of the Himalayas, ~0.35 in the Karakoram along the western edge of the Tibetan Plateau, and ~0.3 for the Andes. Values of δ are lower in the intraorogen plateaus (<0.2) and are higher across the incised orogen flanks (~0.5). The values of δ for the highest peaks are slightly lower than those determined in this study for the European Alps and imply χ values of ~1.35–1.4 for these larger orogens. Gravity data across the Himalayas suggest that the lithosphere in this terrain has a significant degree of flexural rigidity (Karner and Watts, 1983; Lyon-Caen and Molnar, 1985), so values of χ would be reduced since $C < 1$. Therefore, ~2,000 m would seem to be the upper limit of isostatic uplift due to dissection associated with the high peaks of the Himalayas.

Discussion and conclusions

The simple model of landscape dissection outlined above envisages that an elevated plateau generated by tectonic uplift and initially of minimal local relief is subsequently incised by surface processes and that mountain peaks are isostatically uplifted as a consequence. Our analysis demonstrates that the effect of isostatic uplift is limited, largely because of the relatively low depths of incision, relative to base level, along the highest, central parts of mountain belts. It appears that the geomorphic system is unable to respond in a manner that reaches the theoretical limit of the isostatic response to dissection. We conclude, therefore, that the altitudes of the highest peaks in orogens are primarily a consequence of tectonic processes and are not simply a result of the dissection of a moderately elevated plateau.

It is important to appreciate that the model of mountain-belt topographic evolution considered above in which deeply dissected terrain results from the incision of a plateau does not represent the only, or indeed most likely, mode of landscape development. It is a long-standing observation that neighboring peaks attain a similar altitude in many mountain ranges (Penck, 1887). Following the ideas of W.M. Davis, such accordant summits have been regarded as remnants of elevated peneplains, such as in De Sitter's (1952) interpretation of the development of the Alps. This interpretation was challenged from the beginning by Penck (1887), among others. He regarded such accordant summits as being the inevitable consequence of the incision of regularly spaced valleys. Where valley-side slopes are at, or close to, their threshold angle of stability, then progressive valley downcutting as the mountain mass rises will be accompanied by the continuous erosion of crests. Thermochronologic data now demonstrate beyond doubt that many accordant summits have undergone significant erosion because the depths of denudation revealed are well in excess of the existing local relief (Hurford, 1991). For example, an estimated 3–4 km of crustal section has been eroded from the summit of Mount Cook in the Southern Alps of New Zealand (Kamp *et al.*, 1989) and also from the top of K2 in the Karakoram (Foster *et al.*, 1994) during the late Cenozoic. The significance of such data is that in at least some orogens the currently observed peaks do not represent remnants of some initial low-relief plateau surface. Consequently, the assumption that high-relief Alpine topography arises from the dissection of such a preexisting surface is oversimplified.

A model of landscape evolution in which dissection occurs concurrently with mean elevation increase due to tectonic uplift appears to be more realistic for most orogens. Plate convergence causes crustal thickening and tectonic uplift of a linear mountain range near the suture between the two plates. Drainage of the rising mountain range leads to the development of regularly spaced valleys that are incised into the orogen. Initially, rates of slope transport lag behind those of incision (rivers and glaciers) so that peaks undergo little erosion and the isostatic response to dissection acts in concert with tectonic processes to raise peaks. As slopes steepen, mass-movement processes begin to dominate and slope-denudation rates increase, becoming comparable with rates of incision. Ultimately, rates of valley incision and summit lowering become equal, and, in the absence of tectonic uplift, the elevations of both valleys and crests decline if constant local relief is maintained. Should tectonic uplift continue, it is likely that an approximate steady topographic state will be attained in which the tectonic mass flux entering the orogen by plate convergence is balanced by the mass flux of denuded material transported to the foreland. Thus, crustal material is advected through the orogen to the surface, although the morphology of the orogen remains roughly constant, as is apparently the case in the Southern Alps of New Zealand (Adams, 1980).

Acknowledgements

Supported by a NATO fellowship (to Gilchrist) administered by the Natural Environment Research Council and by a grant from the Carnegie Trust for the Universities of Scotland (to Summerfield). We thank C.G. Chase, P. Molnar, and an anonymous reader for reviews of the original manuscript.

References cited

Adams, J., 1980, Contemporary uplift and erosion of the Southern Alps, New Zealand: Geological Society of America Bulletin, Part II, v. 91, p. 1–114.

Ahnert, F., 1984, Local relief and the height limits of mountain ranges: American Journal of Science, v. 284, p. 1035–1055.

Burbank, D.W., 1992a, Causes of recent Himalayan uplift deduced from deposited patterns in the Ganges basin: Nature, v. 357, p. 680–683.

Burbank, D.W., 1992b, Characteristic size of relief: Nature, v. 359, p. 483–484.

De Sitter, L.U., 1952, Pliocene uplift of Tertiary mountain chains: American Journal of Science, v. 250, p. 297–307.

Fielding, E., Isacks, B., Barazangi, M., and Duncan, C., 1994, How flat is Tibet?: Geology, v. 22, p. 163–167.

Foster, D.A., Gleadow, A.J.W., and Mortimer, G., 1994, Rapid Pliocene exhumation in the Karakoram (Pakistan), revealed by fissiontrack thermochronology of the K2 gneiss: Geology, v. 22, p. 19–22.

Holmes, A., 1965, Principles of physical geology: London, Nelson, 1288 p.

Hurford, A.J., 1991, Uplift and cooling pathways derived from fission track analysis and mica dating: A review: Geologische Rundschau, v. 80, p. 349–368.

Kamp, P.J.J., Green, P.F., and White, S.H., 1989, Fission track analysis reveals character of collisional tectonics in New Zealand: Tectonics, v. 8, p. 169–195.

Karner, G.D., and Watts, A.B., 1983, Gravity anomalies and flexure of the lithosphere at mountain ranges: Journal of Geophysical Research, v. 88, p. 10449–10477.

Lyon-Caen, H., and Molnar, P., 1985, Gravity anomalies, flexure of the Indian plate, and the structure, support and evolution of the Himalaya and Ganga basin: Tectonics, v. 4, p. 513–538.

Lyon-Caen, H., and Molnar, P., 1989, Constraints on the deep structure and dynamic processes beneath the Alps and adjacent regions from an analysis of gravity anomalies: Geophysical Journal International, v. 99, p. 19–32.

Molnar, P., and England, P., 1990, Late Cenozoic uplift of mountain ranges and global climate change: Chicken or egg?: Nature, v. 346, p. 29–34.

Penck, A., 1887, Ueber denudation der Erdoberflache: Schriften des Vereines zur Verbreitung Naturwissenschaftlicher Kenntnisse in Wien, v. 27, p. 431–457.

Trümpy, R., 1980, An outline of the geology of Switzerland, in Trümpy, R., ed., Geology of Switzerland, a guide book: Basel, Switzerland, Wepf, 140 p.

Turcotte, D.L., and Schubert, G., 1982, Geodynamics: Applications of continuum physics to geological problems: New York, Wiley, 450 p.

Valasek, P., Mueller, St., Frei, W., and Holliger, K., 1991, Results of NFP 20 seismic reflection profiling along the Alpine section of the European Geotraverse (EGT): Geophysical Journal International, v. 105, p. 85–102.

Wager, L.R., 1933, The rise of the Himalayas: Nature, v. 132, p. 28.

Watts, A.B., 1992, The effective elastic thickness of the lithosphere and the evolution of foreland basins: Basin Research, v. 4, p. 169–178.

9

A THEORY OF
MOUNTAIN-BUILDING

D. Griggs

Source: *American Journal of Science* 237 (1939): 611–50.

The fact that a natural phenomenon admits of one mechanical explanation is proof that there is an infinity of such explanations.

(Paraphrase from Poincaré.)

Abstract

A theory of mountain-building by cyclic convection, thermal and not chemical in origin, is synthesized from: (1) the suggestions of Holmes, (2) the mathematical analyses of Pekeris, Vening Meinesz, and Hales, (3) the writer's experiments on solid flow of rocks, (4) thermal experiments and calculations, and (5) a dynamically similar model to demonstrate the action of cyclic convection currents. The way in which this theory predicts the intermittence of mountain-building is discussed, and its ability to explain the diastrophic cycle.

Previous theories of orogenesis are briefly reviewed and some of their points of inadequacy are discussed.

Introduction

In the following, some arguments are presented which tend to make more plausible the suggestion that convection currents may have operated in the earth's substratum to cause the development of our mountain systems. It is clearly recognized that any such attempt must be highly speculative because of the enormous and indeterminate complexities of the situation. The reasoning is based on a few credible assumptions as to the composition and behavior of the earth's shells. The validity of the argument, therefore, depends entirely on geological and geophysical verification of these assumptions.

236

Any adequate hypothesis of mountain-building must satisfy the three following conditions:

1. It must provide a tangential component of force sufficiently great to fold and thicken the continental crust.
2. It must provide locally sufficient contraction to account for the shortening of the crust in the regions of mountain-building inferred from geological evidence.
3. It must explain the intermittent nature of orogenic processes and the succession of three phases in the mountain-building cycle:
    a. Geosynclinal subsidence and sedimentation.
    b. Compression of the crust and folding of the geosynclinal filling.
    c. Later elevation of the compressed and folded mass.

The literature of geology contains abundant criticisms of all the prevalent theories of orogenesis, based mainly on their inability to satisfy one or more of these conditions. It is not the purpose of this paper to evaluate critically the theories or the objections to them, but to suggest a mechanism which seems physically competent to meet these three conditions.

The basic physical principle on which the present hypothesis depends is not new to geology. Thermal convection currents in the earth's substratum were suggested a century ago by Hopkins (1839).[1] Since that time, several types of convection have been invoked to explain various features of the earth's history. The theory that our mountain systems are the result of convection currents has been the subject of much discussion by European geologists, but has received relatively little attention in this country.

Holmes (1932) has been led by his study of the thermal history of the earth to conclude that cooling of the earth has been inadequate to produce the contraction necessary for orogenesis. He further concludes that thermal convection is to be expected in the earth's substratum, and that the convection currents may be adequate to cause mountain-building. Vening Meinesz (1934) suggests that subcrustal convection currents compress and fold the crust, developing the "Tectogenes" which are thought to underlie the negative-anomaly bands observed in the East and West Indies. He develops simplified equations of convection and calculates the potential stress which they may exert on the crust. Pekeris (1936) develops more rigorously the convection equations, and shows that on the most probable assumptions of temperature, viscosity, and strength of the substratum, convection is to be expected if, in addition, the substratum is homogeneous enough to permit convection. He also calculates the tangential stress developed by such currents, using an entirely different method from Vening Meinesz's. Hales (1936) develops still another mathematical analysis of convection, which indicates the temperature gradient necessary for the maintenance of the currents. This analysis shows that the temperature gradient most commonly assumed by seismologists to explain the observed increase in seismic velocities will lead to the development of convection currents in a homogeneous substratum. Bull (1929) performed an interesting experiment which suggests that the action of convection currents is one of the few mechanisms which will produce nappe structure.

Many other papers have been published by European geologists dealing with various aspects of the convection-current theory of mountain-building, but the papers above referred to are among the most complete, and afford a representative discussion of the hypotheses of the convection mechanism. Most of the discussion of convection has involved the tacit assumption that a molten condition is necessary for convection. The mathematical analyses have been based on the assumption that steady-state convection has been established. The present paper presents some new ideas concerning the way in which subcrustal convection may cause orogenesis, and in particular, suggests the existence of a convection-current cycle, correlated with the mountain-building cycle.

Fundamental orogenic phenomena

The diastrophic cycle

It has long been recognized that in spite of the numerous irregularities of the earth movements culminating in the formation of mountain chains, this activity follows a cycle which has been intermittently repeated during the history of the earth. Bucher (1933) expresses this as Law 20 in his study of the "Deformation of the Earth's Crust": "The typical orogenic cycle begins with a geosynclinal depression and ends with a major uplift. The interval between these limiting events comprises two phases. The first phase is one essentially of quiet sinking, only occasionally interrupted by uplifts; the second phase consists of crustal foldings separated by diminishing epochs of renewed geosynclinal sinking."

For the purposes of the present discussion, it is convenient to divide this generalized cycle of diastrophism into three phases:

1. The geosynclinal phase, in which a strip of the earth's crust is gradually depressed, receiving sediments from the adjacent, slightly elevated land masses, on one or both sides of the trough.
2. The period of crustal folding, in which the contents of the geosyncline are compressed and folded, without major elevation of the area above sealevel. The forces during this period of folding act mainly in the horizontal direction, but with a sufficiently great downward component to prevent the thickened mass from attaining isostatic equilibrium.
3. The elevation of this folded and thickened mass until it reaches isostatic adjustment. The forces acting during this period are dominantly upward; compressive folding is secondary.

There are minor deviations from the average trend in all of these phases in the mountain ranges of the world, but it seems to be the consensus of geological opinion that the history of all the major mountain systems has followed this broad outline of events.

The crustal downfold, or tectogene

One of the greatest contributions to the understanding of tectonics during the twentieth century has been Vening Meinesz' discovery of the great bands of gravity deficiency in the East and West Indies. His careful and painstaking measurements of gravity over the broad expanse of the Pacific Ocean basin disclosed no deviation from normal gravity of more than 50 milligals, as shown in Fig. 1. Peripheral to the East Indian archipelago, however, a persistent gravity deficiency of much greater magnitude was discovered.

Fig. 2 shows the distribution of the gravity anomalies observed by Vening Meinesz (1934) from the Malay Peninsula around the north coast of Australia to the Philippines. Fig. 3 shows the results of a similar set of

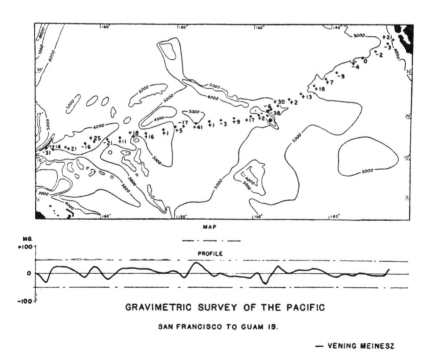

Figure 1 Series of Measurements of Gravity Anomalies Across the Pacific Ocean after Isostatic Reduction. Figure reproduced through the kindness of R.A. Daly.

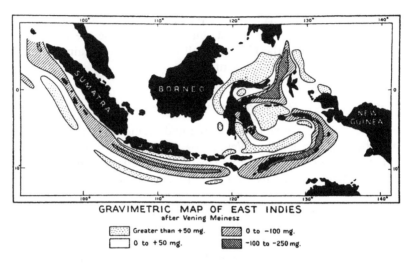

Figure 2 Gravity Anomalies in the East Indies, Showing Belt of Gravity Deficiency Peripheral to the East Indian Archipelago, Flanked by Bands of Positive Anomalies.

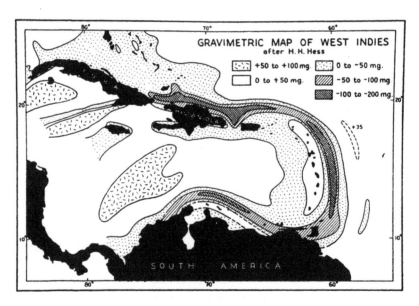

Figure 3 Gravity Anomalies in the West Indies, Showing Similarity to the East Indies.

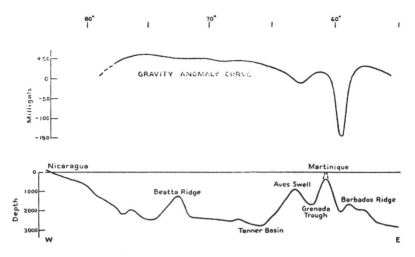

Figure 4 Gravity and Topography Profiles Along Latitude 15° N, in the West Indies.

measurements made in the West Indies by the coöperative enterprise of Vening Meinesz, Hess, Brown, Ewing, and Hoskinson. In each of these island arcs, there is a narrow strip of strong negative anomalies just outside the island festoon. Also in each case, this strip is flanked by irregular patches of less strong positive anomalies. Fig. 4 shows a profile across latitude 15° N. in the West Indies, which illustrates strikingly the intensity of the gravity deficiency and its relation to the topography of the ocean floor.

These negative anomaly bands can be explained only on the assumption of a mass deficiency in the earth's crust at this point. Further, the narrowness of the bands makes it imperative to assume that this mass deficiency is located in the outer 100 km. of the earth's crust. The only explanation that seems geologically possible is that the lighter crust of the earth has been downfolded into the heavier substratum. Fig. 5 (prepared by Hess) shows diagrammatically the way in which such a downfold may account for the observed gravity deficiency.

Both in the West and the East Indies, the distribution of these negative-anomaly strips coincides with the distribution of the most intense late Tertiary folding and thrusting (Vening Meinesz, 1934; Hess, 1938). Each island arc is a branch of the great Tertiary mountain system. It seems fair to conclude that the observations indicate crustal downfolds which were formed as a terminal event of the Laramide-Alpine revolution. It would appear that these regions have just passed through the second phase of the mountain-building cycle. On this assumption it is to be expected that they will, in the geologically near future, be uplifted to form minor mountain masses.

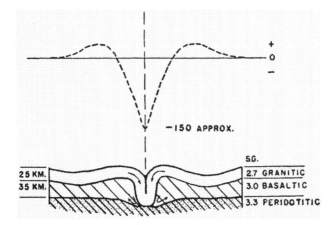

Figure 5 Crustal buckle, specific gravity distribution and resultant anomaly curve. From H.H. Hess.

It is suggested that the crustal downfold postulated in these two areas is a universal development of the second phase of the diastrophic cycle. Kuenen (1936) has termed the downfold a "Tectogene." If conditions during the first phase of the orogenic cycle are such that geosynclinal sedimentation takes place, then this sedimentary mass will be carried on the back of the folding crust and will itself be much more intricately folded. Such a relation has been suggested as the primary movement of mountain-building, as illustrated in Fig. 6 from Hess). If this relationship be accepted, one must next search for a force adequate to compress the crust and cause this great downfold and the concomitant folding of the sedimentary filling of the geosyncline.

Forces available for mountain-building

Primary and secondary diastrophic forces

The forces most easily observed or inferred from geological observations are the so-called secondary forces which tend toward the attainment of equilibrium. Thus, the forces of erosion would in a short time reduce the surface of the earth to a flat featureless plain, if there were not primary forces of earth deformation which continually distort the earth's face. Similarly, any deviation from isostatic adjustment develops forces which tend to restore equilibrium. All of these secondary forces depend on primary deformative forces for their existence, and hence cannot be considered the ultimate cause of mountains.

Five types of primary forces have been suggested in the various theories of orogenesis:

1. The tidal force.
2. The Polflucht force.
3. The Coriolis force.
4. Compression due to thermal contraction of the earth.
5. Viscous drag of convection currents in the substratum.

Before proceeding to a discussion of the magnitude of these forces, let us consider the probable strength of the earth's crust and estimate the size of the compressive stress which is necessary to fold it as a unit. The normal compressive strength of granitic rocks tested in the laboratory is 2,000 to 3,000 kg./cm.[2] It has been shown that high confining pressure such as exists in the earth's crust will greatly increase this strength, if all other conditions are held constant (Adams, 1917; Karman, 1911; Griggs, 1936). Other factors acting simultaneously—high temperature, solutions, and long periods of stress application—undoubtedly change the strength of rocks, but these effects have not been adequately investigated. It is yet too early to deduce from laboratory experiments the probable average strength of the earth's crust.

Studies of the local and regional departures from isostasy serve as the most reliable basis for estimating the strength of the continental crust and substratum. These observations admit of various interpretations, but in general they seem to indicate that the continental crust has a strength somewhere between 100 and 2,000 kg./cm.[2] For the purposes of the present discussion, it seems that the assumption of an average strength of the continental crust of 1,000 kg./cm.[2] will not lead to serious error, and will probably be as accurate, relatively, as many of the other values of physical properties which will be used in calculations.

In contrast to this strength of the crust, the observed flow of the substratum under the load of the glacial ice-cap indicates that it is either lacking in strength altogether, or possesses a strength of less than 10 kg./cm.[2] (Daly, 1938, p. 408).

The tidal force

Some of the advocates of continental drift hypotheses have invoked the westward tidal pull of the moon and sun on the earth's surface to account for westward drift of the continents and consequent mountain formation (Wegener, Joly, etc.). The most telling criticism of this theory is a calculation of the force which can be developed in this manner. This calculation has been made by several people, notably Jeffreys (1929, p. 304), who states, "Tidal currents at their strongest give a bottom drag of the order of 40 dynes/cm.[2] (4×10^{-5} kg./cm.[2]); but this is abnormal, and is reversed in direction in every tide where it does occur. The mean secular tidal friction producing the slowing down of the earth's rotation corresponds to a west-ward stress of the order of only 10^{-4} dynes/cm.^{2} (10^{-11} kg./cm.^{2}) over the earth's surface."

This westward stress acts tangentially over the whole bottom of the continental mass. Hence the compressive stress in the crust is greater than this: roughly in proportion to the ratio between the area of the continent and its cross-sectional area. This factor would be at a maximum of the order of 100:1, which would make the compressive stress in the continents 10^{-9} kg./cm.^{2}

The fact that this stress of tidal drift is only one trillionth (10^{-12}) of the stress necessary to compress and fold the crust seems to preclude its serious consideration as a cause of mountain-building, or even as a modifying influence.

The Polflucht force

Staub (1928) has been the chief exponent of the hypothesis of mountain formation as a result of equatorward drift of the continents in response to the "Polflucht" force. Because the continents are lighter than the substratum and float in it, they are acted on differentially by the centrifugal force of the earth's rotation. This centrifugal force has a component parallel to the earth's surface which tends to make the center of gravity of the continents move toward the equator. It is easy to calculate this component. Using Lambert's (1921) equations, we find for a continent 100 km. broad and 6 km. deep, a compressive stress of 5×10^{-4} kg./cm.^{2} For a continent 15 km. thick, Jeffreys (1929, p. 301) calculates the stress to be 4×10^{-3} kg./cm.^{2}

Thus we see that the maximum Polflucht force is of the order of one ten-thousandth of that required for mountain-building. Hence it would seem that this force may also be excluded from serious consideration as a primary cause of mountain-building.

The Coriolis force

Because the earth's axis of rotation is inclined to the line of gravitational attraction of the sun, it wobbles, like a gyroscopic top that gets a little out of the vertical. The forces that cause this wobbling, or precession, act differentially on parts of different density, and produce a component of force which tends to make the continents move. Jeffreys calculates this force and concludes (1929, p. 301), "The stress in precession that makes the whole of the earth precess at the same rate, instead of different shells precessing at different rates, is at most of order 60 dynes/cm.^{2} (6×10^{-5} kg./cm.^{2}), and this must be mainly alternating in direction." So we see that this Coriolis force is of the same order of magnitude as the tidal force, and much too small to cause mountain-building deformation of the earth's crust.

Compression due to thermal contraction of the earth

Cooling of the earth will produce a compressive stress in the crust limited only by the strength of the shell resisting the shrinkage so developed. Although there is no question that a force so derived might be sufficient to deform the crust, its adequacy must be examined on other grounds. Two considerations are of prime importance in a discussion of the theory that our mountain systems originated by a wrinkling of the earth's crust resulting from thermal contraction:

1. Is the earth cooling rapidly enough to produce the amount of contraction observed in the Tertiary mountain belt?
2. Can the compressive stress developed by uniform shrinking of the earth's interior be transmitted through the crust for thousands of miles to produce the localized deformation observed in the Tertiary mountain belt?

1. Our present knowledge of temperature gradients and the distribution of radioactive, heat-producing materials renders the answer to the first question indeterminate. Holmes (1932, p. 171) has shown that, if the radioactive elements were distributed throughout the earth in the proportion in which they occur in the average rocks of the crust, and if radioactivity proceeded at all depths in the earth, this in itself would provide a supply of heat fifty times as great as the heat lost from the earth's surface by radiation. The discovery of the amount of radioactivity in rocks makes it impossible to estimate how much the earth is cooling at the present time, and even renders possible the conclusion that the earth may not be cooling at all, but may have reached an equilibrium in which all the heat lost is supplied by radioactivity.

It is of interest to calculate the amount of cooling necessary to account for the observed contraction of the Tertiary mountain belt, assuming that all the contraction of the earth is taken up in the shortening observed in this belt. For this calculation, it is necessary to know the coefficient of thermal expansion for the materials of the earth under the pressures and temperatures which exist at depth. This we do not know, but if we assume that they have the same value as rocks at the surface, we find that an average cooling of the order of 1,500° for the whole earth is necessary. It seems probable that the pressure in the earth would greatly lower the coefficient of expansion, so that this figure would be too low. Such an amount of cooling seems excessive for the last three per cent of the earth's history.

2. The thermal contraction theory necessitates that the compressive stress in the crust be transmitted over a large fraction of the earth's circumference to cause the folding localized in the Tertiary mountain belt. If the substratum be assumed to have low strength compared with the crust, it will

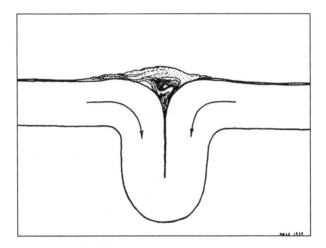

Figure 6 General section of the Alps superimposed on the tectogene. Both features drawn to the same scale with no vertical exaggeration. From Hess.

tend to contract uniformly in a radial direction, and stress differences will be equalized by flow. There will be little tendency to develop tangential motion in the substratum. It follows that the localization of deformation in the crust involves a tangential movement of the crust over the substratum, varying from zero in the center of the shields to a maximum at the mountain abutments. This over-riding motion of the crust is hindered by the viscous drag of the substratum, and it would seem that the condition for transmission of compressive stress through the crust is that this force of viscous drag be small in comparison with the compressive stress.

In order to calculate the viscous drag of the substratum, we must know the amount of crustal motion (shortening of the crust in the mountain regions), the time required for the movement, the viscosity of the substratum, and the distribution of the cooling which is responsible for the motion. The last of these depends on solution of the thermal conditions in the cooling earth, which we have seen is inadequately known. The other three variables are known to a first degree of approximation. From simplifying assumptions of all these conditions, the writer has made a rough calculation that the compressive stress in the crust developed by the viscous drag in the substratum would be of the order of 750 kg./cm.[2] In the light of our poor knowledge of the thermal conditions, however, this must be considered as at best an enlightened guess. If it be true, then it would seem that the drag of the substratum would provide sufficient resistance to the motion of the over-riding crust to prevent transmission of stress over large areas, and would cause more widespread distribution of deformation instead of localization into two mountain belts over the whole face of the earth.

A rough approximation to the conditions to be expected from thermal contraction can be attained in a dynamic model of the earth's outer shells. We shall see later, in the discussion of the dynamic model, that it provides some measure of confirmation of the conclusion reached above.

Many other criticisms of the contraction theory are to be found in the literature. In the opinion of the writer, none of the criticisms rests on sufficient evidence to disprove the hypothesis, but in toto they present significant difficulties which stand in the way of its acceptance.

The force due to viscous drag of subcrustal convection currents

Convection currents are familiar to anyone who has watched a bowl of soup being heated, a pot of coffee, or a maple syrup condenser. When heat is applied to the bottom of any body of liquid in excess of a certain amount, the hotter lower part becomes less dense than the upper part and instability is set up. Convection currents arise at the first disturbance of this equilibrium. In a broad expanse of liquid these currents have the tendency to form definite cells of more or less polygonal outline. Fig. 7 shows a vertical section through some experimental convection cells. The breadth of the cell tends to be about three times the depth, for plane surfaces of heating and cooling.

Any liquid heated from below has a tendency to lose heat both by conduction and by convective transfer of material. Conditions favoring convection are: (1) low viscosity, (2) low conductivity of the material (increase in size has the same effect as lowering the conductivity), and (3) a large temperature gradient. The effectiveness of convective transfer of material depends on the rate of flow. The high viscosity in the earth hinders rapid flow, but the great size favors it. The velocity of convective flow decreases inversely in proportion to the viscosity of the medium, but increases in proportion to the square of the size of the convecting cell. Another factor favoring convection in the earth is the fact that the amount of heat transferred by conduction in unit time decreases in proportion to the square of

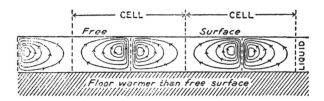

Figure 7 Section through Experimentally Developed Convection Cells. After H. Bénard.

the thickness of the mass through which it travels. A third factor may be of cardinal importance in promoting convection—the substratum may not behave as a viscous liquid, but as a "pseudoviscous" solid. This will be discussed in detail later.

Setting up the equations for convection of the earth's substratum as a viscous liquid, Pekeris (1936) has calculated the drag which the current may exert on the continental crust. His equations are simplified in that the term for adiabatic temperature change is not included, and the effect of variation of the earth's gravitational field with depth is neglected. The corrections due to these effects, however, would probably not change the order of magnitude of the result.

For convection cells of the approximate size of the continents and extending down to the 2,900 km. discontinuity at the earth's core, Pekeris has calculated the velocity of the currents and the tangential stress developed on the continents. Fig. 8 shows the stream lines and the velocities of flow in

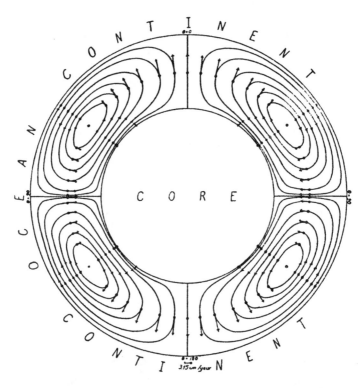

Figure 8. Convection Current Streamlines and Velocities of Flow (Velocity proportional to the length of the arrows). Under a Crust made up of two Polar Continents and an Equatorial Ocean. After C.L. Pekeris.

the case of hypothetical currents rising under polar continents and sinking under equatorial oceans. These velocities were calculated on the basis of Jeffreys' assumption that the viscosity of the substratum decreases from 10^{22} units at the surface to 10^9 at the core. The calculation of stress is independent of the viscosity assumed. Pekeris finds tangential stress (drag) on the continents of the order of 50 kg./cm.^{2} Because this acts on the whole area above the convection cell, and is concentrated into compression at the junction of the descending currents from two adjacent cells, the compressive stress is many times greater than this tangential stress. If the crust behaved as a perfectly rigid shell, but were free to move away from the center of the cell, the ratio between the compressive and tangential stresses would be the same as the ratio between the surface area of the convection cell and the area of cross-section of the crust peripheral to the cell. For Pekeris' case (Fig. 8) this ratio would be 200:1, and the compressive stress 10,000 kg./cm.^{2} Because the crust is not perfectly rigid, and because some force must be expended in thinning it at the center of the cell, the actual force would be considerably less than this. It is estimated that the actual compressive stress would be of the order of 3,000 kg./cm.^{2}

Vening Meinesz (1934, pp. 54–63) used a different mathematical approach to the same problem in connection with the interpretation of the gravity

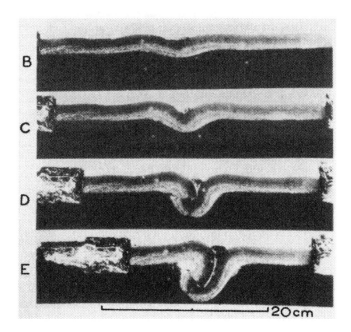

Plate 1
Successive Stages in the Development of a Tectogene During one of Kuenen's Experiments.

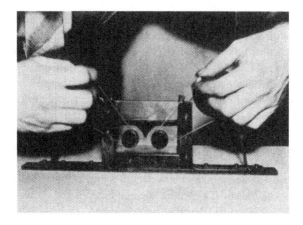

Plate 2
Figure 1 Small Dynamic Model to Simulate the Action of Subcrustal Convection
Currents and the Response of the Plastic Crust. Photograph Shows
Revolving Drums Simulating Convection Currents and the Consequent
Development of a Crustal Downfold.

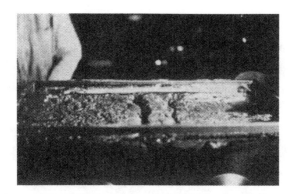

Figure 2 Large Dynamic Model after Development of Crustal Downfold and Two
Underthrusts in the Crust.

anomalies in the East Indies. Assuming convection cells extending only down
to 1,200 km., Vening Meinesz calculates an effective compressive stress in
the crust of the order of 5,000 kg./cm.[2]

Pekeris also calculates the lateral temperature perturbation necessary to
initiate convection on the assumption that the substratum has a threshold
strength below which flow does not occur. He concludes that a lateral

temperature perturbation of a few tens of degrees will start convection even if the substratum has a threshold strength of as much as 50 kg./cm.[2]

Hales (1936) calculates the amount of heat transfer necessary to maintain convection in the substratum, assuming it to have a viscosity of 10^{22} and a coefficient of thermal expansion similar to that at the surface. It appears probable from his calculations that an initial temperature gradient only .1°/km. in excess of the adiabatic gradient will be sufficient to develop convection currents.

In discussing the source of heat for this convection, Hales mentions Holmes' suggestion of deep-seated radioactivity with the comment, "It seems doubtful whether the assumption of a deep-seated layer of radioactivity is necessary. If the conductivity in the core is greatly in excess of that in the shell, then even with the lower temperature gradient the heat brought to the lower surface of the shell by conduction would be more than could be carried away by conduction through the shell. This supply of heat would therefore maintain the convection currents in the shell. The conductivity of the core is probably sufficiently large for this."

Depth of the convecting shell

Among the most fundamental problems of a convection current hypothesis of orogeny is that of the thickness of the convecting layer. Since the convection here in question is due only to the thermal differences in density, any great density discontinuity would act to prevent these density currents from crossing that boundary.

Seismology gives us our only clue as to discontinuities in the earth's shells. Only two discontinuities have stood the test of time in seismological research—the Mohorovičić discontinuity at the bottom of the continental crust, and the first-order discontinuity at the boundary of the central core of the earth (2,900 km. deep). (The discontinuities within the crust are not considered here because it is thought that they act only as secondary modifying influences on the reaction of the crust to the subcrustal convection. It is entirely possible that this is erroneous, and that the crustal discontinuities play a fundamental rôle.)

Various discontinuities have been suggested between the crust and the core: at 475, 1,000, 1,200, 1,700, 2,000 and 2,450 km. The most recent work favors the retention of only the 475 and 1,000 km., and it is interesting to see that one of the masters in this field, Macelwane (1936, p. 231), says, "No one would be more prepared than those who have done the respective pieces of research to admit the tentative and provisional character of the picture of the earth's interior which has been presented . . . Other interpretations may conceivably be developed which will satisfy equally well the data we now have; and tomorrow new facts may be discovered which will sweep away much of our present interpretation."

It must be pointed out that certain types of discontinuities may be no bar to density currents. Thus, if a discontinuity is due to a change in compressibility with pressure, or to polymorphism, it may have little or no effect on a density current crossing it. In the case of polymorphic transitions, the temperature of the material crossing the boundary will be changed slightly by the energy of transition, but this will be the only effect on the current. Jeffreys (1937) suggests that the 475 km. discontinuity is due to a polymorphic transition. The 1,000 km. (Repetti) discontinuity is not a sharp change in properties, but a gradual change in the rate of increase of velocity with depth, and thus might be connected with a change in the compressibility with depth.

The depth of the convecting shell is of primary importance in the distribution of the convection cells over the surface of the earth. Thus, if the cell extends to the core, it is reasonable to expect that the surface extent would be comparable in size to the whole Pacific basin, but if it extended only to the 475 km. discontinuity, the cells would be much smaller. The tangential stresses developed decrease rapidly with the size of the convection cell. Vening Meinesz and Pekeris, however, have calculated that convection in a 1,200 km. shell will be adequate to develop crustal folds. The discussion that follows is based on the arbitrary assumption that the whole substratum down to the core is subject to convection. Other depths of convection will change the magnitude of the calculated effects, but will probably not affect the principles deduced.

The convection-current cycle

Holmes (1932) suggested a hypothesis of thermal cycles in the earth's history based on a chemical fluxing action of the molten substratum. This hypothesis presents thermal difficulties because melting is involved, and necessitates additional hypotheses as to the composition of the rocks of the substratum. Because of the lack of data as to the composition of the substratum, it seems impossible to work out quantitatively any details of this suggested chemically-cyclic convection and so test the theory.

In contrast to this suggested periodicity, the mathematical treatments of Vening Meinesz, Pekeris, and Hales have all been based on the assumption that the equilibrium state of steady transfer of material is attained, as in ordinary laboratory experiments. The convection equations have involved the further assumption that the flow of the substratum follows the laws of viscous flow.

Consideration of the effects of size, the rate of heat conduction, and the nature of rock flow observed in the laboratory have led the writer to the conclusion that convection currents in the substratum will not attain a steady state, but will be periodic in nature. This hypothesis of cyclic convection, thermal in origin, involves no assumption of melting, and the composition of the convecting shell is assumed to be uniform.

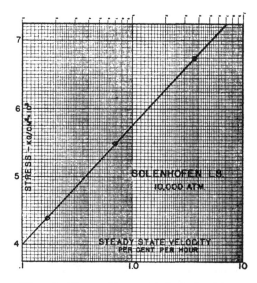

Figure 9 Velocity of Pseudo-Viscous Flow of Solenhofen Limestone as a Function of Compressive Stress from Creep Tests under 10,000 Atmospheres Confining Pressure.

Solid flow of rocks

It is suggested that the substratum does not behave as a viscous liquid, but as a "pseudo-viscous" solid. In a recent paper the writer experimentally demonstrated pseudo-viscous flow in crystalline rocks subject to small stresses for long periods of time in the laboratory (Griggs, 1939). Two types of experiment have now been performed in which it is possible to measure the velocity of this pseudo-viscous flow as a function of shear stress. Figs. 9 and 10 show the results of these two measurements. Fig. 9 presents observations of the pseudo-viscous flow in Solenhofen limestone under a confining pressure of 10,000 atmospheres (equivalent to that at 22 miles deep in the earth's crust). The data are taken from the creep curves published in an earlier paper (Griggs, 1936, p. 562). The shear stresses in this experiment were high and the deformation rapid. Fig. 10 shows a different type of deformation—pseudo-viscous flow of alabaster under small stresses and in the presence of its own saturated solution.[2] It has been suggested that this flow is due to solution and recrystallization (Griggs, 1939, p. 249). Both of these widely different types of experiment show that the velocity of pseudo-viscous flow (v) is related to the stress (σ) as follows:

$$\ln v = k\sigma, \quad \text{or} \quad v = e^{k\sigma} \tag{1}$$

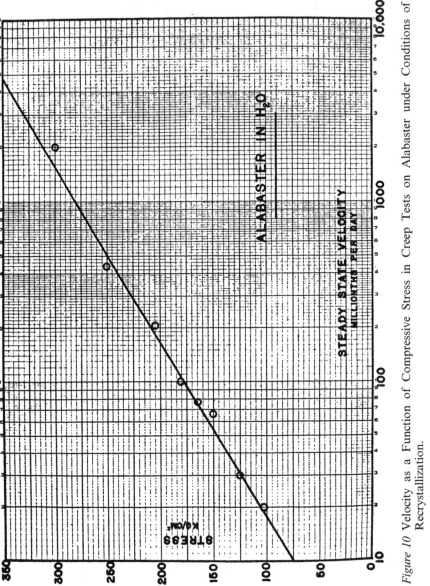

Figure 10 Velocity as a Function of Compressive Stress in Creep Tests on Alabaster under Conditions of Recrystallization.

whereas in viscous flow, the velocity is directly proportional to the stress. True viscosity is invariant with stress, but pseudo-viscosity (η) varies with stress as follows:

$$\eta = \frac{k'\sigma}{v} = \frac{k'\sigma}{e^{k\sigma}} \qquad (2)$$

Present experiments indicate that there may be a threshold stress below which no continuing flow occurs. This would be the "fundamental strength" of the rock under the conditions of pressure, temperature, and solutions which obtained in the experiment (Griggs, 1936, p. 564).

The importance of this type of flow in the present discussion is that as the stress is increased, the velocity of flow increases exponentially. The way in which this affects convection will be suggested in the next section.

Causes of cyclic convection

The condition for the *steady* transfer of material by convection is that the temperature of the transferred mass be materially changed by conduction during the time involved in the transfer. This maintains a temperature gradient which serves as a continuous driving force for the convection currents. Steady convection currents are most commonly observed in the laboratory.

Two factors in the earth, however, tend to violate this condition and to make the currents flow periodically:

1. The great size of the cells which makes possible the development of large forces from small temperature gradients, and permits the transfer of material faster than it can change temperature by conduction.
2. Pseudo-viscous rather than viscous flow of the substratum. This involves an extremely high viscosity for low-stress differences, which makes it possible for a temperature gradient to be set up before convection begins. It further involves a low viscosity at high stress-differences which facilitates rapid transfer of the material.

The driving force for convection is gravitational, due to the difference in density between the rising and sinking columns of a convection cell. It is at once obvious that if the flow of material proceeds more rapidly than cooling by conduction, the hot material of the rising column will be carried over into the sinking column before it has time to cool by conduction, and similarly the cool material of the sinking column will be carried under and into the rising column before it can be heated by conduction. This bodily transfer of hot and cold masses decreases the driving force of convection, and if the transfer is complete enough, it follows that the convection current must stop.

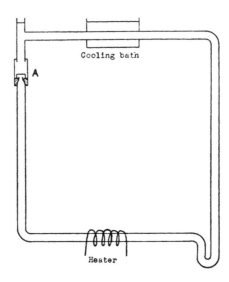

Figure 11 Experimental Cell to Demonstrate Periodic Convection in a System Subject to Constant Heating and Cooling.

Physicists instinctively distrust the idea of developing a periodic convective overturn from a steady supply of heat. In order to demonstrate that it is possible to produce periodic flow from constant heating and cooling sources, the writer has set up an experimental convection cell, illustrated in Fig. 11, in which the conditions are favorable to the production of periodic convection.[3] In this cell the heat source at the bottom and the cooling source at the top are shown, but the boundary conditions are artificial in that the rising and sinking columns are localized in the two side tubes of the cell. The unique feature of this cell is the valve at A. When this valve is removed and a constant temperature-difference applied to the top and bottom of the cell, normal convection may be observed. When the currents first start they accelerate, as the hot material from the region of the heat source is transferred up into the rising column, and simultaneously the material from the region of cooling moves into the sinking column, thus increasing the difference in density between these two columns. As flow proceeds, however, it slows down, and with a few oscillations the velocity approaches a constant rate.

The valve at A—a light inverted cup floating on mercury—opens when the driving force of convection exceeds a certain amount, and closes when the driving force is less than this amount. The introduction of this valve into the system simulates the behavior of a material which has a threshold strength below which flow does not occur, and thus more nearly reproduces the conditions in the earth's crust than convection of a viscous fluid.

When constant temperature-difference of the right magnitude is applied to this new cell, flow does not begin until the left-hand column of the cell has been heated and the right-hand column cooled sufficiently to develop a driving force large enough to open the valve. The valve then opens suddenly; the flow throughout the cell accelerates rapidly for a short period, and then decelerates just as rapidly and stops as the valve closes. After this sudden transfer of material no flow occurs until the temperature difference of the columns is re-established. This period of quiescence is just ten times as long as the period of flow, in the experiment here described. The cycle is repeated continuously with a constant period.

This serves as experimental verification of the suggestion that when flow is rapid enough to transfer hot and cold masses so quickly that they cannot change temperature by conduction, the driving force of convection will vary periodically.

It is instructive to make some rough calculations for the purpose of comparing the rate of convective transfer of material in the substratum with the rate of temperature change in the moving mass by conduction. Both Pekeris and Vening Meinesz have calculated the velocity of convective flow, assuming steady state convection. Pekeris calculates a maximum velocity of the order of 5 cm./yr. for a convection cell 2,900 km. deep; Vening Meinesz calculates 1 cm./yr. for a cell 1,200 km. deep. Two things would make the peak velocity in convection more rapid than that of the steady-state convection assumed in making these calculations: (1) the initial period of acceleration when convective overturn is begun—laboratory experiments develop a peak velocity three times the steady-state velocity; (2) pseudo-viscous flow, which would make the velocity under the maximum stress differences considerably greater than that calculated from the assumption of viscous flow. From these considerations, it would seem likely that the peak velocity would be of the order of 50 cm./yr., and the average velocity of the fastest moving mass during the convective overturn would be of the order of 10 cm./yr.

On the basis of this estimated velocity, the convective transfer of material from the earth's core to the surface would take 30 million years. For comparison with this velocity of transfer, let us calculate the probable percentage temperature change of the transferred mass during this time. Since only the outer shell of the convection cell is moved rapidly, it would seem that we might estimate the order of magnitude of the heat loss by assuming that a layer 200 km. thick were displaced from the bottom to the top instantaneously and held there for 30 million years. Lovering (1935) developed graphically the heat equations applying to the problem of an extensive dike of uniform thickness, which is essentially our problem. If we assume that this 200 km. shell corresponds to the injection of a dike at the base of the continents, then we can use his derivations to calculate the percentage temperature change of this mass. Assuming a diffusivity of .012, a 200 km. dike would retain 75% of its average excess temperature at the end of 40 million years.

If, during convective overturn, the transferred masses lose less than half of their temperature difference, then it follows that the temperature gradient is reversed and the convective driving force drops to zero. For example, if we assume a temperature difference (in excess of the adiabatic temperature gradient) of 500° C. between the core and the surface, and suppose convective transfer of the top and bottom layers within a period of 40 million years, then on the basis of the above rough calculation, the hot mass which rose to the top will still be 375° hotter than its surroundings and the cool mass which sank to the bottom will be 375° cooler than its surroundings. That is to say, the top mass will be 250° warmer than the bottom mass, and we may expect a static equilibrium to be attained and persist until the unstable temperature gradient (hotter at the bottom) is re-established by conduction.

In these calculations the effect of radioactive heating has been omitted. Holmes (1932) has shown that if radioactive elements were distributed throughout the earth in the proportions in which they exist in the crust, the heat so developed would be fifty times as great as that lost from the earth by radiation. Our assumption of convection in the substratum depends on a fairly uniform distribution of radioactivity in the substratum, and so we may conclude that a probable maximum of such activity would be one per cent of that in the rocks of the crust. Taking Holmes' figures for the average of rocks, this gives an annual radioactive heat output of the order of 3×10^{-8} cal./gm. Assuming a specific heat for the rocks of the substratum of .3, this means that in 100 million years the temperature rise due to radioactivity would be only 10° C., which indicates that its omission in the above calculation involves no serious error.

We may now make a rough estimate of the time required to re-establish a temperature gradient which will permit convection to recur. Using the same thermal equations and assumptions as before, we find that the transferred masses will lose 70 per cent of their temperature difference by conduction in about 700 million years. Radioactivity during this period will increase the temperature of the bottom mass by about 70°. From this it seems that an estimate of 500 million years for the period of static equilibrium would be of the right order of magnitude.

These calculations and experiments lead the writer to believe that convection in the substratum must be periodic. Hypothetical phases of the convection current cycle are illustrated in the next section.

Phases of the cycle

Fig. 12 represents successive stages in convection in one cell extending from the surface to the core of the earth. Before convection begins, heat loss from the core sets up a temperature gradient roughly as shown in Fig. 12-1. In a uniform shell this constitutes an unstable equilibrium in which the hot

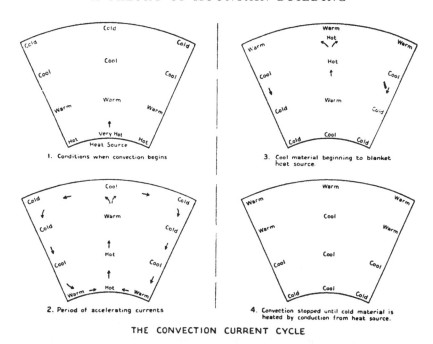

THE CONVECTION CURRENT CYCLE

Figure 12 Successive States in the Development of the Hypothetical Convection-Current Cycle in one Cell Extending from the Surface to the Core of the Earth.

material tends to rise and the cool surface material to sink. When lateral temperature variations become great enough, currents will start. Pekeris has calculated the magnitude of such variations needed to initiate convection and concludes that even in the case of a substratum having a strength of 50 kg./cm.^{2}, "a temperature contrast of a few tens of degrees is sufficient to overcome the above-mentioned strength of the asthenosphere and to start horizontal flow" (Pekeris, 1936, p. 348).

Once started, the temperature instability greatly accelerates the currents. As the hot material rises and the cold material sinks, the central column becomes progressively less dense and the outer column heavier, thus increasing the driving force of the currents (Fig. 12-2). Because of the nature of solid flow as outlined above, this stress greatly accelerates the velocity of the flow, producing an increase in the tangential stress exerted on the crust by the current.

It will be seen that this increase in tangential stress is not proportional to the velocity of the currents, as would be the case in viscous flow, but is proportional to the increase in the driving force due to the greater density difference between the rising and sinking columns of material. Pekeris (1936,

p. 357) shows that the tangential stresses are independent of the viscosity (or "equivalent viscosity" in this case—Griggs, 1939, p. 229).

As the hot material spreads over the surface, the cool material covers the bottom of the cell (Fig. 12-3). This decreases the density difference between the rising and sinking columns and slows down the currents. Finally, when the cool material rises in the central column, and the hot material begins to come down the sinking column, a stage is reached in which density equilibrium is again attained (Fig. 12-4). At this time the currents stop.

The temperature gradient is now the reverse of that before the convection began. No more convection will occur until the original unstable temperature gradient is re-established by conduction from the core and cooling from the surface. The time intervals of these four stages of the convection current cycle are estimated to be of the following order of magnitude:

> First Phase—slowly accelerating currents—25 million years.
> Second Phase—period of rapid currents—5 to 10 million years.
> Third Phase—decelerating currents—25 million years.
> Fourth Phase—quiescence—500 million years.

It would seem probable that during the period of quiescence when the substratum is regaining its unstable gradient, new currents would originate in some other cell in the substratum. For this reason, it seems impossible to set a definite period of time for recurrence of convection in some other cell.

A dynamic model of the earth's outer shells

In order to study the effect of sub-crustal convection currents on the continental crust, a dynamically similar scale model of the earth's outer shells was developed. By applying the laws of dimensional analysis and model theory (see Hubbert, 1937) it is possible to construct a model in which all the important properties of the earth's shells are accurately scaled down, and in which convection currents may be simulated.

Geologic models have progressed through several stages of approximation in the past, and no one of the various attempts has been subjected to a thorough dimensional analysis. The experiments most closely approaching dimensional correctness are those of Kuenen (1936). His apparatus (Fig. 13), utilized for the first time a fluid substratum, so that the crust was free to deform downward as well as upward. His choice of materials for the crust included a mixture of paraffin, vaseline, and oil which had the strength appropriate to a crustal strength of 2,000 kg./cm.[2]

Kuenen developed a compressive stress in the crust by pushing on it with a movable plunger. It was usually necessary to localize deformation of the crust by artificially producing a slight depression. The result of compressing the crust under these conditions was to produce a downfold of the type shown in Plate I. If, during the initial stages of an experiment, the broad

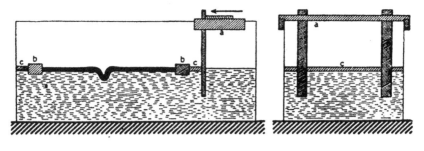

Side view and cross-section of apparatus.
a — pushing mechanism, b — floating beams, c — floating planks.

Figure 13 Schematic Diagram of Kuenen's Model for Developing a Tectogene by Compression of a Crust Floating on a Liquid Substratum.

geosynclinal depression was filled with thin layers of material weaker than the crust itself, it was found that these were intricately folded and thrust in a manner resembling alpine deformations.

In this experiment of Kuenen's the strengths of the crust and substratum were reproduced approximately to scale, and for the first time in experimental geology the geometrical conditions were favorable to the production of a Tectogene. One important factor was neglected, however—dynamical similarity. Kuenen used water for his substratum, which because of its extremely low viscosity did not provide sufficient viscous resistance to the over-riding crust to duplicate conditions in the earth. This viscous resistance of the substratum exerts an important drag on motion of the crust, as shown by the discussion on page 623 of this paper. It is instructive to make a dimensional analysis of Kuenen's model to investigate this point. The principles and symbols for this analysis are those of Hubbert (1937), with Bridgman's (1931) concept of the dimensionless product.

Dimensional analysis of Kuenen's model

a. In order to have the time factor at all manageable, inertial forces must be vanishingly small as in the earth, so that body accelerations may be neglected.
b. The simplest way to avoid complications due to acceleration analysis in this derivation is to use weight (W) instead of mass as one of the fundamental units. The others are time (T) and length (L).
c. Dimensions of the important variables:

$$\text{Length} \quad - \text{L}$$
$$\text{Time} \quad - \text{T}$$
$$\text{Density} \quad - \text{D} = \text{WL}^{-3}$$
$$\text{Viscosity} \quad - \text{DLT} = \text{WL}^{-2}\text{T}$$

d. Model Ratios (Ratio of the magnitude of the property in the model to the magnitude in the earth):

	Symbol	Dimensional Ratio	Value from Kuenen's model
Ratio of length	λ	1	2×10^{-7}
Ratio of density	δ	wl^{-3}	.35
Ratio of viscosity . . .	ψ	$wl^{-2}t$	1×10^{-24}
Ratio of time	τ	t	to be determined

e. From these four ratios it is possible to choose a combination which will be dimensionless. The value of this "Dimensionless Constant" will then be the same in the model as in the earth, and will permit calculation of the unknown ratio.

$$\text{Dimensionless Product} = 1 = \frac{\delta\lambda\tau}{\psi} = \frac{wl^{-3}lt}{wl^{-2}t} \tag{3}$$

$$\text{or,} \quad \tau = \frac{\psi}{\delta\lambda} = \frac{1 \times 10^{-24}}{7 \times 10^{-7}} = 1.4 \times 10^{-17}$$

This means that a process requiring 10 million years in the earth must be reproduced in the model in 1/300 second for dynamical similarity. Such rapidity would violate the basic assumption of this analysis—negligible inertial forces—and so it is impossible for this model to be dynamically similar.

If we are to design a model of the same size and density of materials to be dynamically similar, then we have one variable property—the viscosity—whose value is at our disposal. This analysis shows us that by correctly choosing the value of the viscosity we may make a dynamically similar model to operate at any speed we desire, limited only by the viscosity of the materials available. If, for example, we choose as convenient a model speed such that one minute in the model corresponds to a million years in the earth, we find:

$$\tau = 1.9 \times 10^{-12}$$

Solving equation (3) for the viscosity ratio:

$$\Psi = \delta\lambda\tau = .35 \times 2 \times 10^{-7} \times 1.9 \times 10^{-12} = 1.3 \times 10^{-19}$$

So that for the model, the substratum must have a viscosity of

$$\eta = 1.3 \times 10^{-19} \times 10^{22} = 1,300 \text{ c.g.s. units}$$

In setting up a model to illustrate the action of convection currents, the writer used two sizes—a large model in which an earth process requiring a million years could be reproduced in one minute, and a smaller one in which the same process could be reproduced in two seconds. The properties of the larger one are summarized in the following table:

Table I Properties of Dynamically Similar Model.

Property	Model Ratio	Actual Value in Model
Length	1×10^{-6}	60 cm.
Density	.6	1.8 (substratum)
Viscosity	$.8 \times 10^{-18}$	6,300 c.g.s. units
Strength	$.6 \times 10^{-6}$	.6 gm./cm.^{2} (ca.)
Time	1.3×10^{-12}	1 min. = ca. 1 my.

It is interesting that when the viscosity of the substratum is chosen for dynamical similarity, the model does not behave in the same way as that of Kuenen. When the crust is compressed by a moving plunger in the same manner as in his experiments, it shows no tendency to develop a downfold, but instead the compression is taken up by thickening of the crust immediately in front of the advancing plunger. The viscous drag of the substratum seems sufficient to prevent the transmission of compressive stresses for long distances through the over-riding crust, and causes local thickening of the crust instead. This evidence from the behavior of a scale model provides some measure of confirmation of the conclusion reached on page 623 of this paper which would seem to be a substantial difficulty of the thermal-contraction theory of orogenesis.

The next problem is to produce the effects of subcrustal convection currents in the model. Dimensional analysis shows that it is impossible to produce currents of sufficient velocity by thermal means. This would require a temperature difference between the bottom and top of the model of the same order of magnitude as that existing between the core and the crust of the earth—thousands of degrees. Since the effect on the crust is primarily that of a current passing under it and turning down at some point, however, this current can be simulated by the rotation of a paddle wheel strategically located in the substratum. In the actual model rotating drums were used, to avoid the irregularities of paddle motion. The friction at the surface of the drums sets up currents not greatly different from those observed in convection. This model corresponds to a vertical section across the boundaries of two adjacent cells. The third dimension is not reproduced.

The dynamic convection model

The principles of the dynamic model are best illustrated in Plate II, Fig. 1, which is a photograph of the small model. Here the small glass cell holds a clear liquid (glycerine) on which floats a black plastic crust (cylinder oil and fine sawdust). The drums are rotated by means of small pulleys belted to the larger pulleys which are turned by hand. When the rotation of the drums is slow, the plastic crust is thickened and pulled down into a slight trough (in this model the oil sticks to the glass walls so that it is impossible to see the surface configuration in the photograph). As the currents increase in velocity, the crust is thickened more and more until finally a narrow downfold is formed as shown in the photograph. At this time of rapid currents, the surface of the crust is irregularly folded, and its average level is just a little lower than before the currents began. If the currents are slowed down and stopped, the downward component of force disappears and the thickened mass rises to buoyant equilibrium, so that its surface is considerably above its original level. Thus the reaction of the crust to the three phases of the convection current cycle is demonstrated.

Because the inertial forces were not negligible in this small model, and because it was desired to study in more detail the structures produced, a larger model was built with the physical properties given in Table I. A photograph of this model in operation is shown in Plate II, Fig. 1. Here the substratum was very viscous waterglass and the continental crust was a mixture of heavy oil and sand, which possessed a yield point that could be varied by changing the proportion of oil used, and so made to have the proper strength for the model scale. The rotating drums show in the foreground. At the time the photograph was taken, the drums were in motion, and had developed a downfold and the two thrust faults in the crust shown by the dark lines. The actual development of these structures cannot be satisfactorily depicted short of moving picture portrayal. The writer has a series of movies of the apparatus and is endeavoring to make arrangements so that anyone interested can obtain them at a nominal cost.

The type of thrusts formed at the junction of two down-currents is shown in Fig. 14. This sketch illustrates the relations during the height of current velocity. The similarity in type of structure to that of Kober's Orogen is striking. An interesting feature of these thrusts is that they are formed by passive resistance of the overlying mass and active underthrusting of the foundation.

When only one drum is rotated (corresponding to the development of a single convection cell), the effect on the crust is different—the crustal downfold formed is not so narrow, and is asymmetrical in that it is steeper on the side facing the current. The structures are correspondingly asymmetrical. As the current velocity is increased, the crust is markedly thinned above it, and transported into the thickened part of the crust. Finally, the current sweeps all the crustal cover off and piles it up in a peripheral downfold. This stage in the development is illustrated in Fig. 15, drawn from the model.

This indication that a singly active cell may sweep off the superjacent continental crust opens wide avenues for speculation as to the formation of

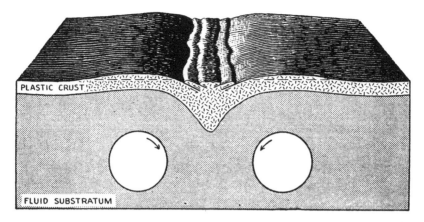

Figure 14 Stereogram of Large Model with Both Drums Rotating, Showing Tectogene and Surface Thrust Masses with Relations Similar to Kober's Orogen.

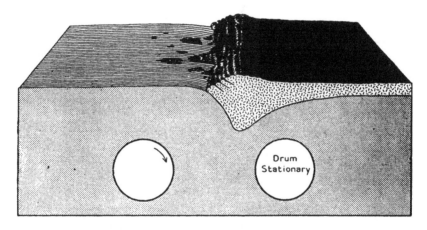

Figure 15 Stereogram of Large Model with only One Drum Rotating, Showing Development of Peripheral Tectogene.

the circum-Pacific mountains and indeed as to the primary segregation of the continental masses themselves. Here is a possible deformative force which could effectively counteract the tendency of erosion to distribute the continental material uniformly over the surface of the globe.

Correlation of the convection and orogenic cycles

We began with a generalized review of the mountain-building cycle; progressed through the hypothesis of thermal convection-current cycles in the

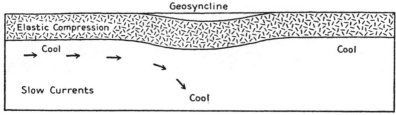

1. First stage in convection cycle – Period of slowly accelerating currents.

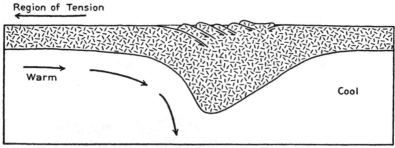

2. Period of fastest currents – Folding of geosynclinal region and formation of the mountain root.

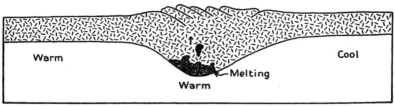

3. End of convection current cycle – Period of emergence. Buoyant rise of thickened crust aided by melting of mountain root.

Figure 16 Hypothetical Correlation between Phases of the Convection-Current Cycle and Phases of the Mountain-Building Cycle. Structural Relations Drawn from the Model.

earth; saw from model experimentation how subcrustal currents may deform the continental crust; and now we proceed to a synthesis of the facts and inferences gleaned from these varying modes of approach.

Fig. 16 shows the suggested correlation between the convection-current cycle and the mountain-building cycle. During the first phase of the convection-current cycle, when the currents are slowly accelerating, they will exert an ever-increasing tendency to compress the crust. The compressive force in the crust reaches a maximum where the convection current dives down at the boundary of a cell. At this point there is a vertical component of stress as well as the tangential drag, and the combination of maximum compressive stress and the vertical stress localizes the crustal deformation which in the first stage of the mountain-building cycle takes the form of a geosyncline (Fig. 16-1). If the currents were constant in velocity, the geosyncline would reach an equilibrium position in which no further sinking would occur. With the slowly accelerating currents, however, the compressive stress is constantly increasing and causes continued depression of the geosynclinal trough, in agreement with the geological evidence for continued subsidence of geosynclines. According to the estimate given above, this period of acceleration would be of the order of 25 million years.

When the currents attain high enough velocity, the compressive stress on the crust is sufficient to deform it violently. This increase in compressive stress is accompanied by an increase in the vertical component at the cell boundary, and between the two the crust is downfolded as shown in Fig. 16-2. The thrusts shown in the diagram were added from a study of the model.

In these diagrams and in the discussion of the model, no attempt has been made to reproduce the sedimentary filling of the geosynclinal trough. The thrusts shown in Fig. 16-2 are foundation thrusts, corresponding to similar features in the Alps described by Swiss geologists, and it is to be expected that, as in the Alps, each foundation thrust would be connected with a nappe or thrust in the sedimentary rocks above. The folding and faulting in the sedimentary filling, however, must necessarily be very much more complicated than in the relatively homogeneous and massive granitic crust. This increase in complexity can certainly be better demonstrated in the region of the "roots of the nappes" in the Alps than in any model.

During this period of orogenic folding by the most rapid currents, which probably lasts from five to ten million years, the thickened mass of the crust is kept submerged by the downtow of the sinking current. The vertical component of stress developed by the current acts to prevent the swollen crust from rising to isostatic equilibrium. This persistent deviation from isostatic adjustment which has been shown in the model is corroborated by the common geological observation that mediterranean sedimentation occurred simultaneously with the peak of diastrophism in the mountain systems of the world.

As the currents decelerate in the third phase of the convection cycle, the compressive force in the crust decreases, the downtow decreases, and the thickened mass rises buoyantly, gradually attaining isostatic equilibrium as the currents slow and stop (Fig. 16-3). This produces the elevation which characterizes the third phase of the mountain-building cycle. Observation of the model shows that during this rise, the thickened part of the crust expands laterally and exhibits a tendency to flow away from the center of the folded portion of the crust under the influence of gravity. The absence of compression which characterizes this phase of the cycle, and the tendency toward lateral expansion are in agreement with geological observations that the period of elevation is not a period of dominant compression, but is one of dominant uplift and normal faulting. This change in character of deformation is difficult to explain by the hypothesis of thermal contraction, but can be readily demonstrated in the cyclic-convection theory.

The persistence of major isostatic disequilibria in the East Indies has puzzled Vening Meinesz, Umbgrove, and Kuenen (Kuenen, 1936, pp. 196–7). If the gravity deficiencies here observed are due to crustal downfolds, it is reasonable to suppose that they were formed at the time of the greatest folding of the sediments on the adjacent islands. This conclusion seems unequivocally established by the exact parallelism of regions of most intense folding and the negative anomaly bands. The age of this folding is Miocene, so that one must explain the persistence of the anomalies for something like 20 million years. This presents great difficulty to the thermal contraction theory of orogeny, but is to be expected from the cyclic-convection theory, since we have seen that the period of deceleration is estimated to last 25 million years.

Application of the cyclic-convection theory to earth structures

Alternative interpretations of major structure

The primary purpose of this paper is to suggest a possible mountain-building mechanism. The application of that mechanism to details of earth structures is beyond the scope of this discussion. It may be of interest, however, to suggest in the most speculative way how this cyclic convection might operate to produce the major trends of the mountain systems.

The most fundamental question in applying this hypothesis to orogeny is the depth of the convective cells. In the absence of conclusive seismological evidence of the existence of a sufficiently sharp density discontinuity in the earth to prevent convective overturn between the crust and the core, it may be assumed for purely speculative purposes that the convection extends throughout this 2,900 km. depth. This would predict a size of the convection cells of the order of 7,000 to 10,000 km. in diameter (at the surface).

Holmes (1932) and Pekeris (1936) have suggested that the blanketing effect of the continents with their high radioactive content will cause sufficiently excess temperature under the continents to initiate rising currents there. These currents would act to spread the continents and to form mountains peripheral to them. Holmes published maps showing the hypothetical effect of this action on continental structures.

It seems conceivable to the writer that the temperature differences within the substratum inherited from the preceding convection cycle may be of more importance in localizing the cells than the blanketing effect of the continents. This opens the attractive possibility of a convection cell covering the whole Pacific basin, comprising sinking peripheral currents localizing the circum-Pacific mountains and rising currents in the center. Such an interpretation would partially explain the sweeping of the Pacific basin clear of continental material, in the manner demonstrated by the model. A minor cell might be suggested with its center in the southwest Indian Ocean, accounting for the Himalayan-Alpine bifurcation.

If this be assumed, then one may carry the speculation further and suppose that the previous cycle occurred as far from this location as possible —namely, about the central Atlantic Ocean. This location is nearly central to the Appalachians, Hercynian mountains, and the Post-Carboniferous mountains of Brazil and Africa.

In favor of the first alternative of Holmes and Pekeris we have the advantage of the thermal explanation for the localization of convection currents. The predominant development of thrusts inclined toward the ocean basins also indicates, by analogy with the model structures, currents rising under the continents. On the other hand, when slightly stronger crusts were used in the model, the direction of thrusting reversed. A tenuous argument in favor of oceanic rising currents is the distribution of deep-focus earthquakes. Visser, Leith and Sharpe, and Gutenberg and Richter all agree that foci of deep earthquakes in the circum-Pacific region seem to lie on planes inclined at about 45° toward the continents. It might be possible that these quakes were caused by slipping along the convection-current surfaces. These flow surfaces would be expected to dip toward the continents on the hypothesis of Pacific up-currents.

In contrast to the rest of the paper, this section is *purely* speculative.

Conclusion

Of the five primary forces which have been invoked by various theorists to explain orogenic deformation of the earth's crust, three are totally inadequate in magnitude. In contrast, thermal contraction may produce abundant force, but is open to other objections as a mountain-building force. Chief among these are: (1) thermal contraction does not seem capable of providing sufficient shortening to produce the Tertiary mountain system; and (2) transmission of compressive stress through the earth's crust over long distances to provide the localized deformation of the mountain systems is greatly hindered by the viscous drag of the substratum on the overriding mass. Calculations indicate that this viscous drag would cause uniform thickening of the crust instead of localized downfolds. Kuenen produced localized downfolds in his model under conditions in which the crust transmitted the compressive stress as it would have to according to the theory of thermal contraction. Dimensional analysis of Kuenen's model shows that it is dynamically incorrect. A dynamically similar model was constructed and, under compression similar to Kuenen's, showed no formation of a down-fold, but thickening of the crust at the point of stress application.

Pekeris', Vening Meinesz' and Hales' calculations show that convection currents can develop adequate compressive stress in the crust to cause orogenesis. A new mechanism of solid flow is suggested for the substratum, with experimental illustrations of this type of flow from two sets of creep tests. Consideration of what seem to be the first order factors leads to the hypothesis of a convection-current cycle. Experimentation with a dynamically similar model which simulates the action of the convection-current cycle produces effects resembling diastrophism of the earth's crust. The structures developed by the action of these currents in the model show striking resemblance to the structures developed in the earth. The phases of the convection-current cycle correlate with the phases of the orogenic cycle.

The hypothesis of orogeny developed from these observations is attractive because it seems to satisfy better than any other the three fundamental conditions of a mountain-building mechanism.

1. Provision of an adequate compressional force.
2. Local provision of sufficient contraction for orogenesis.
3. Explanation of the intermittent nature of orogenic processes and the threefold character of the mountain-building cycle.

Evidence sufficient to establish any theory of this kind can hardly be found within a short time, and has never been put forward in support of any previous theory. This difficulty seems inherent in the very nature of the problem. Accordingly the present theory, necessarily founded on insufficient evidence is here presented because it can be effectively tested only by use and the critical scrutiny of others.

Acknowledgments

The writer wishes to express his deep appreciation to Mr. A. Lawrence Lowell and the Lowell Institute for providing incentive for the present work in the form of an invitation to deliver a series of lectures at the Lowell Institute. The Lowell Institute also financed the construction of the models. The suggestions of Professors L.J. Henderson, P.W. Bridgman, R.A. Daly, M.P. Billings, and Mr. R.W. Vose were most helpful.

Notes

1 Names followed by dates in parentheses refer to references listed at end of this paper.

2 The latter experiments were performed with the aid of a grant from the Pennrose Fund of the G.S.A.

3 This experiment was suggested by Professor L.J. Henderson.

References

Adams, F.D., and Bancroft, J.A.: "Internal Friction during Deformation and the Relative Plasticity of Different Types of Rocks," Jour. Geol., Vol. 25, 597–637, 1917.

Bridgman, P.W.: "Dimensional Analysis," Yale Univ. Press, 1931.

Bucher, W.H.: "The Deformation of the Earth's Crust," Princeton Univ. Press, 1933.

Bull, A.J.: "Further Aspects of Mountain Building," Proc. Geol. Assn. (England), Vol. 40, p. 105, 1929.

Daly, R.A.: "Strength of the Earth's Outer Shells," This Journal, Vol. 35, 401–425, 1938.

Griggs, D.T.: "Deformation of Rocks Under High Confining Pressures," Jour. Geol., Vol. 44, 541–577, 1936.

——: "Creep of Rocks," Jour. Geol., Vol. 47, 225–251, 1939.

Hales, A.L.: "Convection Currents in the Earth," Mon. Not. Roy. Ast. Soc., Geoph. Sup., Vol. 3, 372–379, 1936.

Hess, H.H.: "Gravity Anomalies and Island Arc Structure . . ." Proc. Am. Phil. Soc., Vol. 79, 71–95, 1938.

Holmes: "The Thermal History of the Earth," Jour. Wash. Acad. Sci., Vol. 23, pp. 169–195, 1932.

Hopkins, W.: "Researches in Physical Geology," Phil. Trans. Roy. Soc. London, 381–385, 1839.

Hubbert, M.K.: "Theory of Scale Models as Applied to the Study of Geologic Structures," Bull. G.S.A., Vol. 48, 1459–1520, 1937.

Ingersoll and Zobel: "Mathematical Theory of Heat Conduction," Ginn and Co., Boston, 1913.

Jeffreys, H.: "The Earth," Macmillan, New York, 2nd Ed., 1929.

——: "On the Materials and Density of the Earth's Crust," Monthly Notices of the Royal Ast. Soc., Vol. 4, 50–61, 1937.

Kuenen, P.H.: "The Negative Isostatic Anomalies in the East Indies (with Experiments)," Leidsche Geo. Mededeelingen, Vol. 8, 169–214, 1936.

Lambert, W.D.: "Mechanical Curiosities of the Earth's Field of Force," This Journal, Vol. 2, 129–158. 1921.

Lovering, T.S.: "Theory of Heat Conduction Applied to Geological Problems," Bull. G.S.A., Vol. 46, 69–94, 1935.

Pekeris, C.L.: "Thermal Convection in the Interior of the Earth," Mon. Not. Roy. Ast. Soc., Geophysical Sup., Vol. 3, 343–368, 1936.

Staub, R.: "Der Bewegungsmechanismus der Erde," Geb. Borntraeger, Berlin, 1928.

Vening Meinesz, F.A.: "Gravity Expeditions at Sea," Vol. II Netherlands Geodetic Commission, Delft, 1934.

von Karman, Th.: "Festigkeitversuche unter allseitigem Druck," Zeitschr. des Vereins deutscher Ingenieure, Vol. 55, 1749–1757, 1911.

10

SIGNIFICANCE AND ORIGIN
OF BIG RIVERS

P.E. Potter

Source: *Journal of Geology* 86 (1978): 13–33.

Abstract

Ancient big river systems can be identified by a combination of facies mapping, paleocurrent study, analysis of
unconformities, and the careful reconstruction of the tectonic
history of a region. Ancient or modern, the location of big
river systems on cratons largely follows structural lows such as
deep-seated rifts, aulacogens, and geofracture systems; on
a continent-wide scale many big rivers debouch on trailing
continental margins and into marginal seas, where they tend to
be localized by deep geofracture systems intersecting the coastline. Big river systems have their greatest longevity on cratons
where some have persisted as long as one sixteenth of earth
history. A major factor in any river system is the marine history of its drainage basin, a history that in turn is dependent
upon the region's tectonic history.

Introduction

Sandstone forms some of the earth's earliest preserved rocks in Greenland,
rocks 3.7 b.y. old (Bridgewater *et al.* 1976); is genetically associated with the
erosion of continents and their margins; is found in all depositional environments in abundance except possibly on abyssal plains far removed from
coast lines; and, petrographically and chemically, has been studied much
more in ancient rather than modern deposits.

The initial object of this study was to establish a baseline, both petrologic
and chemical, for a better interpretation of ancient sandstones by studying
modern sands from the world's larger rivers. I emphasize the sands from big
rivers for several reasons. First, most big rivers have large, long-lived deltas
which have played a major role in basin filling in both deep and shallow

waters. In addition, sampling big river sands, difficult as it can be, is much simpler than trying to collect samples from the vast number of the world's smaller rivers that empty into the ocean. Modern eolian sands were excluded, because they are much less abundant than the deposits of modern big rivers, and additionally, it seems much more probable that there is a relationship between the sands of big rivers and plate tectonics than between the sands of continental deserts and plate tectonics.

What might such a study achieve? Petrologic and chemical study of the modern sands of big rivers can:

(1) provide a standard of comparison for ancient sandstones,
(2) contribute to a better understanding—perhaps even a more comprehensive theory—about the mineralogical and chemical composition of ancient sandstones because, with few exceptions, present theories are largely based on the study of ancient sandstones rather than modern sands,
(3) help geochemists better estimate the chemical mass balance between continental erosion and oceanic sedimentation, and
(4) provide metamorphic petrologists with a useful benchmark about the premetamorphic composition of high rank quartzofeldspathic rocks.

Certainly, in comparison to the vast effort expended on the study of sedimentary structures, bedding, vertical sequences, and shapes of modern sand bodies, the petrologic and chemical composition of sands from modern big rivers has been much ignored. This of itself suggests that such study could be most fruitful. And as more and more ancient basins become better known, the role of long-lived, large river systems as their principal agents of supply is being more widely recognized.

But what is meant by the term big river? From a geological point of view a big river is one with either a large drainage basin, a long course, one that transports a large volume of detritus, or has a large fluid discharge. For a sedimentologist interested in past earth history, the fluid discharge of ancient big rivers is the most elusive of all of these to estimate, especially for a large river, where it is unrealistic to think of finding a cross section and slope to determine relevant hydrologic parameters. Hence here I have stressed drainage area, sand composition, and resulting facies patterns leaving to others questions of their paleodischarge, recognizing, nonetheless, that their discharge is probably the most fundamental, *single quantity* by which a river is characterized.

Because estimates of bed-load discharge for many of the world's big rivers are either lacking or are very poor, a better appreciation for a big river is had by considering length and drainage basin area. A plot of drainage basin area versus length for the world's 50 largest rivers, ranked by drainage basin area, shows that all but one has a length greater than 1,000 km and suggests that on the average a stream 1,000 km long would have a drainage basin of

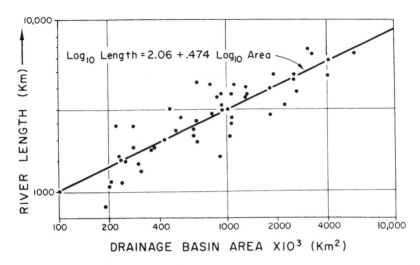

Figure 1 Log–log plot of length versus drainage basin area and regression equation for the world's 50 largest rivers, ranked by drainage basin. Data from Inman and Nordstrom (1971, table 2).

about 100,000 km² (fig. 1). These 50 streams also collectively drain a very significant portion of the continents—about 47% excluding Greenland and Antarctica. From still another point of view, rivers with the five largest drainage basins—the Amazon, Zaire (Congo), Mississippi, Nile, and Yenisei—alone account for about 10%. Or consider only the Amazon, which alone accounts for about 5% of present continental area.

As I studied the sands of big rivers, I became more conscious of big rivers themselves, how they were formed, and their role in geological history so that now I can argue with conviction that big ancient river systems are indeed an overlooked, under exploited theme in earth history. Consider for a moment just a few of the provocative questions that can be asked:

(1) What are the geologic controls—tectonic, climatic, and geographic— that determine the development of big rivers?

(2) How well can big river development be related to plate tectonics?

(3) How well do we really know the size relationships between *depositional basins* and their *river systems*; i.e., how frequently and under what conditions is a large depositional basin associated with a large river system? If we can do this with profit for alluvial fans, for example, should we not try to explore similar relationships for big river systems? And could not such possible relationships be relevant, perhaps, to the economic study of sedimentary basins?

(4) What has been the occurrence of big rivers in the earth's past history; i.e., is the Amazon the largest ever or were there possible super-Amazons

in the past, perhaps either on a Pangaea or in the very early Precambrian, when the world ocean may have been much smaller than today and/or when at some point in earth history there were no grasses or even lichens?

(5) What conditions are required for a *long-lasting* rather than a *short-lived* big river system? And how commonly is a big river system short- rather than long-lived?

(6) And finally, how many case histories of big river systems can we document in the geologic record? Or expressing this in a different way are the sediments of ancient big river systems mostly preserved in ancient delta systems? Or from still another point of view, how many of the deposits of big rivers have been consumed at plate junctions along continental margins?

Let us begin by first considering some of the above questions about river systems and then, with this knowledge in hand, later consider Potter (1978) the petrology and chemistry of the sand that modern big rivers carry to the ocean and its possible significance for the interpretation of ancient sandstones.

Significance of big rivers

There is a common association of a big river—with or without a delta—with a submarine canyon and an abyssal fan. This raises the possibility that most big basins, ancient or modern, were supplied by sand and mud from a big river system which deposited not only clastics in an alluvial valley but also in a subaerial delta which in turn had marine delta front deposits and possibly even associated deep-water turbidites. Kuenen implied this in his 1957 paper on the longitudinal filling of oblong sedimentary basins.

Good modern examples of the above are the Ganges–Brahmaputra Rivers, their combined deltas, and the submarine fan of the Bay of Bengal, the latter over 2,000 km long. The Ganges–Brahmaputra system emphasizes very well the definition of a delta proposed by Moore and Asquith (1971, p. 2566) as, "a subaerial and submerged contiguous sediment mass deposited in a body of water primarily by the action of a river." Or to see better the significance of a big river to the fill of sedimentary basins, consider a major rift system on a continent. If no big river is present, the rift may be filled by marine sediments possibly including hundreds of meters of evaporites. If, on the other hand, the rift is far inland and its floor is above sea level, it will be filled by marginal alluvial fans and possibly lake deposits. But if a big river, one that erodes a well-watered hinterland with high relief, debouches into the rift, it will be filled by a vast volume of terrigenous debris, most probably connected to the world ocean and certainly form a major delta. Nor should one forget the effects of a large volume of mud transported to a coastline and its consequent inhibition of carbonate-producing organisms.

Another consequence of a vast volume of mud introduced to a shoreline is the reduction of coastal wave power by the development of a shallow, gently sloping, offshore profile. Thus, for many reasons, the presence or absence of a big river debouching into a basin significantly alters the character of its fill.

From another point of view, 11 of the world's largest rivers—the Yellow, Ganges, Brahmaputra, Yangtze, Indus, Amazon, Mississippi, Irrawaddy, Mekong, Colorado, and Red (North Vietnam)—supply about 35% of the suspended load carried to the ocean (Drake 1976, table 2b). It seems safe to assume a comparable role for big rivers in the past—as long as the continents were as well watered and as high standing as today. And from the standpoint of world resources, deltas, submarine fans, and canyon fills provide a very significant fraction of the world's petroleum reserves. In addition, many major coal basins are commonly associated with rivers, both large and small. But what of other mineral deposits? Although here the answer is less clear, it is true that after 50 years of supremacy of the magmatic, hydrothermal concept of ore deposition in rich small lodes, interest has now shifted to the vast volumes of stratiform copper and sulfide deposits in sedimentary basins—a shift of research thinking from a cooling to a sedimentation history (Söhnge 1974). Certainly any comprehensive sedimentary analysis of such deposits, which may well be most common in the marine portion of a delta system, invites a consideration of their paleo-drainage systems.

The literature of ancient river systems, especially those predating the late Tertiary, is very widely scattered and not voluminous. All too commonly, it is the indirect product of other study—usually by many different people who each contribute but one or two facets of a paleodrainage reconstruction. Even rarer are papers discussing the *geologic controls* on big rivers.

The age of a river system

Central to a geological discussion of big river systems is the question of their age. The time of origin of a river system is defined (Mann and Thomas 1968, p. 187) as the earliest date at which a continuing, persistent river occupied the region in question—be it a small, post-Wisconsin stream in a glaciated area or a major river crossing much of a continent, such as the Nile. Thus a river system can date from either the last marine regression, the last significant tectonic warping or uplift, the waning of a continental ice sheet, or perhaps from the cessation of vast outpourings of lava. A river system can be terminated by a marine invasion, one that totally inundates its drainage basin, or by new, differently oriented tectonic deformation, continental glaciation or lava outpouring. Or any of the above events can affect only a part of the drainage system so that its age becomes, in some sense, composite. Still another determinant of its age is a drastic change in rainfall; the system might dry up with a striking increase in aridity or, conversely, be initiated by enhanced rainfall. As we will see, some river systems can be remarkably persistent and long-lived even though they may have experienced some major interruptions from time to time.

Case histories

Below are some selected examples that show diverse paleodrainage reconstructions of ancient river systems and how they were made. While not all of the selected examples are big rivers, all do illustrate the different factors that affect major rivers and also show how they were recognized and mapped.

Delta systems

Most studies of ancient delta systems (including both subaerial and submarine parts) also illustrate the lower portion of a river system, generally showing at least part of its lower alluvial valley. An early example is provided by Gilligan (1920, fig. 21), who studied the Millstone Grit and inferred a river system draining southward from Norway into Great Britain. Sedimentological studies of Carboniferous and other coal measures around the world commonly identify, by facies mapping and crossbedding orientation, the location and source of their supplying rivers even though few draw them as far into a hinterland as did Gilligan. Smith (1975) provided a series of maps, all of the same scale, of ancient deltas, some of which include portions of their alluvial valleys. Another interesting source, but only for a view of small segments of ancient river systems, is Conybeare's (1972) summary of petroleum accumulations in ancient fluvial deposits.

In the Tertiary of Texas, Fisher and McGowen (1967) mapped a series of seven small deltas, all of which formed along a hinge line, where oceanic crust underlies continental crust (McGookey 1975), and consequently a thick terrigenous section progrades basinward from thin Mesozoic carbonate shelves. A remarkable feature of these small, now buried, deltas is their spatial coincidence with present southward draining rivers (fig. 2). This shows then that, where ancient river systems prograded seaward into areas of either deep water or rapid subsidence, the resultant deposit will have a *vertical persistence in time* leading to a *vertical persistence of facies*. In contrast, on a stable craton a long-lived river system can migrate over very long distances and thus perhaps contribute to the fill of *several rather than simply one basin*. On such a long cratonic journey, a major river system will avoid, if possible, regional arches and instead follow basin lows.

Widespread paleogeographic reconstructions

A notable river system that has been reconstructed through studies of facies mapping and crossbedding is that of the ancestral Mississippi, called the

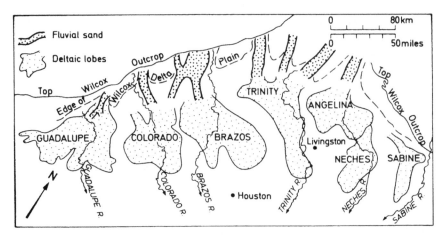

Figure 2 Coincidence of present rivers with their ancient Tertiary deltaic equivalents south of hinge line along the Texas Gulf Coast (Fisher and McGowen 1967, fig. 6).

Michigan River by Swann (1963, fig. 4). This river system (fig. 3) has been oriented south and southwestward across the North American craton since the early Carboniferous. The deltas of this long-lived river system range from thin and widespread with great lateral migration on the craton to thick and geographically stable, where the present Mississippi debouches into the Gulf of Mexico on oceanic crust. On the craton, down-dip delta migration in a single Chesterian cycle in the Mississippian appears to have occurred over distances of as much as 1,000 km in length with lateral shifts up to 350 km.

Part of the evidence for the persistence of this river system is found in the paleo-channel systems of the region's major pre-Pennsylvanian unconformity which has been mapped in great detail, especially in the Illinois Basin (Bristol and Howard 1971). In the Mississippi Embayment the present river rather faithfully follows the axial thickness of the Mesozoic–Cenozoic fill of the embayment, which according to Ervin and McGinnis (1975), is an expression of a reactivated rift that had its beginnings in the late Precambrian (fig. 4).

The most significant fact about the Michigan–Mississippi is its longevity of about 250,000,000 years or about one sixteenth of the earth's history! How many other river systems in the earth's history persisted as long? And how many even longer?

McMillian (1973, fig. 23) made a very widespread reconstruction of a Tertiary paleodrainage system across much of central and eastern Canada, combining available studies of preglacial topography with regional geology

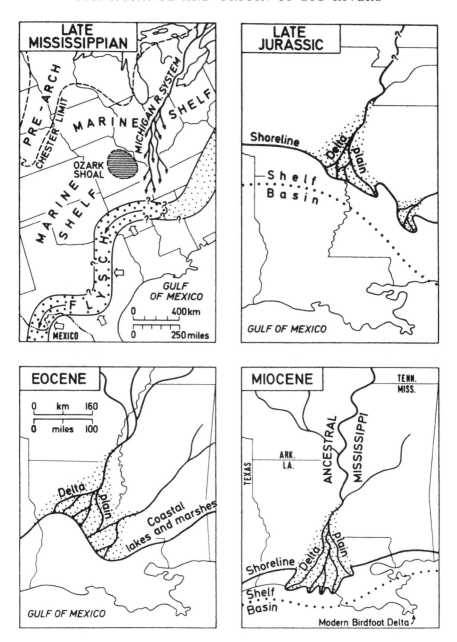

Figure 3 Michigan–Mississippi River system on North American craton has been oriented to the southwest since the early Carboniferous times (Pettijohn *et al.* 1972, figs. 12–16).

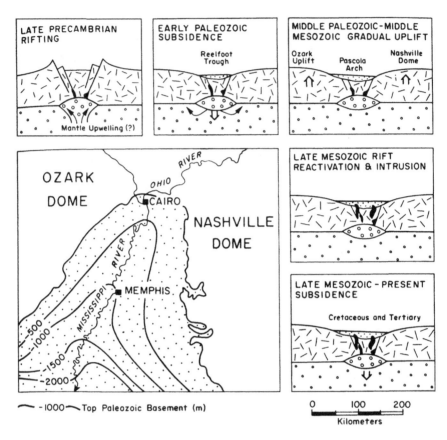

Figure 4 Correspondence between present position of Mississippi River and structural axis of Mississippi Embayment plus inferred geologic history since late Precambrian (redrawn from King 1969, and Erwin and McGinnis 1975, fig. 5).

(fig. 5). This drainage system, now dismembered by Pleistocene glaciation, well illustrates the fact that the river supplying a basin marginal to a craton may have had a source area very, very remote from its basin margins. Consequently, there may be very little relation between the petrography of the sands deposited in the basin and the composition of its neighboring rocks, because the chief source is a well-watered highland perhaps several thousand kilometers or more distant from the depositional basin. Another possibility is the occurrence of large lakes or small inland seas between the chief supplying highland and the final marine basin, a modern example being the flow of the Saskatchewan into Lake Winnipeg, which acts as a sediment trap, with final discharge, over 500 km away, into Hudson Bay. Consequently, sand composition in a marine basin could be almost totally unrelated to the supply from the river's headwaters.

280

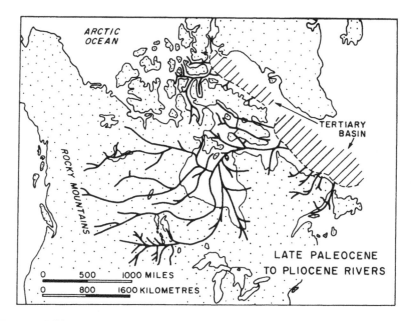

Figure 5 Widespread reconstruction of Tertiary paleoriver in eastern and central Canada (redrawn from McMillian 1973, fig. 23).

Very different is Martin's (1975) paleodrainage study in southwestern Africa, which is based chiefly on outcrop mapping. Here Paleozoic valleys have relief up to 1,000 m, were infilled by Permo–Carboniferous glacial deposits and later by Lower Cretaceous plateau basalts and are now being exhumed. These basalts, which also occur in much of eastern and southern South America, are believed to have resulted from deep tensional fault-ing associated with the separation of South America and Africa. The paleodrainage system mapped by Martin is oriented towards the present Africa coast and thus suggests the presence of a proto-South Atlantic Ocean in later Paleozoic time. Twenty years ago—even ten years ago—how many of us would have thought the mapping of paleo-valley systems, even along an admittedly stable continental margin, could have led to such a conclusion?

The mapping of ancient river systems can also be useful in the tectonic analysis of Mesozoic and Cenozoic mountain chains by helping to identify the uplift of orogenic core zones, as has been nicely demonstrated by Eisbacher *et al.* (1974). Structural reentrants (part of fold belt convex toward foreland), structural salients (part of fold belt concave toward foreland), and longitudinal, intramontane fault systems are major factors controlling late orogenic drainage away from uplifted core zones. The boundary of an uplifted area separates aggradational molasse basins from eroded source areas. Debris from such areas can be carried by rivers flowing in longitudinal, frequently fault-controlled valleys, which emerge from the uplifted source areas in structurally low reentrants, and is deposited in two major ways: in elongate molasse basins, outside of and commonly subparallel to the strike of the orogenic belt of which one good example is the Ganges in India; or deposition may occur in some smaller successor molasse basins within the source area. A good example of both is found in the Cretaceous reconstruction of paleodrainage in parts of western Canada (Eisbacher *et al.* 1974). In addition to facies mapping and paleocurrent studies, petrography is useful in defining specific source regions and their associated river systems. Although post-orogenic river systems such as these are commonly short-lived, perhaps only several million years or less in intensely deforming fold belts, they can be as long as 1,000–2,000 km. Eisbacher *et al.* (1974) also illustrated, discussed, and compared Himalayan drainage with that of the molasse of the Canadian Rockies.

Structural control

In North America the rift valley of the Rio Grande in New Mexico is one of the best documented examples of how a river's course can be controlled by rifting that is most probably related to a major, preexisting flaw in the crust (fig. 6). The summary below is taken from Chapin and Seager (1975) supplemented by the more regional view of Belcher (1975).

The Rio Grande rift is a pull-apart structure formed by the separation of the Colorado Plateau from the craton to the east, which probably began in the early Miocene and most probably produced a series of shallow basins, ones dominantly filled with terrestrial conglomerates, sands and muds, plus volcani-clastic debris and lavas, all of which accumulated in a series of semi-arid and arid bolsons. Continued extension yielded a series of horsts and grabens some of which formed fault block mountains that shed additional fanglomerates. About 3 m.y. ago, judging by the age of the first axial river deposits, through drainage was first established with possibly a change to a wetter climate and/or uplift in the headwaters initiating the through drainage.

The above history was reconstructed by careful field mapping of complex structures and deciphering the sedimentology of the rift fill, including its age dating both by vertebrate paleontology and fission track methods. More important, however, the Rio Grande in New Mexico may be a fairly typical example of the history of a river in a very early phase of rifting in a continental interior—long before subsidence and the localization of a major delta system.

Two studies somewhat different from the foregoing concern hypotheses about the origin of three famous rivers—the Colorado, the Amazon, and Paraná.

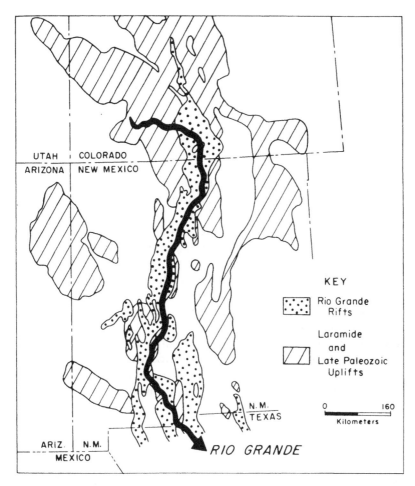

KEY

Rio Grande
Rifts

Laramide
and
Late Paleozoic
Uplifts

0   160
Kilometers

RIO GRANDE

Figure 6 Rio Grande River in New Mexico closely follows rift of Miocene age (redrawn from Chapin and Seager 1975, fig. 2).

McKee *et al.* (1967) reviewed the diverse hypotheses for the origin of the Colorado and defined five broad stages of development of its drainage, although their exact time is still subject to some discussion:

(1) withdrawal of the Cretaceous sea,
(2) early Cenozoic (Laramide) deformation and erosion,
(3) later re-elevation,
(4) mid-Cenozoic volcanism, post lava faulting and additional deposition, and

(5) eastward headward erosion to establish the present course. Notable here are the effects of volcanism, faulting, folding and broad uplift plus the initial withdrawal of an epeiric sea. Withdrawal of an epeiric sea, or what is virtually the same, the formation on a low coastal plain, is probably the most common of all, another good example being the suggested origin of the Tennessee River (Milici 1968).

Because the area is larger and less well known, Grabert's (1971) hypothesis about the Amazon and Paraná drainage systems has fewer specific details but still is very exciting. Citing a widespread lateritic cover up to 40 m thick over much of the central part of the Brazilian shield, he assumed that since the Triassic or even since the late Paleozoic most of this shield has been uplifted and exposed to weathering and inferred that its paleodrainage was chiefly radial toward surrounding seas and oceans. Thus its western rivers should have flowed into the Andean geosyncline. But the mid-Tertiary uplift of the Andes reversed this pattern causing the Amazon to flow eastward (occupying the position of a structural low) and the Paraná to flow southward subparallel to the strike of the Andes. Most probably a widespread inland lake developed during the early stages of this major drainage reorganization.

Another interpretation of the present positions of the Amazon and the Paraná—one that is very imaginative and provocative—is that of de Rezende (1972), who has emphasized the tectonic control of two gigantic *geofracture systems* related to the separation of South America and Africa (fig. 7).

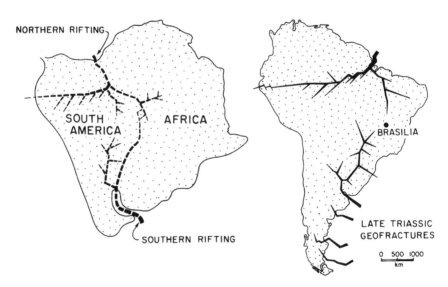

Figure 7 Continental separation and its related geofracture system (redrawn from de Rezende 1972, figs. 4 and 6).

284

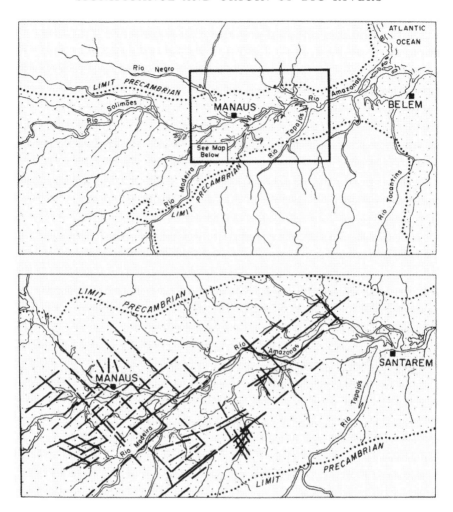

Figure 8 Distribution of Phanerozoic sediments and fault systems in central part of Amazon Basin (redrawn from Cezar de Andrade and Bezerra de Cunha 1969, figs. 1 and 3).

Throughout much of its lower course the Amazon does occupy a combined graben and/or linear structural low which preserves lower Paleozoic sediments as a long, narrow belt extending across the Brazilian Precambrian shield with notable parallelism of rivers and lineaments within it (fig. 8). It has also been suggested that the Paraná in far southwestern Brazil and adjacent Uruguay and Paraguay follows the axis of greatest thickness of Cretaceous basalts simply because the crust is depressed by their extra weight (Cordani and Vandoros 1967, p. 214–215).

Table 1 Methods for Mapping Paleo-River Systems.

FACIES MAPPING

Primarily the recognition of fluvial and deltaic systems plus related shallow and deep water slope deposits including turbidites of associated submarine fans. Low energy delta systems most widely recognized in the ancient.

UNCONFORMITIES

Identified by structure maps of unconformity surfaces, cross sections, paleo-geologic maps, valley fill isopach maps, and isopach maps of interval from a lower datum plane to overlying unconformity surface. See Anderson (1962) for discussion of relative merits of each as well as Martin (1966) and Chenoweth (1967). Best used in the subsurface, but also applicable to the outcrop when good exposures are available.

PALEOCURRENTS

Crossbedding orientation most commonly used and unambiguously indicates paleoslope except in some barrier systems of high-energy deltas. Excellent correlations between unidirectional paleo currents, valley fills of unconformity entrenched river systems, and paleoslope except where there are pronounced buried escarpments that may have locally produced trellised paleodrainage and a complicated pattern of littoral sands (McCubbin 1969). Paleocurrents of last deposited regressive unit in a basin indicate flow of the resultant consequent paleodrainage.

TECTONIC HISTORY

The final integration and "pay off" of many different activities including all of the above plus outcrop mapping, regional structure, and regional geophysics. Virtually all big river systems tend to follow structural lows such as rifts, abortive aulacogens, geofracture systems, broad lows between regional arches, intermontane longitudinal faults, structural entrants and salients, and major trends of folding.

Summary

The above examples—and there are no doubt many, many more—show that the tools required to identify and map ancient river systems are available to virtually everyone and are simply a combination of facies mapping, paleocurrent study, the mapping of unconformities, and a study of regional tectonics—even plate tectonics if we think back to de Rezende's geofracture systems and the location of the Amazon and Paraná Rivers in South America (table 1). Because these methods are applicable anywhere that sedimentary basins remain substantially intact, as they do on most cratons, many more ancient river systems can be identified and mapped throughout much of geologic time, certainly well back into at least the middle Precambrian and perhaps earlier.

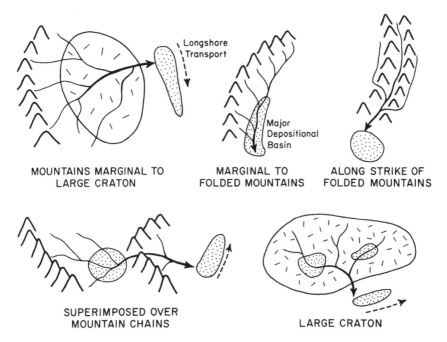

Figure 9 Simplified morphologic assessment of major types of big rivers.

Controlling factors

I found it instructive to examine modern river systems on a world relief map and make a simplified morphologic summary of existing patterns (fig. 9). Four major patterns explain most of the different types of big rivers of the present day. Thus a big river can derive most of its debris from mountains marginal to a large craton (Amazon and Mississippi), may be marginal to a fold belt and flow parallel to it (Ganges and Paraná), may flow along the strike of a mountain chain (Mekong and Magdalena), and may be super-imposed across several mountain chains (Columbia and Danube). The fifth possibility, a river system on a large craton without bordering mountains, was possibly more common in the past when low rather than high relief prevailed.

There is, moreover, a relation between the type of continental coastline and its density of major rivers, the type of coastline being defined as (1) a collision coast (the leading edge of a continent and of island arcs), (2) a trailing-edge coast without compressional deformation, and (3) marginal sea coasts protected by island arcs. Inman and Nordstrom (1971, p. 15) found, in a worldwide study of coastlines, that the distribution of mouths of major rivers is closely related to trailing edge and marginal sea coasts: the 28 largest rivers in the world all discharge across trailing and marginal

287

coastlines where are found 25 of the world's largest deltas. Although there are some exceptions—for example, the Columbia, which is 29th in size and drains across a collision coast—they believe that the above results reflect the continental asymmetry of water sheds—the presence of high orogenic belts on the collision side of continents. An additional factor that favors the formation of large subaerial deltas in small marginal seas is the reduced wave power of that sea. Another possible consequence of this asymmetry is that trailing continental coasts will normally receive more sediment than collision coasts, because they have larger drainage areas. Thus, in broadest view, rivers respond to the megageomorphology of continents, which is relatable to plate tectonics. If, in addition to rifting and downfaulting parallel to a coastline, there is also a fault or rift system at some angle to the coast, it tends to collect and focus major paleodrainage, as is true for the Niger, Amazon (fig. 10), Paraná, and the Mississippi. Where this has continued to the present, there is a Tertiary delta.

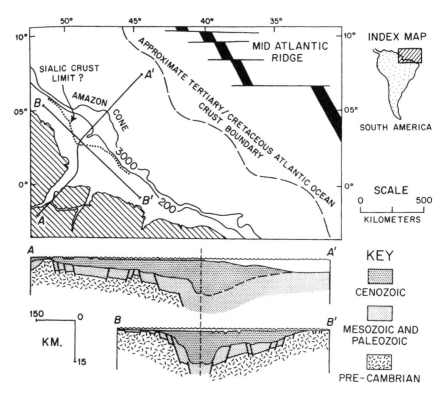

Figure 10 Rift control on lower part of Amazon River (redrawn from Bacoccoli and Texeira 1973, fig. 4).

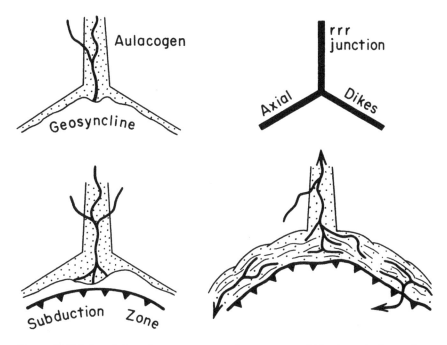

Figure 11 Triple points, aulacogens, mountain systems, and big rivers (redrawn from Burke and Dewey 1973, fig. 2).

Another aspect of plate tectonics that has been suggested as a controlling factor on river systems—in addition to de Rezende's geofractures—are plume generated triple junctions (Burke and Dewey 1973). Mantle plumes—rising mantle material under a continent—are believed to be the precursors of uplift, rifting and uplift-generated triple junctions that may lead eventually to continental break up (fig. 11). Dewey and Bird suggest that the failed arm of the triple junction—the one that does not become a trailing edge of a continent—is a zone of weakness, becomes a rift system, and consequently localizes a major river with a delta at its mouth, the classic example being the Niger River and its delta. This rifting is believed to be the cause of most aulacogens—the long, narrow and deeply filled troughs that extend into cratons. Thus the tectonic history of a region is one of the key factors in its river systems, affecting both their size and longevity.

It seems clear from the foregoing that most large rivers require a large, well-watered continent on which location and orientation of the drainage system depends on the tectonic history—the asymmetry of many continental watersheds which is directly relatable to plate tectonics as well as more local differential subsidence and uplift within a continent. Thus in very broad terms, a major river system depends—for its longevity, size, and

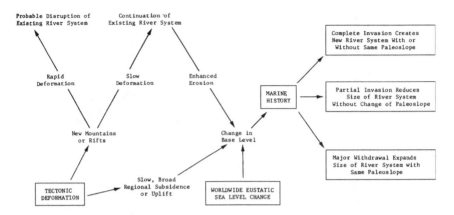

Figure 12 Effects of kind and rate of tectonic deformation and eustatic sea level change on marine history which is the immediate control on a river system. See text for additional effects of climatic change, glaciation and outpouring of lavas.

orientation—mostly on the tectonic and marine history of the region in which it flows (fig. 12), as has been pointed out by Mann and Thomas (1968, p. 187). If tectonic reshaping is rapid, new mountains and basins are created and the old drainage system is dismembered and destroyed. If, on the other hand, tectonic reshaping by local uplifts (or even the development of folded or faulted mountains) of an already established river system is slow, the river system, especially in a humid climate, will not significantly change, because erosion can keep pace with deformation as is probably true for the Danube in eastern Europe. Conversely, if the Danube's watershed had been arid, could it have maintained a course through the fold belt of western Romania? And, of course, a major worldwide eustatic change of sea level can markedly alter the size of river basins, a small rise temporarily eliminating many smaller ones and a major rise, of several hundred meters, perhaps even much of some of the major ones.

Two additional factors that can play a role are continental glaciation and outpouring of lavas, the former possibly completely disrupting and destroying the system (cf., fig. 5) and the latter less likely to do so, at least through most of the earth's history except possibly for its very earliest about 4 b.y. ago.

And finally, a major climatic change through progressive change in rainfall—perhaps induced by the breakup of a very large continent resulting in more rainfall because of greater proximity to the ocean, the rise of a mountain chain, or the drifting of a continent such as Australia into an arid zone, are possibilities that deserve attention, especially when we remember that some long-lasting cratonic based rivers may have persisted for one sixteenth or more of the earth's history. And, of course, in a cratonic setting a major river system could easily be reintegrated, after a period of aridity, through a system of overflowing ponds and lakes in much the same manner as occurs in glacial terrains after the retreat of a continental ice sheet.

Speculations

Below I comment on several intriguing questions that deserve additional consideration by sedimentologists.

What can be said of the possibility of some ancient rivers surpassing the Amazon in size—either measured by length, area, volume of detritus, or discharge? One view of this question is that Pangaea—the hypothetical pre-Gondwanaland megacontinent and possible earlier equivalents—was sufficiently well watered to have had mostly external drainage and thus would have provided many such opportunities. In short, a very large well-watered land mass consisting of welded cratonic blocks with one or more active fold belts would have been a perfect setting for super-Amazons. An opposing view, however, is that of Meyerhoff (1970, p. 32 and 40), who has suggested that such a large continent, because so much of it would have been very far from the ocean, would have been largely arid or semi-arid and thus unable to support super-large rivers. More to the point, however, is not simply its size, but where its mountain ranges occurred in relation to marine air masses; if relief were everywhere low and highlands far inland, aridity could be minimal. In any case, one should recognize the possibility that on such a super-continent a very long river system might diminish in discharge as it flowed from well-watered highland to the ocean—or even possibly disappear in a large inland, saline sea such as the Caspian and thus never reach the world ocean.

And what of the size relationships between depositional basins and the drainage area of their associated rivers? This question is very relevant when a large delta system, one in which both its subaerial and submarine parts are preserved, occurs with its proximal subaerial part next to a continental margin and its submarine and more distal part landward and overlapping a craton. Such an example is the Middle and Upper Devonian delta system in the Appalachian Basin in Pennsylvania and the nearby portions of New York and New Jersey (Walker and Harms 1971), where over 12,000 feet of sediment has accumulated, yet its supplying river system has never been mapped and delineated, because it may have been beheaded and dismembered by an early opening of the Atlantic (Dinley 1975, fig. 4). Hence possible relationships between depositional basins and the size of their supplying drainage basins can be very relevant for paleogeographers and others who reconstruct earth history. At least some data from the modern world can be used to help formulate an answer as shown by plots of the area of the world's 50 largest river basins and the area covered by 269 sedimentary

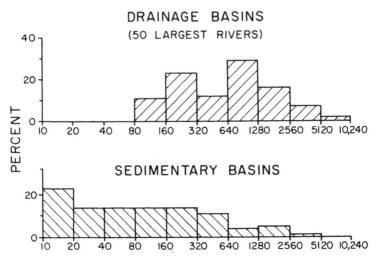

Figure 13 Distribution of the world's sedimentary basins derived from data in Vassoyevich *et al.* (1972, table 2) and areas of the world's 50 largest river basins (Inman and Nordstrom 1971, table 2).

basins (fig. 13). Clearly the area of most sedimentary basins is much smaller than that of the 31 largest watersheds of world rivers even though the west Siberian Basin, the world's largest onshore sedimentary basin, covers an area of close to 3,000,000 km^2 and is surpassed only in the watersheds of the Amazon, Congo, and Mississippi. Again the tectonic setting of the coupled drainage and depositional basins is most probably the determining factor. For example, many small molasse, successor basins appear to have been filled by but one or two small rivers with watersheds of comparable size, the watersheds having, however, moderate to even high relief. On the other hand, many intracratonic basins, when supplied by big rivers, were probably only fractionally as large as their watersheds. And, because cratonic rivers can be very long-lived, one river system can contribute to a series of basins as it migrates across a craton as has the Mississippi and its ancestors in North America. Still another possibility is a large river on a craton whose delta is on oceanic crust and depositing in deep water at the shelf edge such as that of the Mississippi or even beyond it as in the subsea fan of the Amazon. Under these conditions, the depositional basin will be a very small fractional area of the supplying watershed. Thus, in terms of area, large rivers draining a craton and emptying onto a continental margin underlain by oceanic crust most probably will fill thick elongate basins whose area will perhaps be but between 5–10% of the drainage basin of the river.

Almost 20 years ago Kuenen (1958, p. 329) commented on the size of flysch basins in relation to the size of their supplying rivers. He noted that as a rule most flysch basins are generally filled by large volumes of fine to medium sands, interbedded with shales, in contrast to the coarser detritus, especially gravels and conglomerates, of many molasse basins. From this he implied a distant source of moderate relief for most flysch basins whereas he envisioned a proximal highland shedding coarse debris for the majority of molasse basins. A contributing factor, unmentioned by Kuenen is, of course, that a subareal delta commonly intervenes between the site of flysch deposition and its supplying river thus providing a sediment trap for coarser debris. And, if not a subareal delta, possibly an estuary.

Another problem is to consider the role of vegetation on river systems. The two key questions are, "What effect, if any, did an absence of forest cover in pre-Devonian time have on the depositional facies of alluvial and subaerial deltaic deposits?" and "What effect did this absence have on the compositional maturity of big river sands—the kind of sandsize detritus they carried?"

Schumm (1968, p. 1583) has given an appraisal of how pre-Devonian rivers and erosion in the pre-Devonian probably differed from those of today.

During the time before significant terrestrial vegetation appeared, the hydrologic regimen of arid regions would have been much as today, and even in humid regions the products of weathering were swept from hillslopes and divides into well-developed networks of stream channels. Unlike the humid regions of today where weathering products are stored on hillslopes as well as in alluvial valleys, the bare prevegetation hillslopes would have lost material as rapidly as it weathered to a size that could be transported. These immature sediments accumulated in valleys and on pediment surfaces. Sediment production proceeded at a rapid rate in hot humid climates, but an arid-appearing landscape existed everywhere. Bedload channels were ubiquitous and wide braided streams occupied the entire floor of alluvial valleys. Upon leaving the sediment source areas the channels would not have been confined, and vast alluvial-plain piedmont deposits formed. Depending upon the environment of deposition and tectonics, either great thicknesses of clastics accumulated or sheetlike deposits formed, as floods spread across piedmont areas, reworking and sorting the sediments.

This view suggests mostly braided deposits with a minimum of interbedded muds with the contained sandstones generally lacking well-defined shoaling upward, point bar sequences. At present, unfortunately, there are an insufficient number of sedimentologic studies to permit a statistical comparison of depositional facies of pre- and post-Devonian alluvial deposits to test Schumm's vivid picture, even though some fining upward cycles have been recognized as far back as the middle Precambrian (Hoffman 1969, p. 453–455). Certainly, a comparison of the relative abundance of fluvial subfacies throughout geologic history would seem to be most rewarding. Schumm's view also implies a predominance of compositionally immature sands, a view which is totally at variance with what is found in the pre-Devonian record, where mineralogically mature and supermature sandstones abound. Considering only North America, think of the supermature St. Peter and Simpson sandstones of Ordovician quartz arenites described by Ketner (1966) in the western Cordillera, and the many thick, clean quartz arenites of the middle Precambrian in Canada. Hence, either there were plants such as lichens, the partial pressure of CO_2 was very much greater than today causing greatly accelerated weathering of feldspar and rock fragments resulting in compositionally mature and supermature sandstones, or tidal currents were stronger and high energy beaches more prevalent than today (cf. Swett *et al.* 1971, p. 411–412), the combination tending to eliminate sandsized rock fragments.

Schumm's idea also has some implications for both discharge and area. Clearly discharge would be more episodic with a much greater proportion of flood to normal discharge. Size of the drainage basin would seem to be but little affected, however.

The last question—and the most speculative of all—is to me the most intriguing. What can be said of the question of the very earliest rivers in the earth's history, when the crust was all volcanic, well before the first cratons developed? Anaheusser (1975, p. 15–16) has described some early metamorphosed Archean sediments from South Africa, which he believes are precratonic in age. An outstanding analysis of this problem was made almost 50 years ago by a Swedish geologist, Assar Hadding (1929) in his paper, "The First Rains and Their Geological Significance." Hadding's vivid perception (p. 21) is well worth reading today.

> An ash-covered earth, no traces of water, no traces of life. A hot surface of earth under a hot atmosphere, rich in water-vapour and no doubt also in ash-dust. Volcanic cones and folding mountains of greater height than the present ones, tablelands (future continental platforms) and wide depressions (future oceans), equally void and ash coloured. Thus may the earth be pictured before the condensation of water.

Then the first rains fall.

> What a wonderful impression is made on us by that which now takes place! Falling drops hit a ground never touched by water, a ground so hot that it has not been able to hold any liquid water. What salts were not to be leached out! How rapidly could not the

rills cut deep grooves in the ash! Never have the rivers been so full of mud as during this first condensation. Never have depressions been so quickly silted up, never have the oceanic basins had such a supply of salts, and never has the circulation of water by condensation and evaporation been greater than in this early period.

And later Hadding then asks (p. 27),

Did the first condensation occur simultaneously all over the globe or did it start in certain places, for example, the poles?

Could these first rivers—and no doubt they were big ones—have been born, comparatively speaking, as gigantic flash floods leaving behind on lowland areas a vast rubble-strewn, poorly sorted deposit of volcanic debris ranging from the finest volcanic silts to gigantic volcanic, angular boulders? Following Hadding, did such rivers first develop in polar regions, where condensation may have occurred first? Although we may never know whether or not the earth's early river system first developed in polar regions, are comparable deposits preserved not in the earth's crust, as Hadding thought, but rather on the Martian landscape where Viking I and II have revealed a vast system of now abandoned, gigantic braided streams (Anonymous 1976 and Casados 1976). Recently, Sharp and Malin (1975) discussed these Martian channels, based on the photography of Mariner 9 (1971–1972) and concluded they were most probably produced by fluvial erosion and deposition occurring perhaps as long as $3–3.5 \times 10^9$ years ago. Twenty years ago, even five years ago, how many of us would have ventured to predict that Hadding's perceptive deductions about the earth's primitive landscape could be more readily seen on the surface of Mars than in the sedimentary record of the earth?

Acknowledgements

I am greatly indebted to Farouk El-Baz; Francis J. Pettijohn; G.N. Rassam; Raymond Siever; Richard Williams; and M. Gordon Wolman, for their thoughtful help and suggestions. Mr. Richard Spohn very kindly and patiently helped me with many of the references and Mrs. Ruth Scott made all the illustrations. Mrs. Jean Carrol patiently and carefully typed and improved the manuscript.

References cited

ANAHAEUSSER, C.R., 1975, The geological evolution of the primitive earth: evidence from the Barberton Mountain Land: Univ. Witwatersrand, Econ. Geol. Research Unit, Inf. Cir. 98, 22 p.

ANDERSON, MARVIN J., 1962, Paleodrainage patterns: their mapping from subsurface data and their paleogeographic value: Am. Assoc. Petrol. Geologists Bull. 46, p. 398–405.

ANONYMOUS, 1976, Mars: An informal photographic essay: EOS Trans. Amer. Geophysical Union, v. 57, p. 708–711.

BACOCCOLI, GUISEPPE, and TEIXEIRA, ALVARO ALVES, 1973, Tendencias futuras da exploracao petrolifera na margem continental Atlantica: B. Tec. Petrobras, v. 16, p. 17–76.

BELCHER, ROBERT C., 1975, The geomorphic evolution of the Rio Grande: Baylor Geol. Studies Bull. 29, 64 p.

BRIDGEWATER, D.; KETO, L.; MCGREGOR, V.R.; and MYERS, J.S., 1976, Archaean gneiss complex of Greenland, in ESCHER, A., and WATT, W.S., eds., Geology of Greenland: Geological Sur. of Greenland (in press).

BRISTOL, H.M., and HOWARD, R.H., 1971, Paleogeologic map of the sub-Pennsylvanian Chesterian (Upper Mississippian) surface in the Illinois Basin: Ill. Geol. Sur. Circ. 458, 14 p.

BURKE, J.F.; KEVIN, J.F.; and DEWEY, J.F., 1973, Plume-generated triple junctions: Key indicators in applying plate tectonics to old rocks: J. Geol., v. 81, p. 406–433.

CEZAR DE ANDRADE, C.A., and BEZERRA DE CUNHA, F.M., 197 1, Revisao geologia da bacia Paleozoacia do Amazonas: Annis XXV Congresso Brasileiro de Geologia, p. 93–112.

CHENOWETH, P.A., 1967, Unconformity analyses: Bull. Am. Assoc. Petrol. Geologists, v. 51, p. 4–27.

CONYBEARE, C.E.B., 1972, Petroleum accumulations in ancient river sands: A paleogeographic view: Inter. Geol. Congress 24th Montreal Session 5, p. 59–73.

CHAPIN, C.E., and SEAGER, W.R., 1975, Evolution of the Rio Grande Rift in the Socorro and Las Cruces areas, in SEAGER, W.R. et al., eds., Guidebook of the Las Cruces Country: New Mexico Geol. Soc. 26th Field Conf., Nov. 13, 14, and 15, p. 297–321.

CORDANI, U.G., and VANDOROS, P., 1967, Basaltic rocks of the Paraná Basin, in Problems in Brazilian Gondwana Geology, First Int. Symposium Gondwana Stratigraphy and Paleontology, Curitiba, p. 208–231.

DE REZENDE, W.M., 1972, Post Paleozoic geotectonics of South America related to plate tectonics and continental drift: Sociedade Brasileira de Geologia, Anais do XXVI, Congresso Brasileiro de Geologia, p. 205–210.

DINLEY, D.L., 1975, North Atlantic Old Red Sandstone—some implications for Devonian paleogeography, in YORATH, C.J.; PARKER, E.R.; and GLASS, D.J., eds., Canada's continental margins and offshore petroleum exploration: Canadian Soc. Petroleum Geologists Mem. 4, p. 773–790.

DRAKE, D.E., 1976, Suspended sediment transport and mud deposition on continental shelves, in STANLEY, D.J., and SWIFT, D.J.P., eds., Marine sediment transport and environment management: New York, Wiley, p. 127–158.

EISBACHER, G.H.; CARRIGY, M.A.; and CAMPBELL, R.B., 1974, Paleodrainage pattern and late orogenic basins of the Canadian Cordillera, in DICKINSON, W.R., ed., Tectonics and sedimentation: Soc. Econ. Paleon. Mineral., Sp. Pub., v. 22, p. 143–166.

ERVIN, C. PATRICK, and McGINNIS, L.D., 1975, Reelfoot rift: reactivated precursor of the Mississippi Embayment: Geol. Soc. Amer. Bull., v. 86, p. 1287–1295.

FISHER, W.L., and McGOWEN, J.H., 1967, Depositional systems in the Wilcox Group of Texas and their relationship to occurrence of oil and gas: Trans. Gulf Coast Assoc. Geol. Soc., v. 27, p. 105–125.

GILLIGAN, A., 1920, The petrography of the Millstone Grit of Yorkshire: Geol. Soc. London Quart. Jour., v. 75, p. 251–294.

GRABERT, H., 1971, Die Prae-Andine Drainage des Amazonas Stromsystems: Muenster Forsch. Geol. Palaeontol., no. 20–21, p. 51–60.

HADDING, ASSAR, 1929, The first rains and their geological significance: Geol. Foren. Forhandlingen, v. 51, p. 19–29.

HOFFMAN, PAUL, 1969, Proterozoic paleocurrents and depositional history of the East Arm fold belt, Great Slave Lake, Northwest Territories: Canadian J. Earth Sci., v. 6, p. 441–462.

INMAN, D.L., and NORDSTROM, C.E., 1971, On the tectonic and morphologic classification of coasts: Jour. Geology, v. 79, p. 1–21.

KING, P.B. (Compiler), 1969, Tectonic map of North America: U.S. Geological Sur., scale 1 : 5,000,000.

KUENEN, P.H., 1957, Longitudinal filling of oblong sedimentary basins: Kon. Nederl. Geol. Mijnbouwkundig Genootschap, Geol. Ser., v. 18, p. 189–195.

—— 1958, Problems concerning source and transportation of flysch sediments: Geol. en Mijnbouw (N.W. Ser.), v. 20, p. 329–339.

—— 1966, Geosynclinal sedimentation: Geologische Rundschau, v. 56, p. 1–19.

MANN, C.J., and THOMAS, W.A., 1968, The ancient Mississippi River: Trans. Gulf Coast Assoc. Geol. Soc., v. 18, p. 187–204.

MARTIN, H., 1975, Structural and palaeogeographical evidence for an Upper Paleozoic sea between southern Africa and South America, in K.S.W. ed., Gondwana geology: Canberra Australian Nat. Univ. Press, p. 37–59.

MARTIN, R., 1966, Paleogeomorphology and its application to exploration for oil and gas (with examples from western Canada): Bull. Am. Assoc. Petrol. Geologists, v. 50, p. 2277–2311.

McCUBBIN, D.G., 1969, Cretaceous strike-valley sandstone reservoirs, northwestern New Mexico: Bull. Am. Assoc. Petrol. Geologists, v. 52, p. 2114–2140.

McGOOKEY, P.P., 1975, Gulf coast Cenozoic sediments and structure: An excellent example of extra-continental sedimentation: Trans. Gulf Coast Geol. Societies, v. 25, p. 104–120.

McKEE, E.D.; WILSON, R.F.; BREED, W.J.; and BREED, C.S., 1967, Evolution of the Colorado River in Arizona: Museum of Northern Arizona, Flagstaff, 44 p.

McMILLIAN, N.J., 1973, Shelves of Labrador Sea and Baffin Bay, Canada, in McCROSSAN, R.G., ed., The future petroleum provinces of Canada—their geology and potential: Canadian Society Petroleum Geologists, Mem. 1, p. 473–517.

MEYERHOFF, A.A., 1970, Continental drift: Implications of paleomagnetic studies, meteorology, physical oceanography and climatology: J. Geol., v. 78, p. 1–51.

MILICI, R.C., 1968, Mesozoic and Cenozoic physiographic development of the lower Tennessee River: In terms of the dynamic equilibrium concept: J. Geol., v. 76, p. 472–479.

MOORE, G.T., and ASQUITH, D.O., 1971, Delta: Term and concept: Geol. Soc. Amer. Bull. 82, p. 2563–2568.

PETTIJOHN, F.J.; POTTER, P.E.; SIEVER, R., 1972, Sand and sandstone: Berlin–Heidelberg–New York, Springer-Verlag, 618 p.

SCHUMM, S.A., 1968, Speculations concerning paleohydrologic controls of terrestrial sedimentation: Geol. Soc. Amer. Bull. 79, p. 1573–1588.

SHARP, R.P., and MALIN, M.C., 1975, Channels on Mars: Geol. Soc. Amer. Bull. 86, p. 593–609.

SMITH, A.E., JR., 1975, Ancient deltas: Comparison maps, in BROUSSARD, M.L., ed., Deltas, models for exploration: Houston Geol. Soc., p. 531–555.

SÖHNGE, P.G., 1974, Sedimentary ore deposits: A review of research trends: Trans. Geol. Soc. S. Africa, v. 77, p. 159–168.

STAFF, 1976, Mars in formal photographic essay: EOS, Trans. Amer. Geophysical Union, v. 57, p. 713–724.

SWANN, D.H., 1963, Classification of Genevievian and Chesterian (Late Mississippian) rocks of Illinois: Ill. Geol. Sur. Rept. Inv. 216, 91 p.

SWETT, K.; KLEIN, G. DEV.; and SMIT, D.E., 1971, A Cambrian tidal sand body—the Ereboll Sandstone of northwest Scotland: An ancient–recent analog: J. Geol., v. 79, p. 400–415.

VASSOYEVICH, N.B. et al., 1972, Sedimentary basins: Int. Geol. Congress, 24th Session (Montreal), Sec. 5, p. 187–194.

WALKER, ROGER G., and HARMS, J.C., 1971, The "Catskill delta": a prograding muddy shoreline in central Pennsylvania: J. Geol., v. 79, p. 381–399.

11

TIME, SPACE, AND CAUSALITY IN GEOMORPHOLOGY

S.A. Schumm and R.W. Lichty

Source: *American Journal of Science* 263 (1965): 110–19.

Abstract

The distinction between cause and effect in the development of landforms is a function of time and space (area) because the factors that determine the character of landforms can be either dependent or independent variables as the limits of time and space change. During moderately long periods of time, for example, river channel morphology is dependent on the geologic and climatic environment, but during a shorter span of time, channel morphology is an independent variable influencing the hydraulics of the channel.

During a long period of time a drainage system or its components can be considered as an open system which is progressively losing potential energy and mass (erosion cycle), but over shorter spans of time self-regulation is important, and components of the system may be graded or in dynamic equilibrium. During an even shorter time span a steady state may exist. Therefore, depending on the temporal and spacial dimensions of the system under consideration, landforms can be considered as either a stage in a cycle of erosion or as a system in dynamic equilibrium.

Introduction

Current emphasis on the operation of erosion processes and their effects on landforms (Strahler, 1950, 1952) not only has opened the way to new avenues of research but also introduces the possibility of misunderstanding the role of time in geomorphic systems. As Von Bertalanffy (1952, p. 109) put it, "In physical systems events are, in general, determined by the momentary conditions only. For example, for a falling body, it does not matter how it has arrived at its momentary position, for a chemical reaction it does not

299

matter in what way the reacting compounds were produced. The past is, so to speak, effaced in physical systems. In contrast to this, organisms appear to be historical beings". From this point of view, although landforms are physical systems and can be studied for the information they afford during the present moment of geologic time, they are also analogous to organisms because they are systems influenced by history. Therefore, a study of process must attempt to relate causality to the evolution of the system.

It is the purpose of this discussion to demonstrate the importance of both time and space (area) to the study of geomorphic systems. We believe that distinctions between cause and effect in the molding of landforms depend on the span of time involved and on the size of the geomorphic system under consideration. Indeed, as the dimensions of time and space change, cause-effect relationships may be obscured or even reversed, and the system itself may be described differently.

Acknowledgments

The writers wish to thank several colleagues who reviewed this paper. The comments and criticism of the following were most helpful: M. Morisawa, Antioch College; A.N. Strahler, Columbia University; R.J. Chorley, University of Cambridge; and John Hack, R.F. Hadley, and H.E. Malde of the U.S. Geological Survey. Malde not only made numerous suggestions for improvement of the paper, but he also suggested the format for tables 1 and 2.

Time, space, and the fluvial cycle of erosion

The description of the changes occurring in a landscape with time, according to the cycle of erosion as propounded by Davis, is encountered less frequently in current geomorphic writings. In the study of geomorphic processes earth scientists are applying themselves to modern problems, and the spatial-temporal range of their research is considerably curtailed. This is necessary if the knowledge of processes is to be developed; however, even in this work the historical aspect of landscape evolution or the time dimension should not be neglected. The neglect of time leads to confusion and needless controversy. For example, recent papers by Hack (1960) and Chorley (1962) may startle some geomorphologists by the rejection of the time dimension, which is a major concern of the geologist. The discussion that follows is an attempt to show that what Hack and Chorley suggest need not be a break with tradition but is simply a method of considering the landscape within narrow temporal limits.

Hack (1960, p. 85) suggests that many elements of the landscape are in dynamic equilibrium with the processes acting upon them; that is, "The forms and processes are in a steady state of balance and may be considered as time independent". He compares this condition with that of a soil undergoing erosion at the surface at the same rate as the lower boundaries of the soil horizons move downward into the regolith (Nikiforoff, 1959). Hack (1960, p. 94) continues his argument as follows: "The theory of dynamic equilibrium explains topographic forms and the differences between them in a manner that may be said to be independent of time. The theory is concerned with the relations between rocks and processes as they exist in space. The forms can change only as the energy applied to the system changes".

This concept is synonymous to that of the physical system described by Von Bertalanffy in which, "The past is, so to speak, effaced". Nevertheless, after excluding time from his system, Hack considers it in the following qualification, "It is obvious, however, that erosional energy changes through time and hence forms must change". A change of erosional energy can be initiated by many factors, of which diastrophism or climate change are the most obvious. In addition, with the passage of time erosional modification of the landforms themselves will affect erosional energy. Therefore, it appears impossible to exclude time and history from a consideration of landforms except during the study of purely empirical relations among variables, which may or may not reflect causality.

Chorley likewise feels that freedom from the historical approach is desirable, because research efforts are then directed toward a study of the rate and manner of operation of erosional processes, the empirical relations that exist between a landscape and its components, and the relations between the erosion processes and the landform. Chorley (1962, p. 3) illustrates the difficulty of reconciling the two approaches, the Davisian cycle of erosion and dynamic equilibrium, as follows: "In the former, the useful concept of dynamic equilibrium or grade rests most uncomfortably; in the latter . . . the progressive loss of a component of potential energy due to relief reduction imposes an unwelcome historical parameter".

To resolve the controversy resulting from these two viewpoints it may be necessary to think only in terms of large and small areas or of long and short spans of time. A choice must be made whether only components of a landscape are to be considered or whether the system is to be considered as a whole. Also, a choice must be made as to whether the relations between landforms and modern erosion processes are to be considered or whether the origin and subsequent erosional history of the system is to be considered. In table 1 an attempt is made, using a hypothetical drainage basin as an example, to demonstrate that the concepts of cyclic erosion with time and timeless dynamic equilibrium are not mutually exclusive.

The variables listed in table 1 are arranged in a hierarchy we believe approximates the increasing degrees of dependence of the variables considered. For example, time, initial relief, geology, and climate are obviously the dominant independent variables that influence the cycle of erosion. Vegetational type and density depend on lithology and climate. As time passes the relief of the drainage system or mass remaining above base level

Table 1 The status of drainage basin variables during time spans of decreasing duration.

Drainage basin variables	Status of variables during designated time spans		
	Cyclic	Graded	Steady
1. Time	Independent	Not relevant	Not relevant
2. Initial relief	Independent	Not relevant	Not relevant
3. Geology (lithology, structure)	Independent	Independent	Independent
4. Climate	Independent	Independent	Independent
5. Vegetation (type and density)	Dependent	Independent	Independent
6. Relief or volume of system above base level	Dependent	Independent	Independent
7. Hydrology (runoff and sediment yield per unit area within system)	Dependent	Independent	Independent
8. Drainage network morphology	Dependent	Dependent	Independent
9. Hillslope morphology	Dependent	Dependent	Independent
10. Hydrology (discharge of water and sediment from system)	Dependent	Dependent	Dependent

is determined by the factors above it in the table, and it, in turn, strongly influences the runoff and sediment yield per unit area within the drainage basin. The runoff and sediment yield within the system establish the characteristic drainage network morphology (drainage density, channel shape, gradient, and pattern) and hillslope morphology (angle of inclination and profile form) within the constraints of relief, climate, lithology, and time. The morphologic variables, in turn, strongly influence the volumes of runoff and sediment yield which leave the system as water and sediment discharge.

Among the variables listed on table 1, every cause appears to be an effect and every effect a cause (Mackin, 1963, p. 149); therefore, it is necessary to set limits to the system that is considered. Obviously neither the causes of geology, climate, and initial relief nor the effects of water and sediment discharge concern us here.

The three major divisions of table 1 are time spans which are termed cyclic, graded, and steady. The absolute length of these time spans is not important. Rather, the significant concept is that the system and its variables may be considered in relation to time spans of different duration.

Cyclic time, of course, represents a long span of time. It might better be referred to as geologic time, but in order to keep the terminology of the table consistent, cyclic is used because it refers to a time span encompassing

302

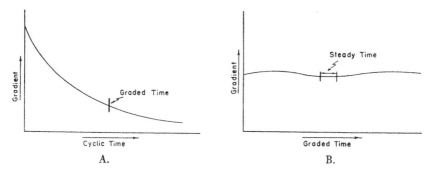

Figure 1 Diagrams illustrating the time spans of table 1. Channel gradient is used as the dependent variable in these examples.
a. Progressive reduction of channel gradient during cyclic time. During graded time, a small fraction of cyclic time, the gradient remains relatively constant.
b. Fluctuations of gradient above and below a mean during graded time. Gradient is constant during the brief span of steady time.

an erosion cycle. Cyclic time would extend from the present back in time to the beginning of an erosion cycle.

Consider a landscape that has been tectonically stable for a long time. A certain potential energy exists in the system because of relief, and energy enters the system through the agency of climate. Over the long span of cyclic time a continual removal of material (that is, expenditure of potential energy) occurs and the characteristics of the system change. A fluvial system when viewed from this perspective is an open system undergoing continued change, and there are no specific or constant relations between the dependent and independent variables as they change with time (fig. 1a).

During this time span only time, geology, initial relief, and climate are independent variables. Time itself is perhaps the most important independent variable of a cyclic time span. It is simply the passage of time since the beginning of the erosion cycle, but it determines the accomplishments of the erosional agents and, therefore, the progressive changes in the morphology of the system. Vegetational type and density are largely dependent on climate and lithology, but they significantly influence the hydrology and erosional history of a drainage basin. If all the independent variables are constant except time, then as time passes the average relief and mass, volume of material remaining within the drainage system, will decrease. As the relief or mass of the system changes so will the other dependent morphologic and hydrologic variables.

With regard to space or the area considered, it is possible to consider an entire drainage system or any of its component parts during a cyclic time span. For example, the reduction of an entire drainage system or only the decrease in gradient of a single stream may be considered (fig. 1a) during cyclic time.

303

The graded time span (table 1) refers to a short span of cyclic time during which a graded condition or dynamic equilibrium exists. That is, the landforms have reached a dynamic equilibrium with respect to processes acting on them. When viewed from this perspective one sees a continual adjustment between elements of the system, for events occur in which negative feedback (self-regulation) dominates. In other words the progressive change during cyclic time is seen to be, during a shorter span of time, a series of fluctuations about or approaches to a steady state (fig. 1b). This time division is analogous to the "period of years" used by Mackin (1948, p. 470) in his definition of a graded stream by which he rules out seasonal and other short-term fluctuations, as well as the slow changes that accompany the erosion cycle.

As an erosion cycle progresses, more and more of the landscape may approach dynamic equilibrium. That is, the proportion of graded landforms may increase, and it seems likely that temporary graded conditions become more frequent as time goes on. However, it is apparent that during this time span the graded condition can apply only to components of the drainage basin. The entire system cannot be graded because of the progressive reduction of relief or volume of the system above base level, which occurs through export of sediment from the system. A hillslope profile or river reach, however, may be graded. Therefore, unlike cyclic time when no restriction was placed on space or area considered, graded time is restricted to components of the systems or to smaller areas.

During a graded time span, the status of some of the variables listed on table 1 changes. For example, time has been eliminated as an independent variable, for although the system as a whole may be undergoing a progressive change of very small magnitude, some of the components of the system will show no progressive change (that is, graded streams and hillslopes). Initial relief also has no significance because the landform components are considered with respect to their climatic, hydrologic, and geologic environment (Hack, 1960), and initial relief with time has been designated as not relevant on table 1.

In addition, some of the variables that are dependent during a long period of progressive erosion become independent during the shorter span of graded time. The newly independent hydrologic variables, runoff and sediment yield, are especially important because during a graded time span they take on a statistical significance and define the specific character of the drainage channels and hillslopes, whereas during a cyclic time span there is a progressive change in these morphologic variables.

The geomorphic variables of hillslope and drainage network morphology of graded time may be considered as "time-independent" in the meaning of Hack (1963, written communication). That is, relict features may not be present, and the landforms may be explained with regard to the independent variables without regard to time.

During a steady time span (table 1) a true steady state may exist in contrast to the dynamic equilibria of graded time (fig. 1b). These brief periods of time are referred to as a steady time span because in hydraulics steady flow occurs when none of the variables involved at a section change with time. The landforms, during this time span, are truly time independent because they do not change, and time and initial relief have again been eliminated as independent variables. During this time span only water and sediment discharge from the system are dependent variables.

Obviously the steady state condition is not applicable to the entire drainage basin. Although an entire drainage basin cannot be considered to be in a steady state over even the shortest time span, yet certain components of the basin may be. For example, a stream over short reaches may export as much water and sediment as introduced into the reach, yet the river as a whole is reducing its gradient in the headwaters (cyclic erosion). In addition, the entire drainage basin may be losing relief as hillslopes are lowered (cyclic erosion); however, segments of the hillslopes may remain at the same angle of inclination and act as slopes of transportation (steady state), or they may retreat parallel, maintaining their form (dynamic equilibrium), but the volume of the drainage basin is being reduced nevertheless. Thus over short periods of time and in small areas the steady state may be maintained. Over large areas progressive reduction of the system occurs, and this is true over long periods of time.

The preceding discussion and the relations presented in table 1 and figure 1 have the sole object of demonstrating that, depending on the time span involved, time may be either an extremely important independent variable or of relatively little significance to a study of landforms.

Fluvial morphology and hydraulics

In this section a specific example of river channel morphology and hydraulics will be cited to illustrate how, as time spans are shortened, there is a shift from dependence to independence among the variables and how, during the shortest span of time, an apparent reversal of cause and effect may occur.

In table 2 an attempt has been made to illustrate the effect of time span on the interrelations between dependent and independent variables of a river system. A similar table has been presented by Kennedy and Brooks (in press) to compare independent and dependent variables in the flume and field situations. As in the preceding discussion, it is the time span or duration of a time period that is important, but Pleistocene and Recent geomorphic history is such that in many areas it is possible to divide the time involved into three spans, termed geologic, modern, and present, each shorter and more recent than the span that precedes it.

It is almost impossible to assign temporal boundaries to these time spans because their duration will vary with each example considered. Nevertheless,

Table 2 The status of river variables during time spans of decreasing duration.

River Variables	Status of variables during designated time spans		
	Geologic	Modern	Present
1. Time	Independent	Not relevant	Not relevant
2. Geology (lithology and structure)	Independent	Independent	Independent
3. Climate	Independent	Independent	Independent
4. Vegetation (type and density)	Dependent	Independent	Independent
5. Relief	Dependent	Independent	Independent
6. Paleohydrology (long-term discharge of water and sediment)	Dependent	Independent	Independent
7. Valley dimension (width, depth, and slope)	Dependent	Independent	Independent
8. Mean discharge of water and sediment	Indeterminate	Independent	Independent
9. Channel morphology (width, depth, slope, shape, and pattern)	Indeterminate	Dependent	Independent
10. Observed discharge of water and sediment	Indeterminate	Indeterminate	Dependent
11. Observed flow characteristics (depth, velocity, turbulence, et cetera)	Indeterminate	Indeterminate	Dependent

geologic time in this sense begins during the Pleistocene Epoch, because it is during the higher discharges of the glacial stages that the width and depth of many valleys were established (Dury, 1962). Since the Pleistocene there undoubtedly have been some modifications of these valleys, but their major characteristics were determined both by the higher discharge of the Pleistocene and the post-Pleistocene adjustments to a changed hydrologic regimen. For this reason geologic time probably should end at perhaps 5,000 or 10,000 years ago. However, to keep table 2 consistent with table 1 geologic time is defined as beginning 1,000,000 years ago and extending to the present.

During geologic time (table 2), time, geology, and climate are independent variables. Initial relief is not included because geologic time does not extend back to the origin of the system. Vegetation and relief are considered to be dependent variables, as is the paleohydrology of the system, which controlled the dimensions of the valley during geologic time. For geologic time we can know little or nothing about the dependent variables in the

306

hierarchy below valley dimensions (table 2), and these variables are classed as indeterminate.

In the time span termed modern (table 2) (arbitrarily defined as the last 1,000 years) the number of independent variables increases, and some previously indeterminate variables become measurable. For example, valley dimensions become independent during modern time because they were defined by the paleohydrology of geologic time and inherited from geologic time. The mean discharge of water and sediment during modern time is also considered an independent variable, because it determines the morphology of the modern channel. Only channel morphology is dependent during modern time. Modern time is 1,000 years in duration; therefore, the observed discharge and flow characteristics, which can be measured only during a brief span of time, are indeterminate.

During the short span of present time (defined as 1 year or less), channel morphology assumes an independent status because it has been inherited from modern time. The present or observed discharge of water and sediment and flow characteristics can be measured at any moment during present time, and these variables are no longer indeterminate.

It is during the brief span of present time that the possibility of an apparent reversal of cause and effect may occur, due to feedback from the dependent to the independent variables. For example, a major flood during this brief span of time might so alter the flow characteristics that a modification of channel dimensions and shape could occur. Just as water depth and velocity can be adjusted in a flume to modify sediment transport, so there is a feedback from flow velocity to sediment discharge and channel morphology. That is, as discharge momentarily increases, sediment that was previously stationary on the channel floor may be set in motion. The resulting scour, albeit minor, will influence channel depth, gradient, and shape. Thus, short term changes in velocity can cause modification of some of the independent variables.

These modifications are usually brief and temporary, and the mean values of channel dimensions and sediment discharge are not permanently affected. Nevertheless, a temporary reversal of cause and effect can occur, which, when documented quantitatively, may be a source of confusion in the interpretation of geomorphic processes. This is best demonstrated by comparing the conflicting conclusions that could result from studying fluvial processes in the hydraulic laboratory and in a natural stream. The measured quantity of sediment transported in a flume is dependent on the velocity and depth of the flowing water and on flume shape and slope. An increase in sediment transport will result from an increase in the slope of the flume or an increase in discharge. In a natural stream, however, over longer periods of time, it is apparent that mean water and sediment discharge are independent variables, which determine the morphologic characteristics of the stream and, therefore, the flow characteristics (table 2, modern time). Furthermore, over very long periods of time (geologic) the independent variables of geology, relief, and climate determine the discharge of water and sediment with all other morphologic and hydraulic variables dependent. Both Mackin (1963) and Kennedy and Brooks (in press) used this identical example to illustrate the need to consider how time spans are relevant to the explanation of fluvial phenomena.

Kennedy and Brooks (in press) state it thusly (Q and Q_s are water and sediment discharge),

> Streams are seldom if ever in a steady state (because of finite time required to change bed forms and depth) and transitory adjustments are accomplished by storage of water and sediment. Water storage is relatively short (hours and days) and occurs simply by the increasing of river stage or overbank flooding; sediment storage (+ or −) occurs by deposition or scour. Thus for the short term, Q_s may be considered a dependent variable, with departures of the sediment inflow from the equilibrium transport rate being absorbed in temporary storage (for months or years). But in the long term the river must assume a profile and other characteristics for which on the average the inflow of water and sediment equals the outflow; consequently for this case (called a "graded" stream by geologists), Q and Q_s are . . . independent variables.

Table 2 is more than a scholarly exercise in sorting variables, for although, as our knowledge increases, it will require modification, it can be of immediate use in the consideration of problems of fluvial morphology. For example, assuming we have arranged the variables in correct order, if flow characteristics are dependent variables for a modern river then currents or the helicoidal flow measured at river bends should not be the cause of meanders. In other words, sinuosity of the river (the ratio of valley slope to channel gradient) influences the flow character not the converse. Specifically, helicoidal flow exists because of a meander, which in turn may exist partly because of past conditions of flow (paleohydrology).

As another example, in a set of data collected for Great Plains rivers it was found that a highly significant correlation exists between valley slope and stream gradient. At first, this appears trivial for slope is correlated with slope. However, if table 2 is correct, the slope of modern valleys is an independent variable dependent on the paleohydrology of geologic time, and the existing, modern channel slope is a dependent variable.

Variations in the sinuosity of the Great Plains rivers have been explained (Schumm, 1963) by the decrease in post-Pleistocene discharge and a change in the amount and type of sediment transported by the rivers. Depending on the changes in sediment load and discharge, the present stream may require a slope identical with the valley slope (sinuosity is 1) or much less than the valley slope (sinuosity is greater than 1 but usually less than 2.5). The valley floor, therefore, is the surface upon which the present river flows, and depending on changes in sediment and discharge, the river may flow at the slope of the valley or at a slope considerably less. Valley slope, as indicated on table 1, is an independent variable exerting a control on stream gradient.

Conclusions

The distinction of cause and effect among geomorphic variables varies with the size of a landscape and with time. Landscapes can be considered either as a whole or in terms of their components, or they can be considered either as a result of past events or as a result of modern erosive agents. Depending on one's viewpoint the landform is one stage in a cycle of erosion or a feature in dynamic equilibrium with the forces operative. These views are not mutually exclusive. It is just that the more specific we become the shorter is the time span with which we deal and the smaller is the space we can consider. Conversely when dealing with geologic time we generalize. The steady state concept can fit into the cycle of erosion when it is realized that steady states can be maintained only for fractions of the total time involved.

The time span considered also influences causality, as the sets of independent and dependent variables of tables 1 and 2 show. If the variables were not considered with respect to the time span involved, in many cases it would be difficult to determine which variables are independent. Mackin's (1963) and Kennedy and Brooks' (in press) suggestions forestall any arguments between workers in the laboratory and workers in the field. In the same manner the disparate points of view of the historically oriented geomorphologist and the student of process can be reconciled.

References

Chorley, R.J., 1962, Geomorphology and general systems theory: U.S. Geol. Survey Prof. Paper 500-B, 10 p.

Dury, G.H., 1962, Results of seismic exploration of meandering valleys: Am. Jour. Sci., v. 260, p. 691–706.

Hack, J.T., 1960, Interpretation of erosional topography in humid temperate regions: Am. Jour. Sci., v. 258-A, Bradley v., p. 80–97.

Kennedy, J.F., and Brooks, N.H., in press, Laboratory study of an alluvial stream at constant discharge: Federal Interagency Sedimentation Conf. Proc., Jackson, Miss., 1963, in press.

Mackin, J.H., 1948, Concept of the graded river: Geol. Soc. America Bull., v. 59, p. 463–511.

—— 1963, Rational and empirical methods of investigation in geology, in Albritton, C.C., Jr., ed., The fabric of geology: Reading, Mass., Addison-Wesley Publishing Co., p. 135–163.

Nikiforoff, C.C., 1959, Reappraisal of the soil: Science, v. 129, p. 186–196.

Schumm, S.A., 1963, Sinuosity of alluvial rivers on the Great Plains: Geol. Soc. America Bull., v. 74, p. 1089–1100.

Strahler, A.N., 1950, Equilibrium theory of erosional slopes approached by frequency distribution analysis: Am. Jour. Sci., v. 248, p. 673–696, p. 800–814.

—— 1952, Dynamic basis of geomorphology: Geol. Soc. America Bull., v. 63, p. 923–938.

Von Bertalanffy, Ludwig, 1952, Problems of life: London, Watts and Co., 216 p.

12

CANONS OF LANDSCAPE EVOLUTION

L.C. King

Source: *Bulletin of the Geological Society of America* 64 (1953): 721–52.

Abstract

The manner in which epigene landscapes evolve is examined and discussed. Slope flattening as a general process of landscape evolution is rejected, and with it Davis' concept of the peneplain. Landscape evolution by scarp retreat and pedimentation is accepted, and several of its consequences are examined.

The opinions derived are expressed at the end of the paper as a series of canons of landscape development

Historical approach

The scientific study of landscape began in the latter half of the nineteenth century and is associated especially with the names of J.W. Powell, G.K. Gilbert, and W.M. Davis in America and Albrecht Penck in Europe. The work of the American trio has always overshadowed in the public regard that of Penck senior, and modern thought still follows closely the pronouncements and opinions of the last of the trio—William Morris Davis. Davis had a remarkably analytical mind, he reasoned lucidly and logically; moreover he wrote with facility, grace, and copiousness, so that his ideas were easily and pleasantly assimilated, became widely disseminated, and carried conviction. He was an inspired teacher, and his ideas found a ready audience and ready acceptance.

Particularly was this so with his concept of the "Cycle of Erosion" under which a landscape uplifted and subjected to the forces of subaerial erosion was first vigorously dissected by the deeply cut valleys of the rejuvenated river systems, with consequent increase of relief, and later reduced by weathering and lowering of the interfluves to a lowland of faint relief which he termed a *peneplain*. The type areas cited for these changes were the areas in which Davis worked in his early professional life, chiefly from Pennsylvania to New England; and thus something of a parochial standard of comparison was instituted for other regions of the globe. Landscapes which did not conform to this standard, such as desert landscapes, were regarded as "abnormal", "accidental", or exceptional.

In his later years, Davis wrote many fine papers, yet none to my mind was more truly philosophical nor more important than his contribution of 1930 wherein the earlier view was revised and many true homologies demonstrated between the landscapes of arid and of humid regions. This harmony calls for revision of a number of the earlier concepts, and particularly the standard of "normality" in landscape types, but the necessary revisions have not been undertaken, with any degree of completeness, either by Davis or by others.

The review of what Davis called "the Cycle of Erosion" is thus broadly the theme of the present paper; but before proceeding with our argument we shall adduce a little more of the historical background.

The cyclic concept did not originate with Davis; it was stated first in Powell's *Exploration of the Colorado River of the West* (1875), in the following terms (we quote from *Geographical Essays*)

> "aerial forces carried away 10,000 feet of rocks by a process slow yet unrelenting until the sea again rolled over the land" and the evenly denuded surface is referred to as "the record of a long time when the region was land".

Powell recognized very clearly that the later stages of reduction in the landscape would be very slow:

> The degradation of the last few inches of a broad area of land above the level of the sea would require a longer time than all the thousands of feet that might have been above it, so far as this degradation depends on mechanical process—that is by driving or flotation; but here the disintegration by solution and the transportation of material by the agency of fluidity come in to assist the slow processes of mechanical disintegration, and finally perform the chief part of the task.

The role of solution as a significant agent of reduction in the later stages of landscape development is open to question. Many senile land surfaces bear a residuum of insoluble matter as a subsurface crust—laterite, silcrete, caliche are examples—but all these hinder degradation in a marked degree, and it is doubtful whether solution is of any real significance in the final reduction of noncalcareous terrains. We shall ignore it in our further discussion.

Davis' specific addition to Powell's theory was to visualize and describe the landforms that, under the action of erosional processes, were evolving in the landscape toward a final stage.

The contribution of Walther Penck (1924), in which the forms of hillslopes are related to continuing earth movements, has not found any enduring acceptance. Nevertheless, Penck junior understood certain classes of landforms (*e.g.*, hillslopes) better than Davis did, and several of his viewpoints such as the parallel retreat of scarps and the difference between senile and initial landscapes are real in nature and make an advance on the earlier Davisian technique.

Landscape studies have now advanced to the stage where statistical analysis is being applied to certain classes of data (Strahler, 1950b). Used critically, so that it does not lead to erroneous conclusions that become established in the minds of scientists, this method should tend to much more exact definition of landforms, processes, and principles. Undoubtedly, if the basic data can be supplied from observation and measurement of landscapes, the method is ideal for disentangling the complexities of forms and processes governed respectively by many factors or variables. What other equally satisfactory approach could there be, for instance, to the complexities of stream flow and grade?

Acknowledgment

The script has benefited from a critical reading by my colleague Dr. T.J.D. Fair, whose contributions are far more important than the few footnotes to his name indicate. Several phrases and sentences have been reframed at his suggestion, and to him I am indebted for the canons numbered (1), (12), and (22). Several of the opinions expressed have also been derived from his careful field studies and published works. All his assistance is gratefully acknowledged.

Statement of the problem

Davis' original conception of landscape evolution under the subaerial processes of rainfall, running water, and weathering has often been summed up in a series of progressive sections in which an uplifted landscape is depicted as first dissected under stream incision with the production of narrow valleys. After the streams have attained grade (defined as the condition in which all major irregularities have been eliminated from the thalweg), the rate of river incision is reduced to negligible proportions, so that the valley bottoms are lowered very little during the remainder of the cycle. According to Davis, the valley sides are then reduced under weathering and surface creep, and perhaps wash, to ever flatter and flatter angles, until they meet upon the interfluves and destroy the last remnants of the initial surface.

Figure 1 Contrasted Slope Profiles.
Widening of valleys and change in slope profile during the cycle. Right, youth to
old age, according to W.M. Davis. Left, with parallel scarp retreat, according to
W. Penck. (From Davis, 1930)

As the slopes continue to flatten, the interfluves are lowered more rapidly
than the river beds until only "a lowland of faint relief remains". To this
ultimate landform Davis assigned the title of *peneplain*. The concept is
illustrated by Davis' own diagram which is here reproduced (Fig. 1). The
progressive lowering of the interfluves under weathering is a vital concept in
the "Normal Cycle" as visualized by Powell, by Davis, and by Cotton.

The Davisian concept has not passed without challenge, notably in his
own Harvard University, where the strong school of Kirk Bryan accepted
parallel retreat of scarps (*q.v.*). Some authorities have indeed rejected the
cyclic concept altogether (Penck, 1924); others, with whom we align ourselves,
have accepted the general concept of a cycle of landforms developed under
erosion, while considering that the detailed forms and sequences depart
considerably from those visualized and adduced by Davis. The essential
differences of viewpoint lie in the interpretation of slope forms and the
manner of hillslope development.

In summary, this opinion, after beginning the cycle of erosion with stream
incision into an uplifted land surface exactly as in the Davisian model,
would place limits to the amount of hillslope flattening, regarding the slopes
as attaining a stable gradient (defined for local conditions) after which
the upper parts of the slopes retreat parallel to themselves. A well-known
example is due to Kirk Bryan, 1922, p. 42–46. At the foot of the slope is left
a *pediment* sloping gently down to the river. This pediment is concave in
profile, so that, when the interfluves are consumed by parallel scarp retreat,
the opposing pediments from adjacent valleys would meet without forming
a broad convexity across the interfluve (Figs. 1, 2).

The ultimate landscape under this philosophy, being composed of many
coalescing pediments, is termed a *pediplain*, and it is distinguishable at
sight from the Davisian peneplain by the multi-concave instead of multi-
convex nature of its surface and the presence of steep-sided rather than
gentle residuals. The fundamental difference between the two is, of course,
not merely a difference of surface form—which indeed may be dubious, for
concavities might occur in relation to peneplains—but of different history
and mode of development.[1]

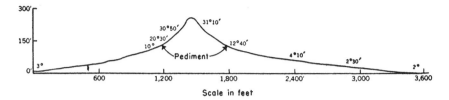

Figure 2 Slopes of a Natal Monadnock.
Surveyed by T.J.D. Fair. On either side is a broad concave pediment, and the hillsides
are steep (30°). Summit convexity is small. This type of monadnock agrees much
better with Penck's doctrine of parallel scarp retreat than with Davis' flattening
slope profiles and widespread convexity (Fig. 1).

The two concepts of an erosion cycle are largely exclusive. Over any given
area, one can be true, but not the other. In the viewpoint of many geologists,
the so-called "Normal Cycle of Erosion" appears under humid climates,
and the Pediplanation cycle under semiarid and arid climates. The viewpoint
is not without absurdity. One can understand that a characteristic cycle and
set of landforms should be generated in frigid zones where water is frozen
into ice and the whole mechanism of abrasion alters, where cirques are
formed by thaw and freeze under specific conditions (D.W. Johnson), and
where roches moutonnées are overridden by masses of solid ice; but that
mere differences in amount and incidence of rainfall, evaporation, and like
factors should have such far-reaching effects as to result in two entirely
distinct cycles requires further investigation. The primary agent moulding
the landscape in both humid and arid cases is water flow, and this should
produce comparable results in both types of region.

This, indeed, was realized by Davis who summarized his position (1930,
p. 145) on landforms of humid and of arid regions:

> The comparison ... will show that, while various contrasted
> features are truly enough unlike in certain respects, their unlikeness
> is rather a matter of degree than of kind. The apparently unlike
> features are really homologous; their resemblances, once recognized,
> are much more striking than their differences. The same may be
> said of the erosional and degradational processes by which appar-
> ently unlike forms are sculptured.

He instituted comparisons between the processes of weathering both
mechanical and chemical, soil creep and rill wash, stream pattern and action,
the respective mantles of waste and types of hillslope (p. 147).

> As the various processes of arid and of humid erosion are thus
> seen to differ in degree and manner of development rather than in

315

nature, and as their differences in degree and manner are wholly due to differences in their climates, so the forms produced by the two erosions may be shown to differ in the degree to which certain elements of form are developed rather than in the essential nature of the elements themselves.

Davis here stood upon the threshold, with the unification of epigene landscapes before him. But the shackles of his earlier thought, nearly 50 years upon him, were not to be broken so easily, and he never entirely accepted real identity of landscape evolution under humid and under arid influences, which is one of our present canons. Thus (1930, p. 147),

> ... although arid erosion resembles humid erosion in many respects the two kinds of erosion nevertheless differ so much in process and in product that they cannot be clearly understood if they are briefly brought together as examples of normal erosion, as has been done by several of the German writers. ... The unlikenesses of the two erosions deserve explicit treatment.

Davis still held, moreover, to his earlier opinion that hillslopes flatten progressively under humid regimes (1930, p. 148):

> A striking difference between humid and arid (granitic) mountains is found in the decreasing angle of slope of the first as the erosion cycle advances, in contrast to the constant angle of slope in the second.

From this we dissent, and regard the manner of hillslope evolution as dependent upon factors such as bedrock, relief, vegetal cover, etc., rather than upon direct climatic influences. We dissent likewise from his separation of monadnocks from inselbergs. The latter are merely a special case of the former, characteristic of granitic and gneissic terrains. Monadnocks are not, as Davis claimed, "pale forms of weakening convexity and lessening slopes" but are truly concave in profile as indeed is the classic example, Mt. Monadnock itself. Davis' theoretical viewpoint of peneplanation obtrudes here and causes misinterpretation of visible landforms.

Still later (1932, p. 408), rejecting Penck's doctrine of parallel retreat of slopes, Davis asserts that:

> ... the retreat of a valley side is usually accompanied by a decrease in the steepness of its slope as well as by the development of a convex profile at its top and a concave profile at its base whatever its original profile may have been. This is largely because a shoulder ... either angular or sharply rounded, at the top of a valley side

yields more rapidly to the attack of weathering and creeping on its two faces than do the more nearly plane surfaces adjoining the shoulder; and at the same time the coarser detritus supplied by the steeper part of the slope and delivered to its base demands a stronger declivity there for its further downhill carriage by wash and creep than the same detritus will require when it has been reduced to finer texture as it advances. Hence even a square-shouldered top edge of a valley-side . . . may soon wear back into a round shouldered top and the . . . round shoulder of short arc and rapid curvature will in later time change into a longer arc of gentler curvature. The concave basal curve is likewise enlarged as it retreats.

It all reads so logically, yet in my experience mapping major scarps and valley sides that have existed, some of them even from Mesozoic time (1944; 1947; 1951a), Penck was right about parallel scarp retreat, and Davis was wrong. Scarp retreat is normally more rapid than pronounced rounding. Otherwise these scarps would have been smoothed out long ago; on the contrary, they are still as steep as is consistent with the rock types of which they are composed (Pl. 1, fig. 1). The final court of appeal is the brutal facts of Nature, as displayed in landscape (Fig. 2).

The nature of hillslopes

Evolution of hillslopes according to W.M. Davis

Davis considered that valley sides behaved rather as rill profiles of indefinite lateral extent, and applied to them the same concept of "grade" that he had used in the development of river profiles. Indeed, the rules governing valley side or hillslope development did not differ, in his view, from those governing flow, transport, and grade in a river system.

Thus Davis said (1909, p. 266) on the development of graded valley sides:

When the migration of divides ceases in late maturity, and the valley floors of the adjusted streams are well graded, even far towards the headwaters, there is still to be completed another and perhaps even more remarkable sequence of systematic changes than any yet described: this is the development of graded waste slopes on the valley sides. It is briefly stated that valleys are eroded by their rivers, yet there is a vast amount of work performed in the erosion of valleys in which rivers have no part. It is true that rivers deepen the valleys in the youth and widen the valley floors during the maturity and old age of a cycle, and that they carry to the sea the waste denuded from the land; it is this work of transportation to the sea that is peculiarly the function of rivers, but the material to be transported is supplied chiefly by the action of the weather on the steeper inconsequent slopes and on the valley sides. The transportation of the weathered material from its source to the stream in the valley bottom is the work of various slow-acting processes, such as the surface-wash of rain, the action of ground water, changes of temperature, freezing and thawing, chemical disintegration and hydration, the growth of plant roots, the activities of burrowing animals. All these cause the weathered rock waste to wash and creep slowly downhill. . . . In the first place a waste sheet moves fastest at the surface and slowest at the bottom, like a water-stream.

In the second place . . . waste sheets normally begin to establish a graded condition at their base and then extend it up the slope of the valley-side whose waste they "drain". Then follows a most important passage (p. 268): "Just as graded rivers slowly degrade their courses after the period of maximum load is past, so graded waste sheets adopt gentler and gentler slopes when the upper ledges are consumed and coarse waste is no longer plentifully shed to the valley sides below. A changing adjustment of a most delicate kind is here discovered. When the graded slopes are first developed they are steep, and the waste that covers them is coarse and of moderate thickness; here the strong agencies of removal have all they can do do dispose of the plentiful supply of coarse waste from the strong ledges above, and the no less plentiful supply of waste that is weathered from the weaker rocks beneath the thin cover of detritus. In a more advanced stage of the cycle, the graded slopes are moderate, and the waste that covers them is of finer texture and greater depth than before; here the weakened agencies of removal are favored by the slower weathering of the rocks beneath the thickened waste cover, and by the greater refinement (reduction to finer texture) of the loose waste during its slow journey. In old age, when all the slopes are very gentle, the agencies of waste removal must everywhere be weak, and their equality with the processes of waste supply can be maintained only by the reduction of the latter to very low values. The waste sheet thus assumes great thickness—even fifty or a hundred feet—so that the progress of weathering is almost *nil*; at the same time the surface waste is reduced to extremely fine texture, so that some of its particles may be moved even on faint slopes. Hence the occurrence of deep soils is an essential feature of old age, just as the occurrence of bare ledges is of youth.

Again, of Old Age: "Maturity is passed and old age fully entered upon when the hilltops and the hillsides, as well as the valley floors, are graded".

It has been deemed necessary to quote fairly fully the older concept of hillslope evolution because I found on a recent tour of Europe and North

America that many geologists in those regions still accepted it as "normal" or standard.

As late as 1950 in a study by Strahler, we encounter this same concept of flattening, graded slopes in an extended form. His equilibrium theory of erosional slopes (1950a) states that:

> slopes maintain an equilibrium angle proportional to the channel gradients of the drainage system and are so adjusted as to permit a steady state to be maintained by the process of erosion and transportation under prevailing conditions of climate, vegetation, soils, bedrock and initial relief or stage. Thus both slopes and streams are graded. . . . As the landmass is reduced both slopes and stream gradients are reduced, being slowly and continuously regraded to maintain approximate equilibrium. As the correlation of stream gradients with slopes suggests, the decline of stream gradient is accompanied by slope reduction. This concept of maintenance of a steady state by slow readjustment is essentially Davis's concept of landmass development in the normal cycle.

But all this fine Davisian armchair philosophizing bears scant semblance of reality to workers from other regions. Its relation to hillslopes and valley sides as observed in the field is often remote. Such features are infinitely more complex than Davis with his single criterion of "grade" would have us believe; and, while the general tenet of "grade" may be ideally true if no other factors existed, certain of Davis' conclusions are demonstrably untrue. We cite but three.

(1) Deep soils do not necessarily, or even characteristically, appear upon old-age surfaces. The plateau landscapes of the high African interior are much older, for instance, than any of the land surfaces of the eastern United States with which Davis was early familiar. Yet soils upon them are universally thin (often only a few inches) and poor in quality. Soils on landscapes a twentieth as old in Europe are deep and fertile. Landscapes of continental extent reverse the Davisian conclusion!

(2) The philosophy calls for progressively flatter slopes, measured from interfluve to stream bed, as the cycle proceeds (flattening of interfluves). While some slopes, or many, may be shown to have flattened in the course of their history, multitudes of hillslopes have clearly not done so. Frequently, in a homogeneous region, the maximum slopes measured, upon large and small hills alike, cluster closely about a mean, affording proof positive that, once a stable angle has been achieved, the slopes have not flattened any more, but have retreated parallel to themselves. Many ancient landscapes, moreover, bear steep-sided residuals (Pl. 3, figs. 1, 3). There have, in all these instances, clearly been no changes in the factors governing slope form, such as is implied under the concept of flattening.

319

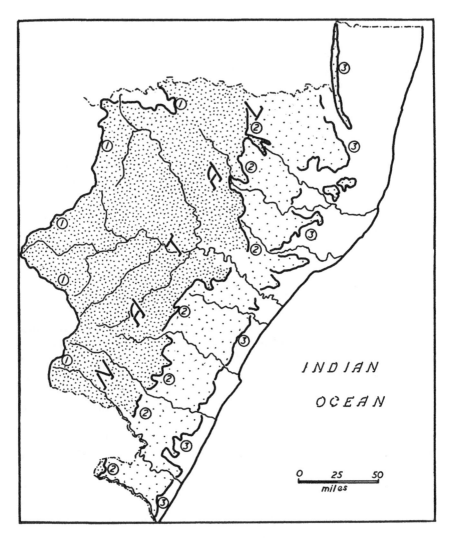

Figure 3 Cyclic Erosion Scarps in Natal, South Africa.
Showing diagrammatically the major river systems and the trends of the major cyclic erosion scarps (in each case over 1000 feet high). These scarps have retreated successively westward scores of miles from an original position near the present coast. The Drakensberg, the oldest, has retreated over 100 miles since the late Mesozoic and is still wall-like and over 4000 feet high (Pl. 1). The scarps are (1) Drakensberg, (2) Ixopo-Pietermaritzburg-Greytown scarp, (3) coastal-hinterland scarp.

And if further proof were needed, what of steep mountain scarps like the wall of the Drakensberg (Pl. 1, fig. 1), which are oriented at right angles to the flow of major river systems. These scarps, generated tectonically in the remote geologic past, in intimate relation to the initiation of new erosion cycles, have retained their steepness unimpaired despite retreat of scores of miles (Fig. 3). The Drakensberg happens to constitute a watershed, but other scarps of similar form run through Natal athwart the middle courses of the major rivers (King, 1947) (Fig. 3). Similar features have been mapped or noted in other parts of the world. Even granting that a majority of these features have been tectonically induced, why have they retreated and why has their steepness not disappeared in the long ages since they were initiated?

(3) The development of a broad summit convexity upon mature and senile divides is a necessary corollary of the Davisian hypothesis. Such convexity is often absent from these situations, a point which we shall discuss under the heading of Peneplain or Pediplain.

The Equilibrium Theory of Slope Development is surely countered by consideration of the processes acting upon hillslopes. Only while the streams are actively incising themselves or undercutting the adjacent spurs and hillslides on meander curves do they exercise control over the forms of adjacent hillslopes. Once the streams cease this activity, the hillslopes are moulded by a set of forces peculiar to themselves and independent of the stream. Hence, with stable base level, hillslope evolution is independent of the later stages of the river cycle. Indeed, in these stages, the river cycle is dependent upon the forms and evolution of the hillsides through the waste—abundant or sparse, coarse or fine—that they supply. In the pedimented landscape of Figure 2 of Plate 1, for instance, how could the river affect the nature of the adjacent hillslopes without re-incision? On the other hand, the effects of hillslope evolution should be immediately reflected in the fluvial regime.

Hillslope elements

Only when hillslopes are analyzed into their several components and the behavior under erosion of each part is examined separately can much progress be made in the study of hillslope evolution. This necessary analysis was undertaken by Wood (1942), and much of the modern study of cyclic erosion begins at this point.

Wood recognized that four distinct elements are sometimes present within the compass of a single hillside. From top to bottom (Fig. 4) these are: (1) the waxing slope (convex), (2) the free face (rock outcrop), (3) the talus or debris slope, (4) the waning slope (including pediments) (concave).

The free face is the outcrop, on the steepest part of the slope, of local bedrock. As the outcrop weathers back, the debris slides clear and builds the

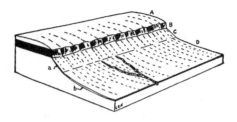

Figure 4 Elements of a Hillside Slope.
A. Waxing slope, B. Free face, C. Debris slope, D. Waning slope (pediment in the sketch). a. talus, b. soil.

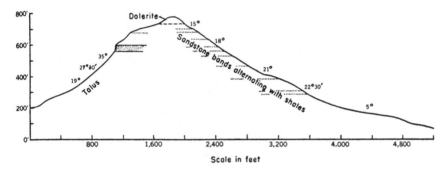

Figure 5 Effect of Structure on Hillslopes.
The single dominating sandstone band (left) generates a steeper slope than several minor bands (right). After Fair.

talus or debris slope below. The declivity assumed by the debris slope is controlled by the size of the fragments, which come to rest at their angle of repose. A talus of boulders 3–4 feet in diameter may have a declivity of as much as 35°. If the talus is not removed with sufficient rapidity, its upper edge encroaches upon the free face, which may ultimately be buried. Debris upon the talus slope weathers to smaller size, which tends to flatten the angle of slope toward the base. But, in general, as the hillslope is worn back, the debris slope maintains the same proportion, so that the hillslope retreats with constant declivity. This it does apparently even though little or no debris be present. In many districts, hillslopes exhibit remarkable likeness of gradient, which could be so only if the hills, large and small, had maintained throughout their erosional history a reasonably constant angle of slope; that is, if the slopes and scarps retreat parallel to themselves[2].

In such districts, where free faces or at least debris slopes abound, the absence of slope flattening is clearly demonstrated. As will be appreciated, hillslopes in the early and very late stages of the erosional cycle depart somewhat from this dictum, but experience and measurement show that, from the time major hillslopes have become stabilized until they lose much of their importance as landscape features, they normally evolve without decided flattening. Parallel retreat of slopes is dependent solely upon forces acting upon the slope. It is independent even of minor changes of local base level.

The agencies which effect the retreat of a hillside scarp are weathering, downhill movement of debris under gravity, sheet wash, and the attack of the gully heads of numerous small streams or rills which correspond, in a measure, to the free face. Of these, the gully heads are usually by far the most important.

Free faces and debris slopes are usually best developed in regions of moderate to high relief. In regions of low relief, these two elements of the hillside slope may be absent, and the waxing slope from above then makes conjunction with the waning slope from below. Under these circumstances, the slope development by parallel retreat is hindered or aborted. Here is an important principle! Large scarps normally exhibit considerable free faces and debris slopes, and hence such scarps undergo parallel retreat. So arises the curious fact, which I have noted in previous publications, that major cyclic erosion scarps in many continents retreat, apparently, at a sensibly uniform rate. This rate is between 1 foot in 150 years and 1 foot in 300 years. Smaller scarps and hillsides, however, may retreat with extreme slowness. Apparently, if the waxing and waning slopes meet so that the free face and debris slope are eliminated, parallel retreat of the scarp may be so slowed as almost to cease. The landscape of countries like the British Isles where concavoconvex slopes are widespread may thus be altering only with extreme slowness compared with landscapes possessing clear scarps with free faces and debris slopes. In this connection, the low relief of most of the British Isles is to be noted.

The waxing slope at the top of a hillside is convex in section. In simple instances it corresponds to the edge of the weathered zone on top of firm bedrock which crops out on the hillside as the free face. As the edge of this zone is unbuttressed, the weathered mantle naturally moves off under soil creep. The curve is convex because more waste must pass a point lower down the slope in a unit of time than passes a point higher up. Convex slopes are thus moulded by soil creep (Davis, 1892; Gilbert, 1909). Lawson's explanation (1932) of the convexity as due to rainwash over saturated soil is not here admitted as a general case. All freely cut hydraulic profiles are concave.

Waning slopes at the base of the hillside exhibit in section the typical hydraulic curve of a surface modelled by water erosion. This is so whether the waning slope merges gently upward into a convex profile of the upper hillside (e.g., the concavo-convex slopes of Southern England) or whether it abuts with more or less sharp angle against the base of the free face or debris slopes (e.g., Southern Rhodesia). All gradations between these two extremes appear in nature.

Beneath the surface wash of the waning slope, either of two subsurface states may exist. First, the subjacent firm rock may pass upward through progressively weathered mantle into soil, in which case the concave surface profile is probably developed only in weathered materials and by surface wash; or, second, the solid rock is sharply truncated by an erosion bevel above which lies an accumulation of transported debris. This latter is the typical condition upon pediments. Of course, landforms that were originally cut-rock pediments may, if the cycle of hillslope development is arrested, deteriorate into the first condition as the transported mantle is slowly removed and weathering proceeds beneath it into the bedrock. Some of the concavities of lower hillslope profiles of northern latitudes may thus have originated as pediments, though their origin as such may no longer be demonstrable.

Normally veneered with detritus, most if not all of which is transported, pediments are essentially cut-rock features, and bare rock is not uncommonly exposed where the veneer thins away toward the base of the commanding hillslope. Here often appears a distinct break in the profile separating the hillslope from the pediment (Pl. 1, fig. 2; Pl. 4, fig. 1). In areas of weak rocks, the break is usually smoothed out into an open curve, but where the bedrock is hard and there is little or no accumulated waste the junction between pediment and free face may be so sharply angular that a boot can scarcely be placed in it (Pl. 2, fig. 1). Sometimes, such abrupt transitions are sharpened by joining, but this is not necessary, and many of the boundaries intersect the joint systems at an angle.

Instances have also been quoted of nicks or inverted pediments. Bryan regarded any explanation of such nicks other than by weathering and rainwash as fantastic. Indeed, many may be shown to be located along semihorizontal joints. They are special features, and are not essential to pediments or pediment theory.

Occasionally, where the inner zone of the pediment is of bare rock or where the pediment extends as a rock fan into a scarp, it is convex over small areas, but such details, though not uncommon, merely represent incomplete stages in the production of the inner pediment.

Often the angular relation in the bedrock is masked by debris fallen or washed from above. Fair (1947, p. 109) has regarded the zone as one in which the coarse waste of the debris slope is broken down under weathering ready for transportation across the lower declivity of the pediment. "Thus the rate at which waste is comminuted determines in no small measure the sharpness of the angle between talus and pediment". Both Lawson and Bryan have also attributed the abrupt change of slope between hillside and pediment to a relatively sudden change from coarse debris on the hillside to finer debris on the pediment. But this can not be the whole truth for the change is even more abrupt, as we have seen, when no debris is present. The essence of the matter seems to be that there is a fundamental change of

Table 1 Typical Gradients upon Pediments*.

(a)	3° to 5°50′ over 4,700 ft:
	5°50′ to 8°50′ over 200 ft.
(b)	2°50′–5°10′ over 2,800 ft:
	5°10′–9°20′ over 300 ft.
(c)	1°–4°50′ over 2,600 ft:
	4°50′–12°40′ over 500 ft.
(d)	2°–5° over 2,250 ft:
	5°–11°30′ over 500 ft.
(e)	4°–4°50′ over 2,150 ft:
	4°50′–13° over 650 ft.
(f)	1°10′–5° over 1350 ft:
	5°–12°30′ over 300 ft.

* From Fair (1947, p. 110), who says:

> Though it rises to a maximum angle of 13 degrees the pediment is essentially a slope
> of less than 5 degrees over by far the greater part of its (profile) length.

geomorphic process acting upon the steep and the gentle slopes respectively to mould them into their present forms. Gravity for instance ceases to be a potent factor, and the activity of running water changes.

Pediments originate and grow by retreat of the hillside elements (free face and debris slope) above them. Thereafter, as their concave profiles guarantee, they are modelled under the influence of running water. The curve so generated is obviously a function of the age and width of the pediment (Table 1).

In certain regions of the globe—*e.g.*, the eastern and southern districts of Southern Rhodesia (King, 1951a)—pediments constitute 60 per cent or more of the landscape. Indeed, King (1949b) has regarded them as possibly the most important of all landforms. The pediment, he has there written, "is the fundamental form to which most, if not all, subaerial landscapes tend to be reduced [under erosion], the world over".

Pedimentation

Much thought has been devoted to the mechanisms by which pediments originate and grow, and a diversity of opinion has been expressed.

McGee (1897, p. 108), attributed the cutting of the wide cut-rock surfaces of the Sonora district, Mexico, to the action of sheet flood. But sheet floods of the magnitude described by him are rare, while pediments can be ubiquitous, so, while both may appear at their best under dry climates, there is not necessarily any genetic relation between them. Sheet floods way mould pediments that already exist, but how do they generate pediments?

Johnson (1931; 1932a; 1932b) laid great emphasis upon lateral planation by streams where they emerge from the mountain front to trim back the scarp and originate the pediment. He pointed to a number of rock fans at the mouths of valleys emerging through the scarp and considered that the scarp was trimmed back when the stream happened to flow along the extreme lateral radius of the fan and impinged against the scarp front. This action should develop a zig-zag outline for the scarp foot, re-entrants of pediment occurring up all the valleys of the scarp face with promontories being left between. Such an outline is, however, not characteristic of the scarps behind pediments; the scarps are straight or only gently sinuous. Likewise the inner edge of pediments should be markedly uneven in the horizontal sense, rising toward and into the valley mouths and falling to intermediate points between them as the successive rock fans are passed. While such features do occur on a minor scale, the inner edge of a normal pediment impresses an observer most because it is level over long distances. Moreover, stream-cut nicks at the foot of scarps are nonexistent.

For the modelling of pediment surfaces also, Johnson appealed to lateral planation by streams. But this must be specifically excluded, for transverse stream beds, alluvia, and allied phenomena are either absent or of minor importance across most pediments. It may be that Johnson, and also Blackwelder (1931), who wrote "Pediments are graded plains due, in the writer's opinion, to the sideways cutting of desert torrents . . . Pediments are essentially compound graded flood-plains excavated by ephemeral streams", were considering special cases, but the broad sweep of pediments outward from the steeper hillslopes toward the axes of valleys is a sufficient assurance that streams do not initiate the pediment landform, nor are they responsible for its continued growth and surface form.

Howard (1942) also supported lateral stream corrasion, stressing the work done by even small tributaries which in arid regions are braided rather than channelled.

Davis (1938, p. 1367) rejected the hypothesis of lateral planation by streams in favor of planation by sheet flood and stream flood, of which he adduced many clear examples from the Basin and Range province.

Berkey and Morris (1927) were of the opinion that the pediments of Mongolia were cut by a network of shallow rills. Rill work is apparent upon virtually all pediments, but as a primary agency of pediment origin and development it is open to the objection that its continued action cuts small but definite channels which become more and more stabilized as the work proceeds. Rills and ephemeral streams are not planing agents *per se*, and pediments are normally developed as planed surfaces at right angles to the courses of the rills from the hillside to the major stream bed. Even admitting a network of rills upon the surface, their action results in a maze of small incisions into the pediment surface, not a broad planing action.

Rich (1935) has conceived of sheet wash, unconcentrated into definite flowlines, playing a leading part in the erosion of pediments:

... that where rock waste is plentiful a surface cannot be lowered below a gradient sufficient to permit the transportation of that waste; that where waste is supplied to an area from outside that area, although it may only be a thin veneer, if it is constantly removed it serves as an effective blanket to protect the underlying rocks from erosion below a gradient sufficient to permit the transportation of the debris across it; and that, other things being equal, the necessary gradient depends upon the size of the particles to be transported.

By regarding both bedrock and alluvia as weathering in their turn, the products being distributed by sheet wash, Rich believes that the form of pediments is largely governed by distribution of waste in transit under the influence of sheet wash. Rich has applied the old Davisian concept of hillslope grade to the agency of sheet wash which, as McGee and Davis have shown, is sometimes present upon pediments. The conception is a valuable one, uniting old ideas and new, and we shall not lose sight of it. But it becomes effective only after the pediment is already in existence and does not explain the origin of pediments, nor the retreat of their commanding hillsides.

Kirk Bryan (1922; 1935; 1936), who wrote much, and authoritatively, upon pediments held that three processes combine to form pediments: (1) lateral planation by streams, (2) rill work, (3) retreat of mountain slopes. In general, also, he lays great emphasis upon the work of ephemeral streamlets in shaping pediments, especially at the inner edges. Thus: "The principal agent in the formation of pediments is the ephemeral stream, which works by incision and lateral planation. However, rainwash, rills and weathering are particularly effective in the later stages of the process".

Fair (1947), in South Africa, opined that, after the retreat of the steeper hillside under weathering, the pediment was moulded under sheet wash.

Present hypothesis of hillslope evolution

The hypothesis here advanced accepts the idea that pediments originate by the retreat of scarps and steeper hillsides. No matter how, or by what agencies, a pediment is later modified; its very existence depends in the first instance upon the process of hillslope or scarp retreat. Thus any hypothesis of pediment formation must explain not only the form and remodelling of the pediment but also the reasons for parallel retreat of hillsides and scarps.

The processes of weathering and the action of gravity (*e.g.*, in forming debris slopes) have already been sufficiently elaborated and are accepted throughout the sequel; but the basis of our argument rests in the manner of water flow, derived from rainfall, over a landscape, and from personal observations and examinations made in the field at times of heavy, and of light, precipitation. The rainfalls studied vary from 3 to 4 inches in an hour to slight drizzles, and the manner of runoff varied from sheet floods down to nothing. The heaviest rainfalls I have personally experienced were 22 inches in 3 days (Zululand, 1940), 13 inches in one day (Durban, 1953), and 11 inches in one night (Knysna district, Cape Province, 1951). The results in all instances were well worth seeing; but probably the most astonishing transformation I have witnessed was in the Kalahari (December 1949) when, for 4 days continuously, the heavens opened, and the contents fell out. At the end, broad lakes stretched across that flat country almost unbroken for miles, and the Great Kalahari Desert had truly become the Great Kalahari Sea. How much rain fell altogether I do not know. The heaviest measured rainfall I have heard of in Southern Africa was in the Zomba district, Nyasaland (1946)—28 inches in 24 hours.

When rainfall is to be discharged from a landscape, the number of variable factors is large: amount and incidence of precipitation; the nature of the soil, porous or clayey, dry or saturated; detailed nature of the surface, smooth or rough; vegetational cover, sparse or close, grass or scrub; terrain, flat or steep; and so forth. These decide the relative proportions of runoff and soakage. The combinations of the various factors are infinite, and the variety of natural conditions is greater than the pen can here describe, but the several standard cases that will be narrated have all been observed over and over again in the African veld and are evidently valid for that type of terrain in which the incidence of rainfall during thunderstorms is high, where the hillslopes display clearly the several elements already discussed, and where the vegetational cover is sparse, leaving exposed notable areas of bare soil.

In what we may describe as "thunderstorms of moderate intensity", the following conditions have been observed on a normal hillside (Fig. 4). As the rain falls upon the upper parts of the slope, runoff slowly increases in volume as it moves downhill. But the volume is small, drainage is often poorly defined, and the amount of work done both by transport of soil and erosional action is not large. In fact, it would appear to be inferior to soil creep, as is indicated by the convex profile of the upper part of the slope. But, as the volume of water increases with distance from the crest of the slope and its speed downhill increases with the steepening declivity, there comes a stage where modification of the surface under the action of running water exceeds the modification due to soil creep. This is the end of the waxing slope.

Under suitable conditions, a free face, washed clean in part by the water, now appears, and, lower down, the debris slope in turn succeeds. Both of these two elements are modified together in the hillslope development. Together they make the steepest part of the hillside, and here the ever-increasing volume of water courses rapidly down in an infinitude of tiny rills, each of which is highly turbulent and is an active eroding agent cutting back into the face of the scarp (Pl. 3, fig. 2). Such rills are normally closely spaced, the "available relief" between them is small, and hence the scarp is eroded back as a whole parallel to itself, rather like a receding wall (Pl. 1, fig. 1). Of course, not all the rills are equally large, and minor irregularities and indentations appear at intervals along the hillside. Some of these may even grow into gullies or dongas, but such irregularities tend to become partly filled with debris, and their further development is hindered. A specialized instance of this process has been termed "gully gravure" by Kirk Bryan.

The rate at which scarps retreat forms one of our later canons (King, 1944, p. 279; 1947, p. xiv).

Retreat of the steeper elements of the hillside leaves at the foot an apron of flat land extending commonly to a stream channel (Pl. 1, fig. 2). This apron of gentle declivity is the primitive pediment.

Storm water coursing rapidly down the steeper face is checked when it reaches the gentle declivity at the bottom. As the volume of runoff still mounts with increasing distance from the crestline of the hill, a much-augmented volume of water has now to be discharged across only a gentle slope. This task is accomplished (King, 1949b) partly by a change in the manner of water flow. The rills spread laterally and join up until they make a thin sheet of water of indefinite lateral extent spreading across the apron in advance of the debris slope. This change in water flow corresponds with the break in profile at the head of the pediment which is so characteristic in pedimented landscape wherever they occur. In areas of strong rocks, as we have noted, the break in profile is abrupt (Pl. 2, fig. 1); in areas of weak rocks, or where detritus and soil are thick, the break is zonal, taking place over a smooth curve. The change from hillside to pediment is sharper also in regions of high incidence of rainfall, where considerable volumes of water have to be rapidly discharged.

The pediment, a smooth landform permitting discharge in sheets over its whole area, is indeed the natural answer to the need for rapid dispersal of storm water: it is the ideal landform to dispose of the maximum volume of water in minimum time, with least erosional damage to the landscape. Pediments, once formed, may thus be expected to be relatively stable landscape features.

The manner of water flow over pediments is of prime importance. The sheets of water, at first only a fraction of an inch deep, flowing over the upper pediment gave rise in the earlier stages of my investigations to much mystification, for, though the rills coming down the steeper slopes were turbid, the thin sheets of water to which they gave rise on amalgamation were often clear, the debris having been dropped temporarily at the foot of the debris slope. The explanation of this curious phenomenon lies in the manner of water flow, for water in thin sheets often moves by nonturbulent, laminar flow. The molecules, instead of pursuing individual, interlacing, looped, and turbulent paths as they do in the thread flow of rills upon the steep hillside, join into laminae, and the laminae slide upon one another in the manner of a pack of cards thrust forward upon a table[3]. The lowest card stays in position while each card above slides progressively farther. In laminar flow, the lowest layer, in contact with the ground, has zero velocity, and the flow as a whole is nonturbulent and nonerosive. So arises the zone of clear water flowing as a thin sheet that may, under favorable circumstances, be observed toward the upper edge of the pediment.

Apparently, a special combination of circumstances is necessary for true laminar flow. The bottom must be smooth, and either of bare rock or fine debris; the sheet of water must be thin; and there must be no restriction upon the lateral spread of the water. Any influence tending to channelize the flow destroys the laminae. Under natural conditions, there is often in thin sheets a mixed flow, partly laminar, partly linear.

Across the pediment, where the volume of water to be discharged continues to increase, the sheet of flowing water becomes deeper. Commonly it also becomes turbid and erosive again. Reynold's Number once more affords the explanation for, when the sheet of water passes a certain critical thickness defined chiefly by the viscosity of water and the smoothness or irregularity of the ground, the sheets of molecules break up, and linear flow takes place again. This *sheet-flood* (as distinct from the laminar flow in sheets) is the agent that erodes and models the pediment into its typical hydraulic cross section. In addition to transporting weathered debris, it is an active eroding agent on its own account.

On the pediment, local circumstances both natural and artificial sometimes cause a concentration of flow lines in the flood, and this produces a channelling of the pediment surface. Such channelling often appears well out from the debris slope, but the gullies so produced often erode headward until they cross even the whole width of the pediment (Fig. 4). These are the typical dongas of badly eroded farmlands. Between the gullies, the pediment surface remains straight in transverse section (King and Fair, 1944).

As the storm passes and the sheet flood subsides, the pediment is covered with an enmeshed pattern of dwindling rivulets, most of them aggrading and leaving a thin surface veneer of temporary accumulation; mud, sand, or grit.

Conditions observed did not always correspond to the scheme outlined above for "storms of moderate intensity". In storms of great intensity (greater than 2 inches per hour perhaps as a generalized estimate), the zone of clear water disappeared from the upper pediment. This was presumably because the volume and discharge speed of water was so great that the rills of the steeper hillside passed over immediately into a thick sheet (sheet flood) upon the pediment, and laminar flow was thereby inhibited. At such times there was no tendency to deposit fine waste at the base of the hillside or top of the pediment. Extreme examples are sheet floods of the type described by McGee (1897, p. 100), but these are exceptional even in a desert environment and reach their maximum torrentiality only in rocky deserts of marked relief.

330

Even in the heavy rains I have witnessed, sheet floods of the depth and violence of those described by McGee were not developed, possibly because the terrains were too absorbent. A suitable physical environment seems to be necessary, as well as concentrated precipitation, to produce extreme sheet floods. Davis has noted (1938, p. 1345) that "the time interval between successive floods on the same area must frequently be extremely long".

In times of less intense precipitation, the rills of the steep slope are prolonged across the upper pediment without joining laterally into a sheet, and tend then to erode gutters across the upper zone of the pediment. In some regions where light rains are frequently interspersed with thunderstorms the upper parts of pediments were consistently irregular and guttered in this way. Parts of Texas such as the Pecos Valley, which receives rains from the Gulf Coast as well as from the interior, may perhaps be cited, though my observations there are too limited to be certain of the example.

Clearly these observations of water flow across African landscapes need to be repeated in many other lands. Such as they are (and they were not comfortably acquired amid sheets of wind-driven rain and vivid strokes of lightning), they offer a sound basis for the understanding of hillslope elements and evolution, agreeing incidentally with the work of Wood. They show why the free face and debris slope retreat rapidly parallel to themselves, why when scarp retreat is rapid the waxing slope is poorly developed, how the pediment is modified under sheet flood, why its upper zone is sometimes irregular or guttered, and in detail how a major pediment may consist of many small pediments graded to shallow channels crossing the major, compound pediment. It explains the paradox noted by King and Fair (1944) of dongas (gullies) occurring most frequently not upon the steeper elements of hillslopes but upon the relatively flat pediments as due to breakdown of the laminar flow conditions into linear flow with the inevitable production of channels. Lastly, it shows why the long concavo-convex slopes of certain northern lands must be regarded as degenerate, evolving no longer by scarp retreat but more or less atrophied. Their active elements, the free face and debris slope, are lost; the waxing slope produced by weathering and soil creep forms apparently at an inordinately slow rate compared with scarp retreat; and the active pedimentation which perhaps formed the original concave slope is now arrested so that this part of the slope now passes downward by gradual transgression through progressively less weathered detritus into solid bedrock. These features are well shown in chalk areas where runoff is minimized by percolation underground.

But more data are necessary before we can carry the conception much further. Before leaving the subject, however, we may add one hint useful for field observations: To distinguish laminar flow from linear flow, it is only necessary to observe the manner in which the water passes an obstacle such as a pebble or the stem of a small bush. If a depressed cone appears in the water surface about the obstacle, the flow is laminar and nonturbulent, if on the contrary the water surface piles up against the face of the obstacle the flow is linear and turbulent.

Convexity of upper slopes

As we have seen, extensive and widespread convexity of upper slopes is a sign of degeneration in the hillslope cycle; it means that soil-creep, slow and smothering, is dominant with weathering over the normally more active process of rain- rill- or sheetwash. This tends to be so more in areas of low than of high relief where the waxing and waning slopes can more easily meet, with elimination of the more actively disposed free face and debris slope. It appears also under climates where rainfall is moderate and evenly distributed, and runoff is consequently at a minimum; and in regions where a turf or sod is developed by the binding action of roots. Lastly, it appears more generally in areas of weak than of strong rocks. In short, bicurved slopes appear in situations where deep weathering is promoted and where the products of weathering are removed but slowly.

Fair (1947), Davis, and Rich, arguing from different premises, have concluded that at the final stage of an erosion cycle the ridge crests should generally become convex in cross section. But we have to record, *e.g.*, from Southern Rhodesia, that convexity, even at the ultimate stage of the cycle is not universal. In the hard, granitic terrains of that land, biconcavity sometimes persists upon pedimented ridges until the opposed pediments meet at a noticeably angular crest. Small castle-koppies (tors) commonly surmount such ridges, and these still maintain angular relations with the pediments at their bases (Pl. 3, fig. 1; Pl. 4, fig. 1). Again, pronounced convexity appears to be a sign of degeneration in the hillslope cycle.

As shown in a later section, pronounced convexity of hillslopes under soil creep is apparently a recent geological phenomenon. Prior to the mid-Tertiary, it exercised but a minor role in landscape evolution. Four general explanations of this new emphasis have been offered:

(1) The usual textbook explanation is that these pronounced convexities are related to a more humid climate. Insofar as more humid climate promotes deeper weathering and a greater thickness of soil mantle, this would allow of more soil creep not only on hilltops but on the lower country as well. *Per contra* the superb hydraulic curve of pediments—the acme of hillslope form under running water—is characteristic of semiarid and arid environments, so that there is a *non sequitur* somewhere. Even consideration of the incidence of rainfall, evenly distributed or episodic, does not remove all the difficulties, and we must conclude that the hypothesis in its present form is inadequate.

(2) Modern French authors have referred summit convexities to a periglacial regime. Pronounced concavo-convex slopes do in truth show a distribution about the areas of Pleistocene glaciation, but this does not of itself explain

the convexity. Professor Hollingworth of University College, London, has privately made the suggestion that the deeper weathering indicated by convexity is connected with deeper frost penetration. It is to be hoped that he will follow this suggestion up.

(3) Late Tertiary time has been punctuated by epeirogenic rise in all continents, with spasmodic uplifts following one another apparently at closer and closer intervals. River incision may thus have been accelerated from stage to stage with the production of steeper gradients on lower (*i.e.*, convex) slopes.

If this were so, then steepening should continue to the bases of hillslopes near the mouths of rivers. Other objections to the hypothesis can also be raised, but it may serve in combination with other factors.

(4) The fourth hypothesis stems from the observation that convexity of hilltops becomes notable from the mid-Tertiary onward, and links this with the great radiation of the carpet- or turf-making grasses during and after the Miocene period. With the production of turf or sod an entirely new factor entered the development of landscape. Surface runoff was hindered, and the development of the water-cut profile was minimized; percolation was enhanced, and deeper weathering and soil creep resulted. So the parallel retreat of slopes with the production of pediments was reduced while the waxing slope was enhanced. In areas of slight relief, or of weak rocks, the free face and debris slope disappeared, smooth concavo-convex slopes became almost universal; and so there came about a fundamental change in the aspect and perhaps the manner of evolution of the landscape.

Homologies between the landscapes of humid temperate, semiarid, and arid regions

As soil removal by wash and creep is still greatest on the steepest part of slopes, namely where the convex and concave slopes meet, it is likely that scarp retreat is not wholly prevented where a sod mantle is present, but the rate at which it occurs must be very much reduced from that upon bare rock and talus.

The manner of scarp retreat, if it exists, in turfed regions is a subject for study. The concave lower hillslopes in such regions do not differ essentially in profile from many pediments, and some which I have examined both in Britain and in northeastern North America do not exhibit a continuous section downward by gradation into bedrock but were apparently at one time cut-rock surfaces, though they seem to have atrophied so that weathering now affects the originally smooth, planed rock surface. Such changes are not inconsistent with the lying of snow during the Ice Age, instead of the continued dominance of running water.

Evidence of associated scarp retreat to form these probable pediments must also be sought. Fair (1948), studying slope form and development in the humid coastal belt of Natal, inquired whether smooth scarps in the Table Mountain Sandstone country, where the free face had been transformed into a waxing slope thinly covered with soil, still retreated in the ordinary way or whether they lost declivity until they were finally obliterated. He concluded that, apart from sections of the scarps which were gashed at intervals by gullies,

> observation suggests that the compound slope, convex above and concave below, assumes a gradient upon which the rate of supply of weathered waste is balanced by its rate of removal by gravity and sheet-wash. Once scarps are rounded off to this gradient, no further loss of declivity occurs as long as the unconsumed (sandstone) remains. It will be seen that the resistance to weathering of the sandstone is partly responsible for this and that slopes on other, less resistant rocks, might continue to flatten with time.

In Britain it has long been customary to speak of the retreat of the Chalk scarp (*e.g.*, west of Salisbury). Much of the British countryside is, indeed, notoriously scarped, and data concerning the lower slopes would be welcome.

In the Appalachian Mountains of North America, transversely concave valley bottoms often occur and not flat flood plains. Do these valley bottoms have concave cut-rock floors? Davis (1930) has quoted valley floors with rock basements "in abundance along the belts of less resistant strata in the folded Appalachians from Pennsylvania and Virginia to Tennessee and Alabama". These seem to differ from pediments only in that nobody has yet called them pediments. This applies also to details of the earlier Appalachian surfaces, as Davis goes on to show[4].

It seemed to me also that certain of the ridge flanks did not coincide with the structure nearly so well as is required by the usual conception of the evolution of Appalachian topography by downwearing under weathering and stream action. Several instances were noted west of Harrisburg where slope angles did not agree with dips of strata, or where the waning slope had transgressed appreciably across the geological boundary from the soft onto the hard strata.

There is likewise the retreat of the Catskill Mountain front from the Hudson Valley which does not seem wholly explained upon structural grounds but requires active scarp retreat. Instances might be multiplied, but my own observations have been too fleeting to command respect, and the research must await the attention of some investigator more leisured in the region.

In the same outstanding paper Davis (1930, p. 136) drew attention generally to the presence of pediments or pedimentlike features in landscapes developed under humid as well as arid climates, thus demonstrating their essential universality in landscapes of subaerial origin. This was an advance toward harmonizing what had been thought of previously as two distinct modes of landscape evolution, and his descriptions of these features and their mode of origin are so important that we quote them *in extenso*. Writing of the valley floors of humid regions, Davis noted that as the valley-side slopes

> ... recede from the banks of the graded stream to which they previously descended ... narrow strips of valley floor ... will be developed at their base, back of whatever flood plain ... is simultaneously formed by the stream. As these strips widen, the concavity of the profile across the valley bottom ... is given broader expression (Figs. 6, 7).
>
> The lateral valley-floor strips are, like the flood plain, underlain with degraded rock which may be called the valley-floor basement. It has a somewhat ragged and vague surface, except that where cut by lateral shifts of the stream it may be more even and better defined. The lateral strips will be everywhere covered by detritus, partly derived from local subsoil weathering, partly washed down from the valley-sides; and the detrital cover will, in time, constitute a large part of the valley floor. .... Each lateral strip of the valley-floor will have, besides its downstream slope, a somewhat stronger

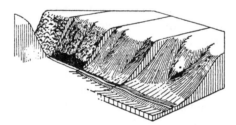

Figure 6 Development of Rock Floors (Pediments) under a Humid Climate. According to Davis (1930).

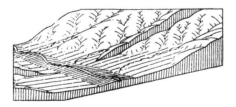

Figure 7 Continued Development of Rock Floors (Pediments) and Incipient Secondary Pediments under Humid Climate.
According to Davis (1930). Note in Figures 6 and 7 the semiparallel retreat assigned to the hillslopes.

335

transverse slope from the valley side towards the stream or its flood plain. . . .

The lateral swinging of the graded stream broadens the flood plain and its rock basement . . . and thus the valley-floor strips may be encroached upon; but in spite of this *they gain width by the retreat of the valley-side slopes.*

(Italics mine)

Davis also noted that

In the meanwhile many brooks, short side branches of the main stream, will excavate ravines in the valley sides (Fig. 7).

After the side-brook ravines are developed to the stage of having their own flood plains,[5] side branches of the main valley-floor strips, each with its underlying rock basement, will be opened along their margins heading farther and farther up the ravines.

Under humid regimes, Davis has here completely admitted scarp retreat with concomitant pedimentation—for what are these rock-floor strips but pediments? (Compare for instance the pedimented landscape of Pl. 2, fig. 2 with Davis' descriptions.) Davis seems to have been prepared to accept them as such, and from long experience in the field I have come quite independently to the same conclusion. They are as homologous with pediments as monadnocks are with inselbergs. Thus I can see no reason for denying to the concave rock floors of humid regions the status and name of pediments. Their development may be less extensive and more imperfect, and they may be more obscured beneath debris than their relatives of semiarid regions; but of their origin by retreat of backing hillsides, and subsequent modification under running water, both sheet and rill flow, I have no doubt.

Fair (1948a, p. 44) has contrasted the pediments of the semiarid interior of Natal with those of the more humid coastal belt. His conclusions may require some modification in the light of the hypothesis of states of water flow enunciated herein, but remain a worthy reference for students of the subject.

Kirk Bryan, too (1940, p. 266), considered hillslope development in arid, tropical, and humid temperate climates and came to the preliminary viewpoint

(a) that slopes of hills are characteristic of the climate and the rock; (b) that these slopes once formed persist in their inclination as they retreat; (c) that they disappear only when all the volume of rock above the encroaching foot-slopes or pediments has been consumed.

In his 1930 analysis, Davis has gone a long way from his simple belief in "graded" hillslopes of the Geographical Essays. His exposition ends indeed

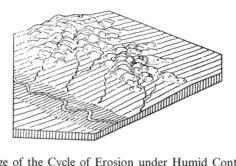

Figure 8 Late Stage of the Cycle of Erosion under Humid Controls According to W.M. David.
Note particularly the wide rock floors, concave upward (pediments), and the steep-sided residuals. These are all typical of pediplain and contrast markedly with Davis' earlier conception of the peneplain with its supposed broad convex divides and faint residuals due to structure only.

with a diagram (Fig. 8) of the ultimate landscape form under the erosion cycle which is far removed from his earlier concept of the classical peneplain. It is none other than the pediplain, typical senile landscape familiar in semiarid environments, utterly unlike the peneplain of broad convexities in that it is composed of numerous broad, concave pediment or pediment-like surfaces from which rise steepsided residuals, not gentle monadnocks. All this accrued from his two-fold recognition of (1) rock basements or pediments and (2) scarp retreat in humid temperate regions.

Yet, with the main principle of the fundamental identity of all epigene landscapes firmly within his grasp, and with the inference that the semiarid landform with widespread pediments is the norm of epigene landscapes clearly before him, Davis relapsed into the old ideas of progressively flattening hillslopes (1932) and the consequent *ideé fixe* of the peneplain. In doing so, we believe, he turned from the gateway to interpretation of the various regional cyclic histories of the earth (King, 1950).

Peneplain versus pediplain

The reader is by now familiar with these two similar, but fundamentally distinct conceptions of the ultimate or senile landscape resulting from epigene erosion (Fig. 1; Table 2).

The distinction between the two types of landscape shows best in tough granitic terrains and is least marked in regions of weak shales and limestones.

Let there be no compounding upon this point: Though both landscapes may conform to the loose definition of "a landscape of low relief resulting from prolonged subaerial erosion" they are not, nor at any stage ever were, the same thing.

Table 2 Characteristics of Mature Peneplain and Pediplain.

Peneplain (Fig. 9)	*Pediplain* (Fig. 8)
Broad flood plains of rivers	Relatively narrow flood plains of rivers
Divides invariably subdued, and broadly convex	Divides concave on both sides or only narrowly convex
Residuals gentle, convex	Residuals may be abrupt, concave in section
Concave profiles only in lower parts	Concave profiles of broad sweeping pediments dominant in the landscape
Origin by slope flattening	Origin by scarp retreat and pedimentation

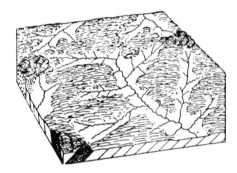

Figure 9 Standard Conception of the Peneplain.
Showing flood plains of rivers, broadly convex divides, and exact correspondence of monadnocks with boundaries of resistant rock formations (After Cotton).

If the prime agents in each case, slope flattening or scarp retreat and pedimentation, operate almost exclusively, then one form is valid and the other is not. If both are regionally operative as, for instance, slope flattening in humid-temperate regions and parallel scarp retreat in semiarid regions, then peneplains may be expected in the former and pediplains in the latter environment.[6] What has Nature to say on this? Do we find both forms, or either exclusively?

Before answering this question we shall consider an opinion of Walther Penck whose concept of senile landforms agrees with the pediplain. This form he called an *Endrumpf*. But Penck recognized also landscapes in which convex slopes appeared owing to gentle uplift and weak incision of the rivers. This landscape type he deemed to be initial in a new phase of erosion and termed it a *Primärrumpf*. He related both landforms to land movement either slower or more rapid than the rate of erosion; I have preferred to think of them in terms of the erosion cycle as a pediplain and a redissected pediplain (two-cycle) respectively—that is, as landforms pure and simple, the former is an end-product of the erosion cycle, the latter, after rejuvenation of the rivers, is the initial stage in a new, second cycle of erosion.

Davis (1932, p. 419) discussed Penck's concept of *Endrumpf* and *Primärrumpf*, but was handicapped in his judgment by his long-held belief in a convexity of old-age landscapes.

My own observations upon old-age surfaces permit very clear and definite conclusions in my own mind. All the truly old-age surfaces I have seen, and I have studied and mapped quite a few, were multi-concave in form, and bore residuals that were steep-sided and concave in profile themselves. The topographic boundaries of such residuals often do not coincide with geological boundaries between resistant and weak rocks. Even if the residuals are of resistant rocks, the topographic discontinuity between the residual and the adjacent plain had almost invariably transgressed from the weak onto the resistant formation. The landforms were pediplains or *Endrumpfe*.

Convexity occasionally appeared along the crests of some of the divides especially in areas of weak rocks such as Karroo shales, but it is clearly a feature induced only as senility advances, and was subordinate to concavity in the landscape observed as a whole. It was always summit convexity only.

Convexity in the neighborhood of the streams upon an old-age land surface, however slight, was found to be an infallible token of incision controlled by the arrival of a new cycle of erosion (Pl. 4, fig. 2), and if the observer follows downstream the amount of incision generally increases, and the second cycle becomes obvious. Below the region of maximum incision, youthful convexity of course decreases again in the second cycle, and concavity appears at the bases of hillslopes.

Judged on these criteria, observed in nature, I must declare that I have seen pediplains and diagnosed dissected pediplains, *i.e.* features akin to Penck's *Endrumpf* and *Primärrumpf* but I have never seen a peneplain. To arrive at this conclusion, I have scrupulously rejected from consideration numerous planed landscapes in the history of which doubt existed (*e.g.*, of resurrection), or where the second cycle had advanced much beyond the incipient stage. I have also neglected many with essentially straight hillslopes where stream incision and hillslope development appear in a delicate state of balance. They do not seem to afford exceptions to the theory. Fair has measured several slopes of this type.

Slow uplift is not essential, in my opinion, to the generation of a *Primärrumpf* from an *Endrumpf*, but slow incision of the streams, from any cause—*e.g.*, gentle tilting toward a basin—brings out the relation.

"Normality" in landscape

Textbooks of geomorphology commonly refer to the following types of erosion cycle: the "Normal Cycle" of Erosion typified in humid-temperate lands, the Semiarid Cycle of Erosion, and the Arid Cycle of Erosion. From this use of the word "Normal" we dissent.

The humid-temperate climates of Europe and North America have been taken as "normal" because they are typical of the home regions of western civilization, they appear abnormal to dwellers of [for instance] semiarid regions who are used to long periods of stable, clear weather and are apt to refer disparagingly to the fickle weather of northern climates. Those in semiarid regions might legitimately point out also that, whereas the action of running water can be plainly seen and studied in landscapes of the semi-arid environment, with its areas of bare ground between tufts of grass, the processes of surface wash are impeded in the northern lands by either a close carpet of grass or by woods which, until recently, were even more widespread. Deforestation has become universal in both Europe and North America only since the Middle Ages. Moreover, the northern parts of both these areas were earlier covered by the greatly extended glaciers of the Pleistocene. Even those areas which were not covered by ice sheets suffered a periglacial regime of permafrost and snowfall instead of rain. A superabundance of waste upon the hillslopes and in the valleys alike was but one of the consequences of this regime.

We ask ourselves therefore, are these regions to be cited as "normal" and serve as a standard of comparison for geomorphic processes and evolution?[7] On the contrary, thoroughly abnormal standards were gratuitously, though not unnaturally, assumed in the classic studies of the subject, and have been perpetuated uncritically since.

An unbiased approach to the use of the word "normal" is now necessary. Davis and Cotton (1941) regarded any departure from the humid-temperate as a "climatic accident", and Davis (1930), even after he appreciated the homologies between landforms under epigene environments and derived the pediplain even under humid regimes (Fig. 8), found himself confirmed in "... an earlier belief that the cycle of arid erosion may be, with all its peculiar consequences, reasonably treated as a variant of humid erosion". Nowadays an unbiased observer, studying the world as a whole, might very well arrive at a very different conclusion.

We seek to establish what is "normal" in epigene landforms. Let us be quite clear upon one initial point: The student who would learn the normal pro-cesses of landscape development, and the normal types of landforms, and the normal manner of scenic evolution *must study them in semiarid regions*. There will he see, in regions of exoreic drainage, the type of rainfall and stable climate which characterizes most of the land area of the globe, and the manner in which this rainfall is dispersed. There generally will he avoid serious com-plications introduced by the climatic fluctuations of the Pleistocene. There will he see best displayed the elements of hillslope form, the most active development of landscape under scarp retreat, and the processes of pedi-mentation, in short the most actively evolving type of landscape (Fig. 10).

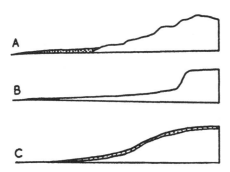

Figure 10 Landscape Profiles under Arid, Semiarid, and Humid Influences.
(A) Broad benches and structure lands with bahadas, (B) Scarps, pediments, and inselbergs, (C) Smooth concavo-convex slopes leading to valley plain and rounded hill outlines.

A deviation from the semiarid mean either toward greater humidity or greater aridity provokes the accumulation of waste in the landscape, symptomatic of a decreased efficiency in landscape evolution. On the one hand is produced a soil blanket that mantles perhaps even the free face and debris slope, with dire results upon the active evolution of the landscape as a whole; on the other abundant waste accumulates in the hollows as fans and bahadas, so preventing the operation of the usual processes of pedimentation which are responsible for the finishing touches put to a landscape. In extreme cases, even the exoreic drainage system may be destroyed, and convex rock surfaces form as base to the alluvia instead of concave pediments. (*Cf.* Lawson, 1915.)

So much for modern landscapes; what was "normal" in the geologic past? All the Mesozoic and early Tertiary erosional landscapes of which I have accurate knowledge were pedimented "semiarid" landscapes, even though they occur in what are now regarded as "humid" regions. Not only regions such as Africa, the Gobi, and Australia were pedimented at those dates but also Europe and North America. Davis has mentioned, for instance, the concave rock floors of the Blue Mountain region of the Appalachians, while Baulig's descriptions of the Oligocene landscape of the Central Plateau of France leave us in no doubt regarding the pedimented nature of that former landscape. Even the small and difficult example of Dartmoor, Southern England, has the characteristic curve of pediments when the oldest ridge crests are viewed in profile, and the famous tors are identical with the castle-koppie residuals of the Rhodesian veld. There, apparent to the trained eye, was anciently a landscape which had reached the penultimate stage in the pediplanation cycle. The dissection of the ancient summit pediplain is far advanced, and indeed a second partial cycle was already developed presumably before the end of the Tertiary. This younger landscape, still older than the main dissection of the region, stands 300–500 feet below the original surface around Postbridge, and itself shows small local pediments.

Only since the middle Tertiary, apparently, have landscapes of "humid" type become prominent, probably as a result of the radiation of carpet grasses. We shall not err if we visualize the surroundings of the Mesozoic dinosaurs and the early Tertiary mammals as landscapes of semiarid type, which were widespread and normal in those times.

We have been considering continental landscapes of erosional type; what evidence is afforded by continental landscapes of depositional type? Therein the same standards regarding climatic normal of the geological past are manifest. Few continental sequences other than those of the Ice Age show evidence of deposition under the "humid" regime. The great majority on the other hand offer abundant indications of deposition under prevailing "semiarid" to "arid" conditions. The Karroo system with the mighty Beaufort Series, reptile-bearing and with correlatives in several continents, or the Devonian and Triassic red beds of the northern hemisphere and the Tertiary sequences of the Midwest of North America are all typical of accumulations in the semiarid areas of to-day.

In seeking fundamental geomorphologic standards, it is therefore necessary to reject almost all existing conceptions about "normal processes" and "normal cycle of erosion". Thoroughly abnormal standards involving the arthritic concavo-convex hillslopes, crippled and reduced to impotence beneath a blanket of sod, were gratuitously assumed half a century ago. These conceptions of normality require revision. It may be difficult, and even painful, for geologists resident in the "humid-temperate" regions to revise their opinions on this topic when all their textbooks tell them otherwise and the landscapes about them appear to confirm this judgment. But over the world as a whole, during the geologic past as well as the present, the facts are inexorable, and the "semiarid" type of landscape must be regarded as the normal, most completely developed, most actively evolving, standard type of landscape.

Epigene cycle of erosion

The secret of landscape evolution lies, evidently, in the mode of development of hillslopes. Now hillslopes may be initiated either as valley sides consequent upon stream incision, or as tectonic features the result of faulting, monoclinal warping, or even gentle tilting in a landscape. This is an important distinction upon which the evolution of the landscape as a whole afterwards depends.

In the first case, after a region has been uniformly uplifted, nickpoints run rapidly up the rivers and tributaries so that stream incision and the production of youthful valleys appear powerfully in the landscape. The interfluves are at first scarcely affected, and, if the initial surface was a plain, there comes into existence a landscape of the dissected plateau type, such as the Appalachian plateaus around Pittsburgh, Pennsylvania. Where the streams were already organized in a previous cycle upon structural controls, an incised trellised pattern develops like that of the folded Appalachians, or the joint-controlled pattern of the Matopo granite area, Southern Rhodesia.

The second case is illustrated by Southern Africa where continental uplifts terminated near the coasts in great monoclines facing toward the sea. These tectonic scarps are directed athwart the drainage lines (*e.g.*, Natal, Fig. 3). Once generated, the scarps that are large and therefore retreat rapidly make a series of risers which separate different cyclic land surfaces ascending step-wise successively away from the coast. The scarps retreat virtually as fast as nick-points advance up the rivers, so that the distribution of successive erosion cycles bears no relation to the drainage pattern whatsoever (Fig. 3). Under erosion, the scarps may not always remain clean and wall-like. Some degenerate into zones of dissected country separating an upper and earlier planation from a lower and later surface; but many of the cyclic scarps of Africa impress precisely because they have not so degenerated but have remained clear and distinct throughout their history. The oldest date even from the Mesozoic, and these are no exception— *e.g.*, the Drakensberg (Pl. 1, fig. 1).

The destruction of earlier cyclic landscapes may be less rapid under the retreat of continental erosion scarps than when the country is first divided up under stream incisions and the proportion of hillslopes available for active retreat is thereby greatly multiplied. In this factor probably lies the explanation for the great ages of many African landscapes as compared for instance with the cyclic landscapes of North America. Smooth erosional landscapes such as the South African highveld or the Rhodesian plateau belong to the Mesozoic cycles, yet standing above them and separated by steep scarps are remnants of even older, conceivably Paleozoic, planations, some of which have apparently been exposed continuously to the weather since that time—*e.g.*, southern Congo.

Nothing like this appears to exist in North America, where few traces of very ancient cycles remain unless their preservation has been aided by burial for a time. A Mesozoic surface is described from Minnesota, and there are one or two other probable remnants of Cretaceous landscapes, but the fundamental surface from which the eastern half of the United States was later carved is the mid-Tertiary land surface, found emerging from beneath the Ogalalla covering formation in the Llano Estacado of the West, and appearing as the summit uplands of the Appalachian Plateaus, Schooley "peneplain", and the New England upland in the east.

We may now follow certain points in the evolution of landscape after the youthful stage. When the streams are already well and widely incised, parallel retreat of hill and valley sides reduces the areas of initial surface upon interfluves. If the relief is low the natural curve of the pediment may soon take it up to meet the waxing slope, so that concavoconvex valley sides appear and slope development becomes moribund. If, on the other hand, the relief is high, shrinkage of the interstream plateaus continues until opposed pediments meet and a biconcave transverse section to the interfluve results. A noteworthy fact emerges here. The upper surface of the interfluve remains in this case virtually without alteration until the hillslope elements of the new cycle encroach upon it. I have seen on outliers of upland surfaces quite unmistakable summit areas of as little as 2 acres surviving virtually without alteration though all the country for miles around had been consumed by scarp retreat in a newer cycle of erosion. The hill called Showe in the Shamwa district of Southern Rhodesia is a superb example of this. Where jointed granites form the terrain, and hillslopes stand steeply, bornhardts are a typical landform under this process (Pl. 3, fig. 3).

Where primary dissection of the country is not by stream incision, but by the retreat of great cyclic scarps (Fig. 3), earlier cyclic land surfaces may survive long after the initiation of the new cycle. They remain practically unaltered until destroyed utterly by the encroachment of the fresh cyclic scarp (Pl. 1, fig. 1). Here is one of the most important canons of landscape development.

Whereas the cyclic erosion of the older Davisian conception, with its emphasis upon slope flattening and universal downwearing under the action of weathering, early obliterated all trace of the initial surface and reduced everything to a peneplain, the newer concept involving cyclic scarp retreat permits the initial surface, itself a record of an earlier cycle of erosion, to remain much longer in the landscape. Inherent in the doctrine is the persistence of the older surface without significant alteration for a long period into the currency of the new cycle. Two cyclic surfaces are for long co-existent, the older above the retreating scarp and the younger below.

The far-reaching consequences of this have already been demonstrated in studies of African and other continental landscapes (King 1949a; 1951b). It forms indeed the primary canon for deciphering the past histories of landscapes, and with its aid the first, tentative correlations of dominant cyclic landscapes have been made from continent to continent. On the downwearing hypothesis, leading to Davisian peneplains, the very possibility of this achievement would be denied.

After scarps have retreated, both the major continental cyclic scarps transverse to the drainage and the minor scarps or valley sides, a wide development of pediments becomes apparent in the landscape. From either side of every stream large and small twin pediments extend laterally and make by coalescence pediment plains upon which stand unconsumed, steep-sided residuals (Pl. 4, fig. 3). In Southern Rhodesia (excluding the sand country of the west) pediments constitute probably 65 per cent of the landscape; the rest is residual hill masses. As the pediments provide the areas for cultivation their social and economic importance is also great.

As they widen they are subject to regrading, especially the steeper part toward the foot of the overlooking scarp. From an initial slope of perhaps 10°, they are reduced commonly to declivities of half a degree or so. This regrading is accomplished chiefly under sheet wash, and even ancient and very flat pediments often show beneath a thin veneer of transported material a smoothly cut, unweathered rock surface. Fair remarked of the general smoothness of pediments in the semiarid Karroo that "sheet-wash and rills are capable of regrading them with little difficulty in response to minor changes of base-level". But such changes are not necessary to regrading as Rich (1935) has also noted: "Drainage diversions and the dissection of abandoned fans and pediments are normal and to be expected. They do not require the intervention of diastrophic or climatic changes".

Conflict of pediments from neighboring streams, in which the stronger (usually the larger) extends its area at the expense of the weaker, is also a very real process causing regrading. The "development of master pediments", as we may call it, has most important consequences when we come to the dating of land surfaces.

In the later stages of the cycle, when the hills are reduced to small, rocky koppies and the pediments stretch perhaps for miles (Pl. 4, fig. 3), a *multi-concave* landscape is characteristic. In Southern Africa, this advanced evolutionary stage is widely developed with respect to stable, or gently falling base levels. Extensive alluvia or even flood plains are seldom present, and the soils may be quite different on opposite sides of even small streams. We may describe as typical the landscape on the east and south sides of Manda (Concession), Southern Rhodesia, where the pediments sweep right down to the swampy area of the streams themselves. The Marodzi valley west from Concession is larger and possesses a strip of flood plain 200 to 400 yards wide. But the bordering pediments are each over a mile in width before they abut against the rocky slopes of koppies. The general slopes of these koppie sides are nearly 30° inclination, but the lower 300 to 500 feet may approach a slope of 45°. The inner, middle, and outer zones of the pediment have slopes of 1°40′, 1°10′, and 30′ respectively, and these estimates are more or less standard for the district. Where the pediments pass as rock fans into the mouths of narrow side valleys they become steeper, sometimes 5° or 6°. The pediments themselves do not seem to exceed slopes of 3°, even when they are narrow and not very concave. In the Glendale-Bindura district of the same colony, the Mazoe Valley, though miles in width, nowhere possesses extensive flood plains but always a narrow river channel with wide, complex bordering pediments. Small residual hills rise from between the various pediment surfaces related to sundry tributaries. And so the descriptions of maturely pedimented landscapes in Africa could be repeated for innumerable instances.

At senility a continental erosion surface is not simple. It consists of a complex assemblage of closely related surfaces (chiefly pediments) referable

Plate 1 Erosional surfaces and scarps in South Africa.
Figure 1 Natal Drakensberg South of Champagne Castle.
View north. Relief over 4,000 feet. This scarp began at the Natal coast as a structural feature, the Natal monocline, in the late Jurassic. It has since retreated westward under erosion the full width of Natal, 150 miles. Its steepness is maintained by innumerable gully heads. The ancient, Mesozoic landscape above has remained virtually without alteration throughout the late Mesozoic and Tertiary. Anon. photo.

Figure 2 Typical Landscape of a Semiarid Region, the Karroo, South Africa.
There is no flood plain, and there are wide pediments backed by steep hillslopes on which are minor structural effects caused by horizontal, resistant sandstones and dolerites. T.J.D. Fair photo.

Plate 2 Erosional features of granite terrains, Southern Africa.
Figure 1 Small Granite Hillock.
Showing waxing slope, free face, and incipient pediment. In the absence of detritus the sharp angle between free face and pediment is clearly shown. Hluhluwe Game Reserve, Zululand.

Figure 2 Retreat of Hillslopes and Growth of Pediments.
In a region of jointed granite (Matopo Hills, Southern Rhodesia). Airphoto from 15,000 feet. Aircraft Operating Co. photo.

Plate 3 Residual hills in Southern Africa.
Figure 1 Steep-Sided Granite Residuals.
Rising abruptly from a smoothly pedimented, erosional plain at the ultimate stage of the cycle of erosion. Eastward from Pietersburg, Transvaal.

Figure 2 Gullied Debris Slope on Loskop, Natal.
The change of gradient from debris slope to pediment is clearly shown. T. J. D. Fair photo.

Figure 3 Bari, a Bornhardt 1,500 Feet High, Chiweshe Reserve, Southern Rhodesia.
The summit of the mountain is closely accordant in level with the Southern Rhodesian plateau, here dissected, though no significant remnant of the plateau surface survives upon the summit. D. Aylen photo.

Plate 4 Pedimented landscapes in Southern Rhodesia.
Figure 1 Castle Koppies (Tors) Rising Abruptly from Smoothly Concave Pediment.
This affords a very clear case of the sharpness of angle which may appear between pediment and free face when there is little waste to form a debris slope. Granite bedrock. Syringa district, Southern Rhodesia.

Figure 2 Broad Pediplain with Steep-Sided Monadnocks.
The convexity of the valley sides in the middle distance shows that the stream is newly incised. The landscape is two-cycle. Amapongokwe River, Southern Rhodesia.

Figure 3 Late Stage in Pediplanation Cycle.
Residuals rise like islands from the extensive pediplain. Sabi Valley, Southern Rhodesia.

to a multitude of drainage lines and basins, large and small, and united only in reference to a single, widespread, long-stable base level.

Even in this stage there may be further, accidental complications such as the multiplication of cyclic surfaces by structure (split nickpoints), local land movement, and so forth; but these factors have been discussed separately (King, 1947; 1951b) and shall not detain us here.

Observational data suggest that even at the senile stage there is little weathering of the firm rock beneath the pediment veneer, but this conclusion may not be valid. In the African theater, Pleistocene climatic changes were expressed in an alternation of arid and relatively more humid phases. As has been remarked by Cooke, the superficial materials overlying rock-cut pediments, when revealed in the sides of dongas, in many places contain Middle Stone Age implements. Following a previous phase of pedimentation there was thus a widespread phase of deposition while the Middle Stone Age folk occupied the scene. Still later came the phase of incision when the dongas were trenched through the depositional blanket. All these changes, covering an enormous area, appear to have been climatically controlled. They obscure the ultimate stage of pedimentation over much of Africa.

Dating of land surfaces

Land surfaces are customarily dated by the discovery of superficial deposits whose age can be defined. But with the regrading and conflict of pediments such deposits, usually thin anyway, may be wholly or partially removed. The net alteration in the landscape produced by one pediment cutting shallowly across another may be small indeed and quite irregular, but from the point of view of dating it becomes highly important.

Let us assume that a pedimented landscape of low relief except for residual koppies and small plateaus acquired a thin cover, partly detrital and partly residual, of Cretaceous age. The original landscape must therefore have been either Cretaceous or pre-Cretaceous. The region is not invaded from outside by incisions or other features of later continental erosion cycles, but continues to develop solely by the processes of pedimentation under which certain struggles for mastery go on, more-favored pediments cutting shallowly across their neighbors and so removing the cover and perhaps a few inches or feet of bedrock. Over the area so replaned, Eocene deposits may accumulate. So from time to time during the Tertiary, portions of the original area are remodelled in this way, having a shaving taken off them as it were and acquiring new and younger superficial deposits.

At no time does the landscape as a whole depart materially from the original planed condition, yet its parts are manifestly of several different "actual" ages,[8] as defined by the various superficial deposits. Unquestionably, the planed landscape is at one point Cretaceous, at another Eocene, at another Pliocene, and at yet another Pleistocene, so that the subjacent bedrock surface is a compound Cretaceous-Eocene-Pliocene-Pleistocene surface. This seems to be essentially the history of the Gobi, and other ancient landscapes *e.g.*, parts of the Kalahari.

But dating land surfaces in this manner is unsatisfactory except for local purposes, and to derive the fundamental or "comparative" age of such a compound landscape we can argue as follows. The landscape, viewed as a whole, was planed first in Mesozoic (Cretaceous) time. From the planed aspect then achieved, it has never since materially departed. What we see now is essentially a Cretaceous surface that has since undergone minor modification only. Its fundamental or "Comparative" age, by which it should be compared with other planed surfaces, is Cretaceous; and so we regard it.

The number and extent of planed Cretaceous land surfaces surviving thus as remnants in landscapes of the present day, especially those of the southern (Gondwana) lands, is remarkable. The survival of these surfaces affords a direct negative to the concept of significant vertical downwearing of landscapes—*i.e.*, to the so-called "Normal Cycle of Erosion".

The rate of retreat of major continental cyclic scarps (*e.g.*, Fig. 3) affords another aspect of the dating of ancient land surfaces. Thus the Natal Drakensberg originated along a coastal flexure which formed the margin of South-East Africa during the Jurassic period (King, 1940; 1944). Wall-like in form, it has retreated over 150 miles westward to its present position. The wall is still 4,000 feet or more high with black, forbidding precipices in its upper part along the borders of Basutoland where the crest rises above 10,000 feet. It is quite unmistakable, and shows little sign of flattening in its 120 million years of existence. The rate at which it has migrated averages about a foot in 200 years.

Knowing the date and place of initiation of the scarps and their present position, similar calculations can be made for several other of the world's major cyclic scarps—*e.g.*, the frontier scarp of Moçambique, the scarp of western Mexico, and so forth. Many of these furnish a rate of retreat between 1 foot in 150 years and 1 foot in 300 years. This shows that over long periods, *ceteris paribus*, the rate of scarp retreat is sensibly constant. In other words the erosive processes responsible for such retreat act at much the same rate the world over.

Only when scarps are small (low relief) and the amount of free face and debris slope is much reduced does the rate of retreat slow up. As we have seen, the rate of parallel retreat upon concavo-convex slopes from which these elements are absent appears to be almost negligible.

Quantitative methods of landscape study

However landscapes originate, it is important that their distribution should be known, and for this there is no better method than routine mapping.

Erosion-cycle maps, or morphological maps exist of many small areas and even of certain countries—France, Belgium, European Russia, Southern Rhodesia, etc.—but many more such maps are necessary. The data they provide are valuable to engineers, soil conservationists, and others as well as geomorphologists. There is no substitute for such maps, even though useful data may be gleaned from the ordinary topographic sheets.

Methods involving frequency of elevations, spot heights on grids, and so forth have been used (Maze, 1944) to bring out the cyclic facets present in multicyclic landscapes, and several methods involving superimposed and generalized profiles have been employed for similar purposes (*e.g.*, Wooldridge and Morgan, 1937); all have yielded helpful results more or less directly.

The application of statistical analysis is still more recent, and the pioneer has been A.N. Strahler (1950a; 1950b). Statistical analysis is essentially the method of the bulk sample, and is admirable for the study of complex phenomena and processes into which enter a large number of variable factors. As yet few geomorphic topics provide data suited directly to statistical treatment, and methods may have to be adapted to the new field of inquiry, so that too facile results should not be expected. The net result must be, however, a greater precision in our thinking.

At present the method of statistical analysis is in danger of lending an air of truth to erroneous conclusions. Decisions concerning type and admissibility of data need to be made most carefully, or the results will be falsified. No amount of statistical work can improve upon dubious assumptions and we must be ever mindful of Huxley's dictum concerning the use of mathematics (Mathematics is a mill which will grind you meal of any degree of fineness, but the quality of the meal depends upon the quality of the grain used). Unless the original data and assumptions are above suspicion, the conclusions lack certainty and perhaps validity. Fundamental errors of judgment made in the sampling process are not eliminated in the method, but they may be corrected in the course of further field work.

Strahler (1950a), using valley sides in the Verdugo hills and assuming that they were no longer undercut by stream action, concluded that hillslopes, in general, flatten progressively with the passage of time. Criticisms here would be: (1) The material adduced is too small in bulk to found a general principle upon. (2) It places a relationship between slope angle and stream gradient which may cease to exist after a certain stage is reached in the landscape cycle (*i.e.* slopes may become stabilized in gradient). (3) The manner of slope development is governed by forces active upon the slope and is independent, except initially, of the river. (4) Even upon the slope, erosional agencies do not act uniformly. (5) There are four different slope elements possible, all of which are modified independently. (6) Equilibrium between various features and factors is assumed whereas such equilibrium patently does not necessarily exist in nature. (7) The general conclusion that slopes flatten progressively is denied in nature by the close agreement of maximum slopes in many districts with a mean angle, and the retention of freshness and activity of major cyclic scarps that have existed from mid-Tertiary or even Mesozoic times. This last, under the Euclidean method, reduces the proposition to an absurdity.

Nor can one accept Strahler's general proposition of the equilibrium theory ("decline of stream gradient is accompanied by slope reduction") for it fails to recognize the great diversity of factors operative in the carving of a landscape. No simple relation can exist between hillside slope and thalweg. After the stage of extreme youth they are modified quite independently. What finer examples need we quote than some of the plates reproduced herewith?

To follow the argument further would profit us nothing. Our object here is not the rebuttal of the Equilibrium Theory of Erosional Slopes or any other variant of Davis' early views upon hillslope evolution; it is merely to emphasize that no amount of statistical analysis will compensate for poor geology. Geologists will be grateful for such practical results as will undoubtedly accrue from the use of the new method, but they will not regard it as a substitute for honest field work, nor will they assign to mathematical expressions an importance equal to facts which can be verified and reverified in the field.

Baulig (1950) has made another point: "Engineers will always use more or less empirical formulas for practical purposes. But there remains to be proved that mathematics has ever revealed in geomorphology an actual relationship that had not been discovered without its aid".

What seems necessary now is a multitude of further observations on the nature of landforms, with special reference to hillslopes and perhaps following the lines indicated by Wood, Fair, and Horton (1945). Quantitative work upon pediments and similar topics is needed, and this will lead to statistical studies. An increased volume of measurement upon topographic forms is necessary and should be undertaken directly in the field rather than from maps in the laboratory.

Also required are extensive observations of the processes operating upon landscape, perhaps stemming from the manner of dispersal of rainfall, through water flow to soil creep and weathering. These need to be correlated with different types of climate and terrain. Studies of stream grade, such as those by Kesseli and others, are of less direct application than processes affecting wide areas of landscape simultaneously. Equally required is a multitude of field studies in which cyclic and noncyclic erosion surfaces are mapped in everincreasing detail. As the number of morphological maps increases, and landscapes are accurately dated, both "actually" and "comparatively", so the erosional histories of the lands will be deciphered and compared, perhaps upon the lines of world-wide correlations.

Canons of landscape evolution

(1) Landscape is a function of process, stage, and structure. The relative importance of these is indicated by their order.

(2) The word *epigene* as applied to landscapes means "at the surface" or "subaerial". It does not include landscapes moulded beneath a solid cover of ice, and certain modifications are understood to be necessary in regions of permafrost.

(3) There is a general homology between all epigene landscapes. The differences between landforms of humid-temperate, semiarid, and arid environments are differences only of degree. Thus, for instance, monadnocks and inselbergs are homologous.

(4) Four elements may occur in a hillside slope. From the top, these are: the waxing slope, the free face, the detrital slope, the waning slope (usually pediment). Each or any element may be suppressed on a given hillslope.

(5) Each of the four elements of hillslope may evolve more or less independently, although each affects the others in some degree.

(6) The most active elements of hillslope evolution are the free face and the debris slope. If these are actively eroded, the hillside will retreat parallel to itself.

(7) In planed landscapes, pediments are the most important features. In stable regions like Southern Africa, pediments may occupy more than half the whole landscape, and locally may exceed nine-tenths of the landscape.

(8) The waxing slope is developed under weathering and soil creep.

(9) When the free face and debris slope are inactive, the waxing slope becomes strongly developed and may extend down to meet the waning slope. Such concavo-convex slopes are degenerate.[9]

(10) Parallel retreat of slopes is aided by (a) high relief, tending to maintain a clear free face and a debris slope; (b) resistant formations, tending to make cliffs (a good free face); (c) horizontal structure in sedimentary rocks; (d) generation originally as a tectonic scarp (*e.g.*, fault or monoclinal scarp).

(11) Erosion of the free face and the debris slope is accomplished chiefly by rill wash forming gully heads.

(12) Whereas the transport of debris upon the waxing slope, free face, and detrital slope is governed by both gravity and water work, that upon the pediment is solely accomplished by water work.

(13) Davis' old deduction of continuous lowering of hillside gradients, a feature also of Strahler's "Equilibrium Theory", is incorrect, and never existed as a general process of landscape development apart from terrains of rocks so weak that they cannot maintain a free face and detrital slope.

(14) Rock floors in epigene landscapes appear commonly between the base of hillslopes and stream channels. These rock floors, which are found

under all three climatic regimens, originate by retreat of the hillslopes behind them and are subsequently modified by the passage of water across them which confers upon them a concave profile. Such rock floors should, in all cases, be called *pediments*.

(15) Stream spacing is closer in humid than in nonhumid regions so that an evenly distributed rainfall is largely discharged by channel flow. Wider spacing of streams and heavier incidence of rainfall favor sheet flow and also allow room for wider pediments.

(16) Pediments are normally veneered with detrital material which is in process of transport across them. But pediments themselves are essentially cut-rock surfaces.

(17) Pediments which have ceased to evolve may show weathering of the bedrock.

(18) A pediment is the ideal landform for the rapid dispersal of surface water, encouraging sheet flow and with a proper hydraulic profile. Pediments are, indeed, moulded under sheet flow.

(19) The pediment is the fundamental landform to which epigene landscapes tend to be reduced the world over.

(20) Gullying may appear upon pediments where laminar flow of water is changed to linear flow.

(21) The break in profile between pediment and hillside may be abrupt if little detritus is supplied from above.

(22) Quantitative study of both slopes and processes provides a sequence of landscape forms different from those propounded by W.M. Davis.

(23) The early studies in the erosion cycle were conducted in Europe and northeastern North America, both of which regions were previously subjected to a glacial or periglacial climate. These areas and landscapes came to be cited as "normal" for the globe, a misconception that should no longer be tolerated.

(24) The standard or "normal" type of landscape, both now and in the geological past, is the semiarid type with broad pediments and parallel scarp retreat.

(25) Processes of erosion and evolution of landforms can, as a consequence of the above, be best observed in semiarid regions.

(26) Semiarid landscapes are the most efficiently developed. Deviation from the semiarid norm results in less efficient transport of waste, seen on the one hand in the broad, alluviated valley floors and smothered hillslopes of humid regions culminating in moraine and till under glaciation; and the abundant fans and bahadas, or even desert dunes of extremely and regions.

(27) Water may flow across landscapes either in threads or in sheets. In thin sheets, water may flow in laminae. Such laminar flow is nonturbulent and nonerosive.

(28) In storms of moderate intensity, the manner of water flow appears to best advantage. Rill flow on the steep hillsides is powerfully erosive; in the upper pediment thin laminar flow (nonerosive) may occur, with deeper sheet flood lower down accompanied by turbulence and erosiveness.

(29) Laminar flow past an obstacle shows a depressed water surface, linear flow banks up the water surface against the obstacle.

(30) Only in late Tertiary time have smooth concavo-convex slopes become common. This is a result of retardation of surface wash by a carpet of grass, and consequent enhancement of soil creep.

(31) Before mid-Tertiary time, landscapes generally were of the semiarid (scarp and pediment) type.

(32) Stream work affects the nature of adjacent hillslopes directly only during the early stages of the landscape cycle. After the streams are graded the dominant agencies are the processes acting directly upon the hillslopes, which evolve in an appropriate manner. The streams are, however, affected by the evolution of hillslopes through the nature of the detritus which these supply for transport by the streams.

(33) New cycles may penetrate inland either by nickpoints and incision of the rivers followed by retreat of the valley sides producing "flanking pediments", or by the retreat of wall-like scarps which are independent of the drainage lines "mountain pediments".

(34) On the whole, a landscape is dissected and reduced in a new cycle more rapidly following widespread river incision than by retreat of cyclic scarps originating tectonically. Conversely, the history of a landscape may be deciphered more readily where cyclic surfaces rise steplike between major cyclic scarps than where the landscape has been gutted by stream dissection.

(35) The ultimate cyclic landform is the *pediplain*, consisting dominantly of broad coalescing pediments. Residuals are steepsided and have concave slopes. Flood plains may or may not be extensive.

(36) A pediplain is multi-concave upward. When the streams begin to incise themselves due to tilting, uplift, or climatic change, convexity enters the landscape adjacent to the stream channels. The interfluves then become transversely convex, and a landform morphologically different from the pediplain is produced. It is an initial stage of the cycle of erosion; the pediplain represents a senile stage. This expresses Penck's *Endrumpf-Primärrumpf* concept insofar as land form is concerned. We do not necessarily follow his further argument relating these differences to decreasing or increasing rates of land movement.

(37) A peneplain in the Davisian sense, resulting from slope reduction and downwearing, does not exist in nature. It should be redefined as "an imaginary landform."

(38) Davis has quoted low granitic domes of arid regions as though their form resulted specifically from long-continued erosion. Observations in Southern Africa show, however, that these forms are normally functions of structure; that, where the vertical systems of jointing are strong, bornhardts and castle koppies (large and small inselbergs) appear, and where flat or gently dipping joint systems are paramount "ruwares" or flat domes of granite appear in the landscape. That such flat domes do not result from the erosion of bornhardts is, for this region, certain (Pl. 2, fig. 1; Pl. 3, fig. 3); they follow the broadly convex joints formed apparently in plutonic rocks by "unloading" as superincumbent rock systems were removed under erosion.

(39) Monadnocks are concave in profile as a rule (Pl. 4, fig. 3) (including Mt. Monadnock itself) and have originated by surface wash rather than by downweathering and soil creep. They are not necessarily sited upon outcrops of more resistant rocks.

(40) Inherent in the pediplanation cycle, with scarp retreat, are *two* cyclic land surfaces, the older above a retreating scarp, the younger below.

(41) Many of the major cyclic erosion scarps originate tectonically as fault or monoclinal scarps, especially along outwardly tilted coast lines.

(42) Major cyclic erosion scarps retreat almost as fast as the nickpoints which travel up the rivers transversely to the scarp. Such scarps therefore remain essentially linear and do not have very pronounced re-entrants where they cross the rivers.

(43) Major continental erosion scarps in many lands retreat at a rate of about a foot in 150 years to a foot in 300 years.

(44) A landscape once reduced to a pediplain may remain in that state for an indefinite time with only minor alteration, until some change, tectonic or climatic, is introduced.

(45) Notwithstanding the above, small changes continually take place by regrading and conflict of pediments. These changes, insignificant in the landscape as a whole, perhaps amounting to the removal of only a few inches or feet of material, produce great differences in the superficial deposits of the pediments.

(46) Land surfaces may be dated by the deposits upon them. "Actual" ages are local ages fixed by directly dating the deposits in any given locality. "Comparative" ages refer to the dates at which land surfaces were originally bevelled, and are obtained from the oldest superficial deposits.

(47) Land surfaces may bear deposits of any age from the oldest, used for "comparative" dating, to the present day.

(48) A tentative approach, using "comparative" datings, has been made toward the correlation of major cyclic landscapes from continent to continent.

(49) More use of quantitative methods is necessary in landscape study; especially needed is more morphological mapping.

(50) When more suitable data are available, statistical analysis may become a useful tool in landscape study.

Notes

1 In discussion Dr. T.J. Fair has commented upon this point: "Though Davis neither measured slopes nor, apparently, appreciated the processes operative upon them, he was nevertheless a master of landscape observation and I find it difficult to believe that he would see on his peneplains dominant convexities when, in reality the form was dominantly concave. I'd feel happier on this point if I could measure low divides in humid-temperate regions but, whatever the conclusion, I don't see that the possibility that peneplains and pediplains have more or less the same form invalidates the pediplanation concept one bit—for it is the process and not the end-form that is the crucial matter here. Can we say (Canon 37) the peneplain is 'an imaginary landform'? I am inclined to believe that Davis *really saw* pediplains, but called them peneplains, and that what is imaginary is not the peneplain itself but the peneplanation process. Did not Davis observe the features of the landscape truly but fail in their genetic analysis?"

2 For "stable angle of slope" there may be composite slopes where contrasted rock types appear in the valley sides—*e.g.*, Grand Canyon of Arizona. Under these circumstances minor free faces alternate with debris slopes. Fair has stated (1947) that it is the outcrops which govern the form of the slope as a whole, and that the slope will be less steep in maturity than with a single cap-rock (Fig. 5).

3 See any textbook on hydraulics, under the title Reynold's Number.

4 *Fair comment!* "See my earlier note that Davis saw pediplains but called them peneplains".

5 I would say even before the flood plains develop.

6 Dr. T.J.D. Fair comments "Davis's peneplains were described mainly from humid temperate climates where, with heavy vegetation and deep soil, creep probably operates more obviously than in semi-arid climates. Thus, just as the pediplain of semi-arid regions probably displays greater summit convexities in humid-temperate regions, so probably do Davis's peneplains of humid-temperate regions display smaller summit convexities in semi-arid areas".

7 Since this was written I have seen the scheme of Morphogenetic Regions based upon climatic type proposed by L.C. Peltier (1950). He lists nine regional types each "normal within its own regime and a 'climatic accident' only when it temporarily encroaches upon the area of another regime". This novel and fundamental point of normality is more far-reaching than that of the text which might then perhaps be termed "mean or average cycle".

8 The age of the surface at any actual spot where datable deposits actually lie upon it.

9 Dr. Fair demurs "Rather does the waning slope extend *upwards* to meet the waxing slope and so give the pediment its predominantly concave form". He is thinking chiefly of African conditions, I of European. Both statements may be true.

References cited

Baulig, H. (1950) *William Morris Davis: Master of Method, Assoc. Am. Geog.*, vol. 40, p. 188–195.

Berkey, C.P., and Morris, F.K. (1927) *The geology of Mongolia*, vol. 2, p. 475.

Blackwelder, E. (1931) *Desert plains*, Jour. Geol., vol. 39, p. 133–140.

Bryan, K. (1922) *Erosion and sedimentation in the Papago Country, Arizona*, U.S. Geol. Survey Bull. 730. p. 19–90.

—— (1935) *The formation of pediments*. XVIth Inter. Geol. Cong., C.R., vol. 2, p. 765–775.

—— (1936) *Processes of formation of pediments at Granite Gap, New Mexico*, Zeit. fur. Geomorph., vol. 9. p. 125–135.

—— (1940) *The retreat of slopes*, Assoc. Am. Geog., Ann. vol. 30, p. 254–268.

Cotton, C.A. (1941) *Landscape as developed by the processes of normal erosion*, Cambridge, p. 300.

Davis, W.M. (1892) *The convex profile of badland divides*, Sci., vol. 20, p. 245.

—— (1909) *Geographical essays*, ed. by D.W. Johnson, New York, p. 734.

—— (1930) *Rock floors in arid and in humid climates*, Jour. Geol., vol. 38, p. 1–27.

—— (1932) *Pediment benchlands and primärrumpfe*, Geol. Soc. Am. Bull., vol. 43, p. 399–440.

—— (1936) *Geomorphology of mountainous deserts*, XVIth Inter. Geol. Cong., C.R., vol. 2, p. 703–714.

—— (1938) *Sheet floods and streamfloods*, Geol. Soc. Am. Bull., vol. 49, p. 1337–1416.

Fair, T.J.D. (1947) *Slope form and development in the interior of Natal, South Africa*, Geol. Soc. S. African Tr., vol. 50, p. 105–118.

—— (1948a) *Slope form and development in the Coastal Hinterland of Natal*, Geol. Soc. S. Africa Tr., vol. 51, p. 33–47.

—— (1948b) *Hillslopes and pediments of the semi-arid Karroo*, S. African Geog. Jour., vol. 30, p. 71–79.

Gilbert, G.K. (1909) *The lowering of hilltops*, Jour. Geol., vol. 17, p. 344–350.

Horton, R.E. (1945) *Erosional development of streams and their drainage basins*, Geol. Soc. Soc. Am. Bull., vol. 56, p. 275–370.

Howard, A.D. (1942) *Pediment passes and the pediment problem.* Jour. Geomorph., vol. 5, p. 1–31, 95–136.

Johnson, D.W. (1931) *Planes of lateral corrasion*, Science, vol. 73, p. 174–177.

—— (1932) *Rock fans of arid regions*, Am. Jour. Sci., 5th ser., vol. 23, p. 389–416.

—— (1932a) *Rock planes of arid regions*, Geog. Rev., vol. 22, p. 656–665.

King, L.C. (1940) *The monoclinal coast of Natal, South Africa*, Jour. Geomorph., vol. 3, p. 144–153.

—— (1944) *Geomorphology of the Natal Drakensberg*, Geol. Soc. S. Africa Tr., vol. 47, p. 255–282.

—— (1947) *Landscape study in Southern Africa*, Geol. Soc. S. Africa Pro., vol. 50, p. xxiii–lii.

—— (1949a) *On the Ages of African landscapes*, Geol. Soc. London Quart. Jour., vol. 104, pp. 439–459.

—— (1949b) *The pediment landform: some current problems*, Geol. Mag., vol. 86, p. 245–250.

—— (1950) *The world's plainlands: a new approach in geomorphology.* Geol. Soc. London, Quart. Jour., vol. 106, p. 101–131.

—— (1951a) *The geomorphology of the Eastern and Southern Districts, Southern Rhodesia*, Geol. Soc. S. Africa, Tr., vol. 54, p. 33–64.

—— (1951b) South African scenery, Edinburgh, 379 p.

—— and Fair, T.J.D. (1944) *Hillslopes and dongas*, Geol. Soc. S. Africa Tr., vol. 47, pl. 4.

Lawson, A.C. (1915) *Epigene profiles of the desert*, Univ. Calif. Publ., Bull. Dept. Geol., vol. 9, p. 23–48.

—— (1932) *Rainwash erosion in humid regions*, Geol. Soc. Am. Bull., vol. 43, p. 703–724.

Maze, W.H. (1944) *The geomorphology of the central eastern area of New South Wales*, Royal Soc. N.S.W., Jour., vol. 78, p. 28–41.

McGee, W.J. (1897) *Sheetflood erosion*, Geol. Soc. Am. Bull., vol. 8, p. 87–112.

Peltier, L.C. (1950) *The geographic cycle in periglacial regions as it is related to climatic geomorphology*, Assoc. Am. Geog. Ann., vol. 40, p. 214–236.

Penck, W. (1924) *Die morphologische Analyse*, Stuttgart, 283 p.

Powell, J.W. (1875) *Explanation of the Colorado River of the West*, Washington, 291 p.

Rich, J.L. (1935) *Origin and evolution of rock fans and pediments*, Geol. Soc. Am. Bull., vol. 46, p. 999–1024.

Strahler, A.N. (1950a) *Equilibrium theory of erosional slopes approached frequency distribution analysis*, Am. Jour. Sci., vol. 248, p. 673–696.

—— (1950b) *Davis' concepts of slope development viewed in the light of recent quantitative investigations*, Assoc. Am. Geog., Ann., vol. 40, p. 209–213.

Wood, A. (1942) *The development of hillside slopes*, Geol. Assoc., Pr., vol. 53, p. 128–138.

Wooldridge, S.W., and Morgan, R.S. (1937) *The physical basis of geography*, London, 445 p.

Printed and bound by CPI Group (UK) Ltd, Croydon, CR0 4YY

01/11/2024

01782632-0018